Zdenko Rengel
Editor

Mineral Nutrition of Crops
Fundamental Mechanisms and Implications

Pre-publication REVIEW

"It is necessary to increase crop yields and maintain good crop quality for the expanding world population. This can only be accomplished if there is a good understanding of the factors affecting the nutrient content of plants. This book discusses mechanisms controlling element uptake, translocation, and accumulation by plants. There is also some discussion of soil factors controlling the supply and availability of nutrients for plants. The effects of good plant nutrition and increasing crop resistance and tolerance to disease are also discussed. This book will be helpful to research scientists, extension workers, teachers, and students."

David L. Grunes, PhD
Soil Scientist (Collaborator), U.S. Plant, Soil, and Nutrition Laboratory, Ithaca, NY

NOTES FOR PROFESSIONAL LIBRARIANS AND LIBRARY USERS

This is an original book title published by Food Products Press, an imprint of The Haworth Press, Inc. Unless otherwise noted in specific chapters with attribution, materials in this book have not been previously published elsewhere in any format or language.

CONSERVATION AND PRESERVATION NOTES

All books published by The Haworth Press, Inc. and its imprints are printed on certified pH neutral, acid free book grade paper. This paper meets the minimum requirements of American National Standard for Information Sciences-Permanence of Paper for Printed Material, ANSI Z39.48-1984.

Mineral Nutrition of Crops
Fundamental Mechanisms and Implications

THE FOOD PRODUCTS PRESS
Crop Science
Amarjit S. Basra, PhD
Senior Editor

New, Recent, and Forthcoming Titles of Related Interest:

Dictionary of Plant Genetics and Molecular Biology by Gurbachan S. Miglani

Advances in Hemp Research by Paolo Ranalli

Wheat: Ecology and Physiology of Yield Determination by Emilio H. Satorre and Gustavo A. Slafer

Mineral Nutrition of Crops: Fundamental Mechanisms and Implications by Zdenko Rengel

Conservation Tillage in U.S. Agriculture: Environmental, Economic, and Policy Issues by Noel D. Uri

Cotton Fibers: Developmental Biology, Quality Improvement, and Textile Processing by Amarjit S. Basra

Intensive Cropping: Efficient Use of Water, Nutrients, and Tillage by S. S. Prihar, P. R. Gajri, D. K. Benbi, and V. K. Arora

Plant Growth Regulators in Agriculture and Horticulture: Role and Commercial Uses edited by Amarjit S. Basra

Crop Responses and Adaptations to Temperature Stress: New Insights and Approaches edited by Amarjit S. Basra

Physiological Bases for Maize Improvement edited by Maria Elena Otegui and Gustavo A. Slafer

Mineral Nutrition of Crops
Fundamental Mechanisms and Implications

Zdenko Rengel
Editor

CRC Press is an imprint of the
Taylor & Francis Group, an informa business

Softcover edition published 2000.

Cover design by Monica L. Seifert.

The Library of Congress has cataloged the hardcover edition of this book as:

Mineral nutrition of crops : fundamental mechanisms and implications / Zdenko Rengel.
p. cm.
Includes bibliographical references and index.
ISBN 1-56022-880-6 (alk. paper).
1. Crops—Nutrition. 2. Crops—Effect of minerals on. I. Rengel, Zdenko.
SB112.5 .M56 1999
631.8'11—dc21

98-48697
CIP

CONTENTS

ABOUT THE EDITOR

Associate Professor **Zdenko (Zed) Rengel, PhD,** Reader at The University of Western Australia, has sixteen years of teaching and research experience in various aspects of plant physiology, soil fertility, and plant nutrition. He has been awarded four prizes and thirteen Fellowships from institutions in Austria, France, Germany, U.K., Japan, U.S.A., and Australia. He is a consultant to various international and Australian institutions and Universities. He has been the invited keynote speaker at twelve international conferences and is currently serving on the Editorial Boards of three international journals. In the last ten years, he has supervised twelve postdoctoral fellows as well as eighteen postgraduate and eighteen Honors students. He has published 103 peer-reviewed articles and thirteen invited book chapters. His current research work spans a wide range of topics from soil chemistry, microbiology, and hydrology, to molecular biology of nutrient transporters, to computer modeling of root growth and nutrient uptake.

ABOUT THE EDITOR

CONTRIBUTORS

Asher Bar-Tal, PhD, is Soil Scientist, Institute of Soils, Weather, and Environmental Science, the Volcani Center, Agricultural Research Organization, Bet-Dagan, Israel.

Ismail Cakmak, PhD, is Professor, Department of Soil Science and Plant Nutrition, Faculty of Agriculture, Cukurova University, Adana, Turkey.

Silvia R. Cianzio, PhD, is Professor, Department of Agronomy, Iowa State University, Ames, Iowa.

Norbert Claassen, PhD, is Professor, Institute of Agricultural Chemistry, University of Göttingen, Göttingen, Germany.

David E. Crowley, PhD, is Associate Professor in Soil Microbiology, Department of Soil and Environmental Sciences, University of California, Riverside, California.

Christof Engels, PhD, is Professor, Section Agroecology, University of Bayreuth, Bayreuth, Germany.

Ian R. P. Fillery, PhD, is Senior Principal Research Scientist, CSIRO Plant Industry, Wembley, Australia.

Robin D. Graham, PhD, is Professor of Plant Science, Department of Plant Science, University of Adelaide, Adelaide, Australia.

Don M. Huber, PhD, is Professor of Plant Pathology, Botany and Plant Pathology Department, Purdue University, West Lafayette, Indiana.

Robert J. Reid, PhD, is Lecturer, Department of Plant Science and Center for Plant Membrane Biology, University of Adelaide, Waite Campus, Glen Osmond, Australia.

Frank W. Smith, PhD, is Senior Principal Research Scientist, CSIRO Tropical Agriculture, St. Lucia, Australia.

Bernd Steingrobe, Dr. rer.hort., is Assistant Professor, Institute of Agricultural Chemistry, University of Göttingen, Göttingen, Germany.

Ross M. Welch, PhD, is Plant Physiologist, U.S. Plant, Soil, and Nutrition Laboratory, Ithaca, New York, and Professor, Department of Soil, Crop, and Atmospheric Sciences, Cornell University, Ithaca, New York.

Pieter Wolswinkel, PhD, is Associate Professor, Transport Physiology Research Group, Department of Plant Ecology and Evolutionary Biology, Faculty of Biology, University of Utrecht, Utrecht, Netherlands.

Preface

Food production will have to increase by 50 percent by the year 2025 to feed a population estimated to be around 8 billion at that time. Judicious fertilizer management based on understanding complex interactions and the multitude of mechanisms underlying nutrient dynamics in the water-soil-plant continuum in the agroecosystems will be paramount in increasing crop and pasture yields and minimizing unwanted losses of nutrients to the environment. Such an understanding can only be brought about by a multidisciplinary approach to mineral nutrition of crops. This book epitomizes such an approach by covering subjects that range from plant molecular biology to soil hydrology. Interactions influencing nutrient availability in the rhizosphere are covered first, followed by descriptions of mechanisms important in uptake of nutrients across the plasma membrane of root cells and long-distance nutrient transport and loading into developing seeds. The role of nutrients in photosynthesis and in tolerance and resistance to diseases and the importance of seed nutrient reserves in plant growth and development are discussed. Physiological mechanisms underlying differential efficiency in uptake and utilization of nutrients form a basis for screening techniques that are the cornerstone of breeding for increased nutrient uptake and utilization efficiency in crops. Measuring movement of water and nutrients down the soil profile is important in understanding nutrient cycling and fertilizer utilization. Finally, our understanding of the complexity of processes involved in nutrient mobilization from soil, uptake by roots, and utilization in plants can be improved by using mechanistic models based on our partial understanding of these processes.

Although I have reviewed all chapters, I also would like to thank Fusuo Zhang (China Agricultural University, Beijing, Peoples Republic of China), Roger Leigh (Rothamsted Experimental Station, Harpenden, United Kingdom), James Fisher (Agriculture Western Australia, Perth, Australia), John Pate and Eugene Diatloff (University of Western Australia, Perth, Australia), and David Crowley (University of California, Riverside, United States) for reviewing selected chapters and offering valuable comments and suggestions to the authors.

I would like to express my thanks to Amarjit S. Basra (Senior Editor of the Crop Science book program of The Haworth Press, Inc.), for advice and encouragement, and the staff of The Haworth Press for care and diligence in producing this book.

Zdenko Rengel
Perth, Western Australia

Chapter 1

Biology and Chemistry of Nutrient Availability in the Rhizosphere

David E. Crowley
Zdenko Rengel

Key Words: iron, manganese, microflora, mycorrhiza, organic acids, phosphorus, rhizosphere, root exudation, siderophore, zinc.

INTRODUCTION

The importance of rhizodeposition and soil microorganisms in plant nutrition is clearly evident in nature, where plant nutrient uptake is directly coupled to nutrient cycling from organic matter, solubilization of metals by root exudates, and the formation of symbiotic relationships with fungi and bacteria. Plant ecologists, for some time, have studied the growth strategies and distribution of plants that are adapted to different climates and soil types and have recognized that there is wide variation in the ability of various plants to thrive in nutrient-limiting environments (Chapin, 1987). However, until recently, the importance of rhizosphere processes that function in plant nutrition has not been a major consideration in modern agriculture, where the practice has been to provide N (nitrogen), P (phosphorus), and K (potassium) in luxurious quantities as synthetic fertilizers (Schaffert, 1994). This has generated concern that the selection and breeding of plant cultivars for agriculture has resulted in the development of cultivars that are highly responsive to fertilizers, but without the traits necessary for growth under nutrient-limiting or adverse soil conditions (Duncan and Baligar, 1990). With the current emphasis on developing better cultivars for sustainable, low-input agriculture, a better knowledge of the processes that contribute to nutrient uptake efficiency has become essential for breeding genotypes that perform well in nutrient-limiting soils.

This chapter considers the influence of root exudates and rhizodeposition on nutrient availability in soils and the importance of root-microbial interactions that affect plant nutrition. Many of these processes, such as rhizosphere acidification, the release of metal chelators, and formation of symbioses with N_2-fixing bacteria and mycorrhizal fungi, have been intensively studied and are fairly well understood. By comparison, much can still be learned about the relative degree of control that plants have in mediating beneficial root-microbial interactions and how this might be manipulated through plant breeding. An ongoing discussion focuses on using plant breeding to tailor the rhizosphere for plant-beneficial microorganisms and microbial communities (O'Connell, Goodman, and Handelsman, 1996). However, rhizosphere ecologists have not yet been able to provide summary information that has direct relevance for plant breeders. In part, this is because the rhizosphere consists of a mosaic of microsite environments that constantly change in response to environmental variables and because almost any process can be shown to occur at some time or location on plant roots. Following a reductionist approach with axenic culture systems, it is possible to study the influence of specific microorganisms on root exudation and cell signaling between plants and microbes. Eventually, such information may be used to select for beneficial microorganisms and to study the factors that enhance their growth and activity on plant roots. In addition, genetic and biochemical methods are now being used to study the structure and species composition of microbial communities on plant roots, which may help to enhance the development of beneficial microbial communities in the rhizosphere.

INFLUENCE OF RHIZODEPOSITION ON MICROBIAL ECOLOGY

When challenged with nutrient deficiency, plant roots exude relatively large amounts of various organic substances into the rhizosphere (see Lynch and Whipps, 1990; Darrah, 1993; Mori, 1994; Pearson and Rengel, 1997). These substances, although presenting an energy and carbon cost to plants, offer ecological and evolutionary advantages because they allow plants to survive and produce new generations under unfavorable conditions. Virtually all of the carbon compounds that are synthesized by plants are at some time deposited into the rhizosphere, either by cell lysis, root leakage, or secretion of root exudates (Rovira, 1979). This general process causes an increase in microbial populations and activity in the rhizosphere, which in turn influences soil pH and redox and increases the mineralization of soil organic matter by soil microorganisms and soil fauna (for reviews, see Uren and Reisenauer, 1988). Microorganisms supported by growth on these substrates can fortuitously influence plant nutrition by causing enhanced release of root exudates (Meharg and Kilham, 1995) and by producing growth factors that influence root growth (Arshad and Frankenberger, 1993). Plants and microorganisms also modify the rhizosphere through respiration, selective uptake of cations and anions, and changes in the physical properties of the soil in the vicinity of roots (for reviews, see Curl and Truelove, 1986; Pepper and Bezdicek, 1990).

Estimates of the quantities of carbon that are released into the rhizosphere vary for different plant species, but range from 30 to 60 percent of the total carbon that is fixed during photosynthesis (Lynch and Whipps, 1990). The proportion of carbon that is allocated to plant root development and to rhizodeposition varies with plant species and developmental stage and is further influenced by abiotic variables such as nutrient availability, aeration, soil texture, and water stress. Rates of root exudate release and the composition of root exudates are also influenced by changes in light intensity, soil temperature, plant age, and the presence and absence of specific microorganisms (for reviews, see Curl and Truelove, 1986; Marschner, 1995a). This complexity precludes broad generalizations about the nature of root exudates and creates an inherent difficulty in extrapolating data from hydroponic studies to actual rhizosphere processes in soil.

To examine plant responses to nutrient limitation or adaptation to adverse conditions such as salinity or heavy metals, most studies on root exudates have been conducted with hydroponically grown plants, in which bulk exudate samples representative of the entire root system are collected. These data provide an indication of plant responses to nutrient deprivation and allow collection of clean samples for chemical analysis. However, the extent to which these data are artifactual of plant growth in hydroponic media, and the manner in which the exudates have been collected, is unknown. In soil, plants release two to four times more carbon into the rhizosphere than hydroponically grown plants (Trofymow, Coleman, and Cambardella, 1987) and have increased rates of rhizodeposition in response to mechanical impedance, anaerobiosis, nutrient deficiency, and the presence of microorganisms. Further complexity is generated by the differential abilities of different species of bacteria to increase root exudation rates three- to tenfold (Meharg and Kilham, 1995). At yet another level of trophic interactions, bacterial-feeding nematodes in the rhizosphere have caused a nearly threefold increase in root exudation in comparison to plants inoculated with bacteria alone (Sundin et al., 1990). Spatial patterns of root exudate accumulation in relation to microbial colonization of actively growing and nongrowing root apexes, as well as temporal considerations such as diurnal pulsing of certain root exudate components to high concentrations in the rhizosphere, are other factors which are eliminated in hydroponic systems, yet which must be considered to understand the influence of rhizodeposition on rhizosphere processes.

Given these differences between hydroponic- and soil-cultured plants, one research challenge has been the development of methodology for determining the composition and quantities of root exudates at localized microsites for soil-grown plants. Redox and pH probes also are now being used to examine root effects on soil chemistry at specific microsites along the roots (Gollany and Schumacher, 1993). Such data can then be incorporated into models that consider localized concentrations of specific root exudate components, the chemical conditions of the rhizosphere, and the effect of these variables on plant nutrient uptake and microbial populations at specific locations in the rhizosphere. This modeling approach begins with a conceptual view of the soil factors influencing nutrient availability, the plant responses to nutrient deprivation that cause changes in exudate release, and the influence of microorganisms that grow on root exudates and further modify soil chemistry and nutrient availability (see Figure 1.1). More complex simulation models, such as those developed by Darrah (1993), consider the actual fluxes of root exudates, microbial growth rates, and rates of root elongation. These data, combined with empirical data for the rate of solubilization of different soil minerals by specific root exudate components, can then be integrated into equations that predict the availability of nutrients in different root zones, for example, for elongating primary roots as opposed to nonelongating short lateral roots. In conjunction with these data, plant physiologists can examine the ion selectivity by, and transport kinetics of plant nutrients and metal complexes across, the root cell membranes.

Practical applications of this integrative modeling approach that are currently driving studies on plant nutrition are the possible use of plants for phytoremediation of heavy metal-contaminated soils and related research that is aimed at understanding the role of root exudates in the bioavailability and transfer into the food chain of heavy metals taken up by crop plants. Certain accumulator plants that are being used for phytoremediation are known to selectively take up and translocate specific heavy metals such as Ni (nickel), Se (selenium), and Pb (lead) (Ernst, 1996). Whether this ability to mobilize heavy metals from the soil solid phase is based strictly on differences in membrane transport capabilities of these particular plant species, or also involves differences in root exudates and the microbial commu-

FIGURE 1.1. Conceptual Model of Rhizosphere Microbial and Soil Chemical Factors Influencing the Mobilization and Uptake of Nutrients by Plants

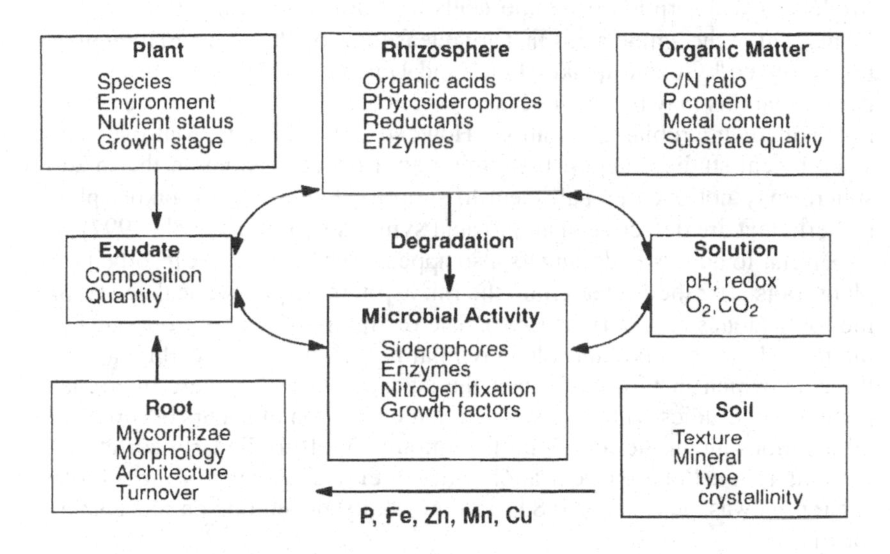

nities that are associated with these plants, has not yet been determined. Undoubtedly, data generated from research on these topics will have spin-offs that will be of importance for designing agricultural plants with improved nutrient uptake efficiency.

Rhizosphere Exudate Components

Carbon materials deposited into the rhizosphere include sloughed cells and root tissues; low-molecular-weight exudates, consisting of soluble organic acids, sugars, phenolics, and amino acids; and high-molecular-weight exudates that comprise the root mucilage secreted at the root apexes (Rovira, 1979; Curl and Truelove, 1986). The low-molecular-weight substances are also contained in cell lysates, which are released when root cortex cells are ruptured either by mechanical damage, grazing, or during the emergence of lateral roots that break through the root epidermis. In addition to these substances, rhizodeposition includes the secretion of enzymes, such as acid phosphatase secreted in response to P deficiency (Helal and Dressler, 1989; Tadano and Sakai, 1991).

Among the different classes of root exudate components, amino acids appear to have little or no role in complexation or mobilization of plant nutrients (Jones et al., 1994). Moreover, amino acids have consistently been found to comprise a small proportion of the root exudate mixture. Models of reabsorption of amino acids by corn roots indicate that up to 90 percent of the amino acids that are released through root exudation may be recovered by plant uptake (Jones and Darrah, 1993). It has been suggested that this may be one means by which plants regulate the size of the rhizosphere microbial population. However, this idea is controversial, since recent studies suggest that amino acid concentrations in the rhizosphere may not even be sufficient to support the growth of auxotrophic bacteria that are defective in amino acid synthesis (Simons et al., 1997).

Similar to amino acids, sugars also appear to be actively reabsorbed by plant roots after their release into the rhizosphere by passive leakage from the roots (Jones and Darrah, 1996). The recapture of sugars from outside the root plays an important role in regulating the amount of carbon lost to the soil, which in turn will affect the size of the rhizosphere microbial population. Studies with corn showed that exudation of sugars occurred at all locations along the root, with the amount of efflux linearly correlated with internal cellular concentration. Moreover, the turnover of the soluble sugar pool was measured at 0.8 to 15 times daily, depending on root spatial location.

Exuded organic acids appear to have a major role both in solubilization of mineral nutrients and as selective growth substrates for growth of microorganisms. Typical organic acids found in root exudate include citric, malic, malonic, acetic, fumaric, succinic, lactic, oxalic, and others. Malate and citrate appear to be major root exudate components for some plant species. For example, in corn grown under sterile conditions, exudation of malate and citrate increases by twelve- and thirty-threefold under nutrient deficiency conditions (Jones and Darrah, 1995). Because organic acids are also excellent substrates for microbial growth (Jones, Prabowo, and Kochian, 1996), these concentrations may only occur temporarily at rapidly growing root apexes that are not yet densely colonized by microorganisms. Nonetheless, models of organic acid accumulation in the rhizosphere suggest that concentrations as high as 50 μM can occur within 1 mm from the root surface (Jones, Darrah, and Kochian, 1996).

Organic acids exuded by roots under P deficiency (an average rate of 0.57 nmol cm^{-1} root h^{-1}; Hoffland, Findenegg, and Nelemans, 1989) may solubilize various P complexes, such as those with Al (aluminum) and Fe (iron) (Gardner, Barber, and Parbery, 1983) and Ca (calcium) (Dinkelaker, Römheld, and Marschner, 1989), as well as rock phosphate (Hoffland et al., 1992). It is interesting to note that canola plants grown in P-deficient soil excrete organic acids (citric and malic) in the apical root zone (1 to 2 cm behind the root tips), but when supplied with the rock phosphate, organic acid exudation occurred at any root part along the root axis, as long as this part was in a direct contact with the rock phosphate particle (Hoffland et al., 1992).

Organic acids are thought to be excreted from the root cells by a H^+-coupled cotransport system (Dinkelaker, Römheld, and Marschner, 1989; Jones and Darrah, 1994; Johnson et al., 1996), even though they may be excreted through anion channels as well, given that at least malate and citrate exist as anions at the cytoplasmic pH (Delhaize, 1995). In contrast to sugars (Jones and Darrah, 1996), organic acids do not leak passively into the rhizosphere of P-deficient plants; Johnson and colleagues (1996) found that succinate represented between 22 and 34 percent of the total organic acid pool in P-deficient roots of white lupin, but was not detectable in root exudates.

Comprehensive Analysis of Root Exudates

To date, few attempts have been made to comprehensively characterize the root exudates of plants. This is due to the necessity of employing a wide array of analytical techniques for quantifying all of the different exudate components, which requires prior knowledge of the compounds that are present in order to devise analytical methods for separation and quantification of the different types of chemicals contained in the exudate mixture. Under axenic conditions, the exudates of hydroponically grown corn were found to be comprised of 65 percent sugars, 33 percent organic acids, and 2 percent amino acids (Kraffczyk, Trolldenier, and Beringer, 1984). However, the relative quantities of these substances, particularly for organic acids, vary among plant species and even among plant cultivars (Cieslinski et al., 1997).

Recently, the utility of ^1H (hydrogen) and ^{13}C (carbon) NMR (nuclear magnetic resonance) techniques for comprehensive analysis of root exudates has been shown for root exudates of barley collected from hydroponically grown plants (Fan et al., 1997). In combination with GC-MS (gas chromatography-mass spectrometry) techniques, a large number of common organic and amino acids were identified, and complete structural information could be obtained even for unknown exudate components. Although this methodology is not practical for routine analyses, the power of this technique lies in its ability to detect all of the predominant exudate components, which then can be routinely quantified by more standard procedures.

A variety of techniques are now being used to characterize the function of root exudates for soil-grown plants. These include root boxes that employ screens to separate soil into compartments at discrete distances from the surface of plant roots (Li, Shinano, and Tadano, 1997). Other systems have been devised that allow measurement of changes in root exudate quantities over time for soil-grown plants (Hodge, Grayston, and Ord, 1996). In our research, plants are routinely grown in root box microcosms that are inclined to direct root growth along a removable plate that permits repeated access to the roots during plant growth (see Figure 1.2). To prevent artifactual movement of microorganisms and soil solutes along

FIGURE 1.2. Root Box Microcosms for Collection of Root Exudates and Sampling of Rhizosphere Microflora in Different Root Zones for Soil-Grown Plants

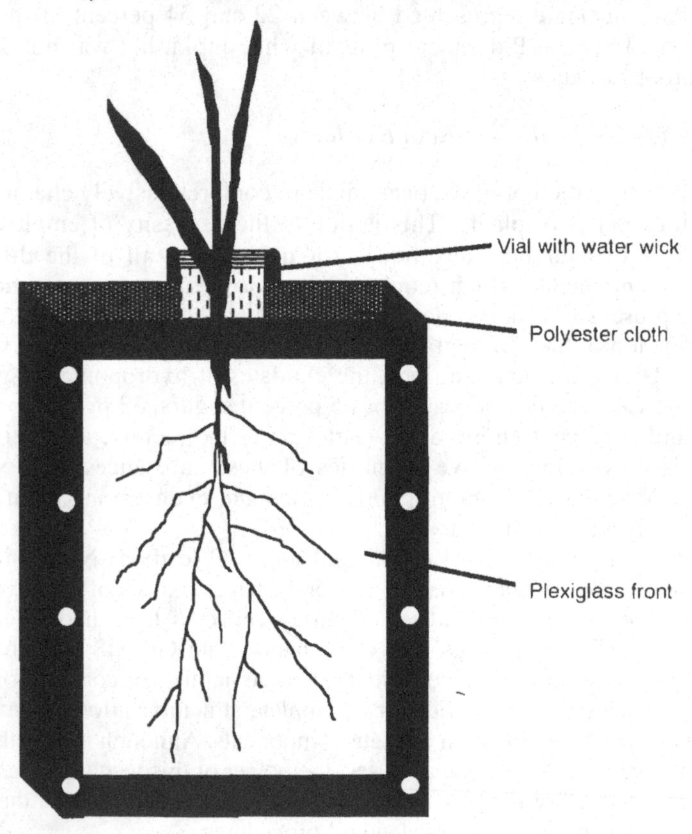

Note: Plexiglass boxes are inclined at a 15° angle to direct root growth along the surface of a removable plate (not shown) that allows direct access to the roots. Plants are watered by means of a wick system that allows maintenance of constant moisture conditions.

the removable front plate, the microcosms are watered by means of a wick cloth placed at the back of the microcosm. The wick extends to a water/nutrient solution reservoir that is replenished to maintain a fixed soil moisture that is monitored gravimetrically. To sample the rhizosphere microflora, the front plate is removed and small strips (measuring 2 mm by 1 cm) of chemically clean, glass filter paper (Pallflex Products, Putnam, CT, United States) are overlaid onto the plant roots to sorb root exudates into the filter. The cover is placed on the box, and after a period of four hours, the filter papers are removed for analysis. Using GC-MS analysis, the organic acid composition of root exudates was determined for cucumber plants grown in a P-deficient soil (see Figure 1.3). GC-MS analysis has been used to quantify changes in root exudate quantities in relation to colonization of roots by mycorrhizal fungi (Marschner, Crowley, and Higashi, 1997). This method has also been used with nitrocellulose filter

FIGURE 1.3. Microsite Sampling Technique Using Quartz Filter Paper Strips for Collection and Analysis of Root Exudate Composition by Gas Chromatography with Flame Ionization and Mass Spectrometry Detection

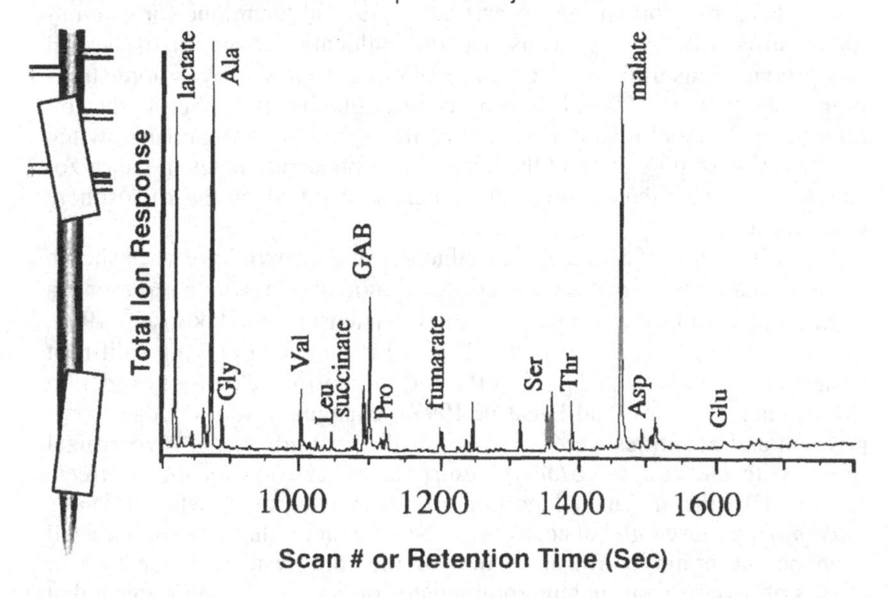

Source: Higashi, Marschner, and Crowley (unpublished data).

Note: Organic acids and amino acids (standard abbreviations) and gamma amino butyrate (GAB) detected in the root exudate are shown adjacent to the corresponding peaks.

papers to sample rhizosphere bacteria associated with the root surface (Marschner and Crowley, 1997). Although still in development for routine analysis of root exudates, this microsite procedure permits nondestructive analysis of the rhizosphere over time and in relation to different environmental factors.

Quantification of Root Exudates

Estimations of carbon allocation to the roots generally are based on measurements using ^{14}C carbon dioxide to pulse-label the root exudates (Meharg, 1994). This allows quantification of root exudation in different zones, the amount of carbon assimilated by rhizosphere microorganisms, and CO_2 release by root and microbial respiration. A recent study with forage rape (*Brassica napus*) showed approximately 17 to 19 percent of fixed $^{14}CO_2$ was translocated to the roots over two weeks, of which 30 to 34 percent was released into the rhizosphere and 23 to 24 percent was respired by the roots (Shepherd and Davies, 1993). Of the ^{14}C released into the rhizosphere, between 35 to 51 percent was assimilated and respired by rhizosphere microorganisms. A particularly useful technique for examining the effect of microorganisms on carbon allocation to the rhizosphere of soil-grown plants utilizes $^{14}CO_2$ pulse labeling, followed by isotopic trapping (Cheng et al., 1994). In this procedure, glucose is applied to the soil after the pulse label is supplied to the plant shoots. Microorganisms switch to utilization of glucose, and the labeled carbon accumulates in the rhizosphere soil rather than being immediately consumed by the rhizosphere microflora.

Quantities of root exudate released to the rhizosphere have been shown to be decreased in response to mycorrhizal inoculation, which represents a significant carbon drain on the plant (Christensen and Jakobsen, 1993; Marschner, Crowley, and Higashi, 1997). In studies employing split-root systems to study local and global effects of mycorrhizae on root exudation (Marschner, Crowley, and Higashi, 1997), changes in root exudate composition and quantities were variable for different arbuscular mycorrhizal fungi, with one fungus, *Glomus deserticulum*, causing localized effects restricted to the inoculated portion of the root system, while *Glomus interradices* caused global changes in root exudation that were manifested even on the nonmycorrhizal portion of the root system. These myriad effects of mycorrhizae and microorganisms on root exudation suggest that it may be difficult, if not impossible, to make general conclusions about the actual composition and quantities of root exudates for soil-grown plants.

Influence of Root Mucilage on Plant Nutrition

One of the least-studied materials secreted by roots is high-molecular-weight mucilage materials that are produced by the apexes of growing roots. Mucilage is composed of a mixture of polygalacturonic acids and polysaccharides that are secreted by the root cap and epidermal cells of actively growing root tips. After secretion, the material is further modified by microorganisms, such that the final substance consists of an amalgamation of high-molecular-weight gelatinous material containing mineral soil particles, soil organic matter, microorganisms, and sloughed root tissues, which together is called mucigel. This material serves to increase contact between the soil and the roots and has been shown to directly influence growth of microorganisms and the solubility and uptake of metal micronutrients by plants. Interestingly, there are chemical differences in metal-binding properties of mucigel from different plants (Morel, Mench, and Guckert, 1986), which results in preferential selective binding of metal cations during their diffusion to the root surface. Differences in metal binding by mucigels of different plants were shown to be correlated with the heavy metal contents of plants, suggesting a direct role in mediating uptake of these and other metal ions (Mench and Martin, 1991).

Mucilage has been shown to be of possible importance for enhancing uptake of Zn (zinc) in dry soils (Nambiar, 1976) through a process involving hydraulic lift of water from the deep soil to dry topsoil (Vetterlein and Marschner, 1993). It has further been speculated that mucigel forms a two-phase system that may directly mobilize P (Matar, Paul, and Jenny, 1967), Fe (Azarabadi and Marschner, 1979), and Mn (manganese) (Uren, 1993) that diffuse from the soil solid phase through the mucigel phase to the plant root. This phenomenon has been termed contact reduction (Uren, 1993) and occurs when mucilage produced by root cells forms a gel that moulds to the surface of soil particles and diffuses into the soil aggregates. In grasses, which do not involve a reductive process to mobilize Fe, enhanced Fe uptake in the mucigel may involve chelator-mediated transport in the mucigel layer (Marschner, 1995a).

PLANT MODIFICATIONS OF RHIZOSPHERE CHEMISTRY

Plants have a number of modifying influences on soil chemistry in the rhizosphere that specifically affect the availability of nutrients. These effects are mediated primarily by root alteration of pH and redox, which control the solubility of metal micronutrients, dissolution of P minerals, and N transformations by soil bacteria.

Acidification of the Rhizosphere

The primary factor controlling rhizosphere pH is the differential uptake of cations and anions from the soil solution, which is balanced, respectively, by the release of protons and hydroxyl or bicarbonate ions into the soil solution. In addition to pH shifts caused by differential uptake of cations, many plants have an inducible proton pump that can acidify the rhizosphere as a direct response to nutrient deficiencies (for a review, see Thibaud, 1994). Induction of rhizosphere acidification occurs as a general response to P deficiency in almost all plants, and to varying degrees in response to Fe, Zn, and Cu (copper) deficiencies in dicotyledonous plants (Fe, Zn; Cohen, Norvell, and Kochian, 1997). In the case of P deficiency, this also involves the secretion of organic acids into the rhizosphere, particularly citric and malic. This response is particularly refined in plant species such as white lupin, which produce proteoid roots that exude large quantities of citric acid under P-limiting conditions (Gardner, Barber, and Parbery, 1983; Dinkelaker, Römheld, and Marschner, 1989).

As N is the predominant macronutrient taken up by plants, the chemical form of N present in the soil as NO_3^- or as NH_4^+ has one of the greatest modifying influences on rhizosphere pH, which can differ by as much as two to three pH units from the bulk soil (Marschner and Römheld, 1983; Tagliavini, Masia, and Quartieri, 1995). For soil-grown plants, these pH effects are localized at the root apexes (Marschner, Römheld, and Kissel, 1986; Gollany and Schumacher, 1993), such that distinct pH differences may occur in different zones along the root axes of the same plant. Manipulation of NO_3^- and NH_4^+ nutrition has been exploited as a possible cultural practice for increasing the availability of P and trace metals to plants (Jungk, Seeling, and Gerke, 1993; Tagliavini, Masia, and Quartieri, 1995; and review by Marschner, 1995b). An increase in rhizosphere pH caused by release of hydroxyl ions during uptake of NO_3^- results in increased solubility of Fe and Al phosphates (Jungk, Seeling, and Gerke, 1993).

Root respiration and accumulation of carboxylate in the rhizosphere soil solution has a similar pH-increase effect in acid soils. In contrast, in alkaline soils where phosphate solubility is controlled by Ca phosphates, NH_4^+ nutrition and concomitant release of protons serve to solubilize these complexes. For example, acidification of the rhizosphere soil of black locust plants increased availability of P from the rock phosphate up to fourfold (Gillespie and Pope, 1990). Such acidification occurs when the N source is NH_4^+, but not when it is NO_3^-, indicating that acidification is due to exudation of H^+ ions that accompany cation influx (Gahoonia, Claassen, and Jungk, 1992; Hoffmann et al., 1994). An increase in P availability and uptake by acidification is accentuated by the ion equilibrium shift from HPO_4^{2-} to $H_2PO_4^-$, with the monovalent form being taken up at a faster rate than the divalent form (Marschner, 1995b).

Rhizosphere acidification by wetland rice plants in response to P deficiency is the result of a more subtle response involving O_2 (oxygen) transport to the roots, which causes oxidation of reduced Fe(II) and release of a proton from the Fe mineral (Kirk and Du, 1997). This is also accompanied by changes in root architecture to a finer-textured root system having greater porosity to O_2 supplied by the shoot, and changes in the cation/anion balance with respect to NH_4^+ and NO_3^- uptake.

Organic Acids

Organic acids have different mechanisms for phosphate solubilization in acid and alkaline soils. In acid soils, sparingly soluble phosphate, such as Fe or Al phosphate, can be mobilized by the lowering of the rhizosphere pH, as well as desorption from sesquioxide surfaces by anion (ligand) exchange (Gerke, 1992). The organic acids form stable complexes with Fe and Al, allowing the phosphate to be less adsorbed to the soil particles (Bolan et al., 1997) and therefore more available for uptake by the root (Otani, Ae, and Tanaka, 1996). In alkaline soils, where P is present primarily as Ca phosphate minerals, the organic acids cause dissolution of these minerals by lowering pH and through chelation of Ca (Bolan et al., 1997).

The organic acids differ in the extent to which they can mobilize P from the soil. The mobilization of P is greatest with citric, followed by oxalic acid, while malic and tartaric acids are moderately effective at mobilizing P. Acetic, succinic, and lactic acids are the least effective at mobilizing P (Nagarajah, Posner, and Quirk, 1970). Malonic acid, the major component of the root exudates of pigeon pea, dissolves $AlPO_4$ better than $FePO_4$, while piscidic acid, present in pigeon pea root exudates in ninefold lower concentration than malonic acid, chelates Fe better than Al and therefore solubilizes $FePO_4$ better than $AlPO_4$ (Ae et al., 1990; Otani, Ae, and Tanaka, 1996).

Phosphatases

Organic P can be hydrolyzed by ectoenzymes, releasing P to be taken up by the root (see Pearson and Rengel, 1997). This was demonstrated experimentally by the mineralization of organic P compounds by enzymes, such as acid and alkaline phosphatases, phosphodiesterase, and phytase, which were added to soil, and extracts were analyzed for orthophosphate (Bishop,

Chang, and Lee, 1994). Phosphatases from various plants may be similar, as shown by purification of phosphatase enzymes exuded from the roots of white lupin and tomato (Li and Tadano, 1996).

Exudation of phosphatases occurs from the apical root region and increases under P-deficient conditions in a number of crop species (Tarafdar and Jungk, 1987). The activity of phosphatases decreased with the distance from the root surface but was still detectable 2.5 mm (Li, Shinano, and Tadano, 1997), and up to 3.1 mm, away (Tarafdar and Jungk, 1987). Exudation of phosphatase by roots of white lupin was confirmed even in sterile soil, thus excluding any possible contribution from microbial phosphatases. However, in wheat, organic P addition enhanced mycorrhizal colonization and exudation of phosphatases from hyphae, thus improving P uptake and growth (Tarafdar and Marschner, 1994).

Phosphatases are not effective in decomposing phytin (inositol hexaphosphate), which represents the majority of stored organic P in many soils. Instead, the enzyme phytase specifically catalyzes the breakdown of phytin. However, plant roots do not exude phytases (which are located in seeds); it is only through microorganisms (e.g., *Aspergillus niger*) excreting phytases that phytin may become a source of P to plants (see Delhaize, 1995).

Cluster Roots

When subjected to P-deficiency stress, some families of native flora (e.g., Proteaceae) can form cluster (proteoid) roots as dense groupings of rootlets along lateral roots. Other nutrient deficiency stresses, including Fe, Mn (Johnson, Allan, and Vance, 1994), and Zn (Dinkelaker, Hengeler, and Marschner, 1995), can also induce formation of cluster roots to a lesser extent. Cluster roots of white lupin have been associated with localized increased concentrations of organic acids, reducing and chelating agents, and H^+ in the rhizosphere; the most prevalent organic acids are citric and malic (Gardner, Barber, and Parbery, 1983; Johnson, Allan, and Vance, 1994; Johnson et al., 1996). Similarly, the cluster roots of *Banksia integrifolia* excreted mostly citric acid (50 percent of the total), followed by malic and aconitic acids, which amounted to 18 and 17 percent, respectively (Grierson, 1992).

The biosynthesis of citrate is increased in cluster roots of white lupin in comparison to normal roots (Johnson, Allan, and Vance, 1994). Increased activity of the enzyme phosphoenolpyruvate carboxylase replenished carbon from the tricarboxylic acid cycle for the synthesis of citrate, indicating that a significant portion of carbon in citric acid is nonphotosynthetically fixed (Johnson et al., 1996). The conversion of malate to citrate may occur in the cluster roots, thus increasing their capacity to supply large amounts of citrate for exudation into the rhizosphere (Johnson, Allan, and Vance, 1994). In white lupin, citrate release is also accompanied by secretion of acid phosphatase, which results in significant depletion of organic P in the rhizosphere (Sakai and Tadano, 1993; Li, Shinano, and Tadano, 1997).

PLANT RESPONSES TO MICRONUTRIENT DEFICIENCIES

Such responses have been studied primarily with respect to Fe and Zn nutrition, which are the two most common trace metal deficiencies in agriculture. These deficiencies are seldom caused by low soil contents of these metals, but rather are due to their poor solubility at neutral to alkaline pH.

Iron

Wet climatic regions tend to have more acidic soils and a greater abundance of partially reduced Fe minerals, such as ferric hydroxide, that support relatively high levels of soluble Fe. At the other extreme, Fe availability is particularly low in semiarid and arid soils that have both high base saturation and highly oxidized and crystallized Fe minerals. Under aerated conditions, Fe minerals that are formed include amorphous and crystalline ferric [Fe(III)] hydroxides that have very poor solubility and slow dissolution kinetics (Schwertmann, 1991).

Plants respond to Fe deficiency by employing a range of mechanisms that may be grouped into strategy I (dicots and nongraminaceous monocots) and strategy II (graminaceous monocots) responses. Strategy I is characterized by increased reduction of Fe(III) to Fe(II) at the root-cell plasma membrane, acidification of the rhizosphere (enhanced net exudation of protons), enhanced exudation of reducing and/or chelating compounds into the rhizosphere, and changes of root histology and morphology (formation of rhizodermal transfer cells, induced root branching, more root hairs, etc.). Strategy II involves increased exudation of metal chelators, phytosiderophores, into the rhizosphere. For more extensive reviews covering various other aspects of the Fe-deficiency stress, the reader is referred to Pearson and Rengel (1997), Jolley and colleagues (1996), Ma and Nomoto (1996), Welch (1995), Marschner and Römheld (1994), Mori (1994), and Römheld (1991).

Strategy I

Acidification of rhizosphere is an important part of the Fe-deficiency response of strategy I plants and functions to increase the solubility of Fe-containing minerals. An increased net extrusion of protons is due to increased activity of the plasma-membrane-bound H^+-ATPase, resulting not only in the rhizosphere acidification but also in an increased transmembrane electrical potential and thus an increased driving force for Fe(II) uptake (Rabotti and Zocchi, 1994; Brancadoro et al., 1995). Proton extrusion, which occurs primarily at the root apexes, causes a 100- and 1000-fold increase in Zn and Fe solubility, respectively, for every unit decrease in pH. Pea (*Pisum sativum*) plants showed an inducible rhizosphere acidification in response to Cu deficiency, which, similarly to Zn, would increase solubility of this metal by 100-fold for every unit decrease in pH (Cohen, Norvell, and Kochian, 1997).

Organic acids, particularly citric and malic (Landsberg, 1981; Brancadoro et al., 1995) and caffeic (Olsen et al., 1981), also have a role in Fe stress-induced acidification of the rhizosphere, which is particularly effective in solubilizing amorphous Fe hydroxide minerals (Jones, Darrah, and Kochian, 1996). The process of net H^+ extrusion assumes a special importance in acquisition of Fe from calcareous soils that are also rich in bicarbonate because bicarbonate binds H^+, prevents rhizosphere acidification, and decreases effectiveness of the membrane-bound Fe(III)-reductase (Romera, Alcantara, and De la Guardia, 1992). In addition to rhizosphere acidification, strategy I plants also modify biological and chemical conditions in the rhizosphere by extruding reducing and/or chelating compounds. This process may not take place in nutrient solution-grown plants (Chaney and Bell, 1987) where relatively high rates of Fe(III) chelates are generally supplied, and therefore, concentration of Fe(III) chelates in the vicinity of the plasma membrane-bound Fe(III)-reductase may be sufficiently high.

In addition to the acidification response, plant-induced changes in the redox status of the rhizosphere also greatly influence Fe availability. Redox shifts in the rhizosphere are associated primarily with depletion of oxygen by root and microbial respiration in water-saturated soils or at microsites of high microbial activity, and in some plants, they represent a specific inducible response to Fe or Cu deficiency (Bienfait, 1988; Cohen, Norvell, and Kochian, 1997).

The inducible reductase in dicotyledonous plants can function in Fe nutrition by reducing Fe(III) to Fe(II) in various Fe(III)-specific chelates from microorganisms or when provided as synthetic chelates (for reviews, see Römheld and Marschner, 1986; Welch, 1995). This results in dissociation of Fe from the chelate, which can then be taken up by a Fe(II) channel or transport system in the root cell membrane. The ability of the reductase to reduce different metals, metal complexes, and ligands can be predicted by standard reduction potentials, such that redox couples having a reduction potential greater than −0.37 volts can be reduced (Bienfait, 1988). Reactivity of the reductase system can be further influenced by changes in pH and O_2 availability in the rhizosphere. The reductase has a broad specificity for many synthetic ferric chelates, but it is not yet certain which natural chelates function as the primary Fe chelators in the rhizosphere. For some time, it was speculated that microbial siderophores may be utilized by means of a chelate reductase, since siderophores are produced by many rhizosphere microorganisms. However, the rates of Fe reduction measured for most siderophores are extremely low in comparison to those measured for synthetic chelates such as EDTA. Possibly, the reductase may function for reduction of citrate- or malate-Fe complexes, or for Fe complexed with heme and fulvic acid.

Iron can exist in soil as ferric phosphate. Alfalfa (*Medicago sativa*) can solubilize ferric phosphate complexes and thus increase availability of Fe (Masaoka et al., 1993). This solubilization of ferric phosphate appears to be mediated by secretion of benzofuran, which also has phytoalexin properties and inhibits the growth of *Fusarium oxysporum* f. sp. *phaseoli*. It is unclear whether the dual action of this particular compound in the root exudates has ecological relevance.

Strategy II

Graminaceous species acquire Fe by releasing phytosiderophores (PS), nonproteinogenic amino acids with a high binding affinity for Fe, and by taking up ferrated PS through a specific transmembrane uptake system (Marschner and Römheld, 1994; Mori, 1994; Welch, 1995). An increased mobilization of Fe from a calcareous soil, even as far away from the root surface as 4 mm, demonstrated a high capacity of PS to mobilize Fe (Awad, Römheld, and Marschner, 1994). In some plant cultivars, PS release is induced only in response to Fe deficiency (Gries et al., 1995), but once the compounds are exuded, they can function in transport of other metals, including Zn and Cu.

The rate of PS exudation from roots (an average exudation rate of 2.9 nmol cm^{-1} root h^{-1}) is positively related to tolerance of different species and genotypes to Fe deficiency (Römheld, 1991), which formed a basis for screening plants for their relative Fe uptake efficiencies (for a review, see Jolley et al., 1996). In general, the broad trend seems to be that plants which release low levels of PS in response to Fe deficiency are adapted poorly to Fe-limiting soils (Gries and Rünge, 1992).

The precursor for biosynthesis of all PS is methionine, which is converted into nicotianamine and then 2'-deoxymugineic acid (DMA) in a series of biosynthetic steps (Mori, 1994). Biosynthesis of DMA is associated with the methionine recycling pathway, which ensures the continued supply of methionine (Ma et al., 1995). From DMA, the biosynthetic pathway may diverge in different plant species (Ma and Nomoto, 1996), resulting in different PS being exuded into the rhizosphere of different species. The PS are synthesized mainly in root tips (Ma and Nomoto, 1996), even though biosynthesis may occur in the meristematic tissue of the shoot as well (Walter et al., 1995). In either case, PS are synthesized continuously and are stored in roots for release into the rhizosphere during a defined period of the day (Ma and Nomoto, 1996).

An increased exudation of PS under Fe deficiency occurs in a distinct diurnal rhythm, with a peak exudation after the onset of illumination, the light ensuring the continuous supply of assimilates from photosynthetically active plant parts (Römheld, 1991). More detailed studies on the diurnal rhythm revealed that it is an increase in temperature during the light period, rather than the onset of light itself, which causes an increase in PS exudation (see Mori, 1994).

Ferrated PS are transported across the plasma membrane of graminaceous species by a transport system that does not carry significant amounts of Zn(II)-PS (Ma and Nomoto, 1993; Ma et al., 1993), but recognizes specifically the stereostructure of the Fe(III)-PS complex (Oida et al., 1989) in preference to similar complexes with Cu(II), Zn(II), Co(II), and Co(III) (Ma et al., 1993). However, these metal cations can decrease the rate of uptake of the Fe(III)-PS complex by competing with Fe(III) for binding to PS in accordance with the stability constants of these metals with PS (Ma and Nomoto, 1993).

It has been well established that the undissociated Fe(III)-PS complex is taken up by corn and rice roots (Von Wiren, Marschner, and Römheld, 1996, and references therein). The uptake of Fe(III)-PS was inhibited by metabolic inhibitors (DCCD or CCCP) and chilling (Mori, 1994), indicating that the transport of the Fe(III)-PS complex across the plasma membrane is an energy-dependent process. Von Wiren, Marschner, and Römheld (1995) found two components of Fe(III)-PS uptake (a saturable, high-affinity component at low concentrations and a linear component at higher concentrations), suggesting that either two transporters are present or that one transporter may assume multiple structural and/or functional forms. Current work in several laboratories around the world is aimed at molecular characterization of the Fe(III)-PS transporter.

Zinc

Exudation of PS from roots increases under Zn deficiency in a range of plant species (Zhang, Römheld, and Marschner, 1989, 1991; Cakmak et al., 1994, 1996; Walter et al., 1994; Rengel, 1997; Rengel, Römheld, and Marschner, 1998). However, an unequivocal experimental proof that PS play a role in mobilization and uptake of Zn from Zn-deficient soils has yet to be reported (see Rengel, Römheld, and Marschner, 1998). This is especially important because PS have a greater affinity for Fe than for Zn (e.g., DMA has a twofold higher stability constant for Fe than for Zn; see Marschner and Römheld, 1994).

Either Zn or Fe deficiency may stimulate production of PS in wheat (Cakmak et al., 1994; Walter et al., 1994; Rengel, Römheld, and Marschner, 1998). However, an increased release of PS under Zn deficiency might be due to an indirect effect, for example, as a response to impaired translocation of Fe from roots to shoots under Zn deficiency (Walter et al., 1994; Rengel and Graham, 1996; Rengel, Römheld, and Marschner, 1998). Such an imbalance in Fe circulation in plants might cause hidden physiological Fe deficiency, resulting in increased PS release.

For wheat, Zhang, Römheld, and Marschner (1991) suggested a model in which PS, released across the plasma membrane, mobilize Zn in the apoplasm of root cells, but dissociation of the Zn(II)-PS complex occurs at the plasma membrane, and only Zn is taken up into the cytoplasm. However, more recent research has indicated that not only splitting of Zn(II)-PS complex at the plasma membrane and uptake of ionic Zn(II) occur but that the Zn(II)-PS complex can also be taken up undissociated, at least by corn roots (Von Wiren, Marschner, and Römheld, 1996). In addition to exudation of PS, Zn deficiency increases root exudation of amino acids, sugars, and phenolics in a range of plant species, including wheat (Zhang, 1993, and references therein). The importance of this exudation has not yet been assessed in terms of increasing plant capacity to acquire Zn from soils with low Zn availability.

Manganese

Environmentally controlled changes in redox occur when oxygen is depleted; NO_3^-, Mn, and Fe then serve as alternative electron acceptors for microbial respiration and are transformed into reduced ionic species. This process greatly increases the solubility and availability of Mn and Fe but is not under the direct control of the plant. In some circumstances, such as in poorly aerated soils, this results in Mn and Fe toxicities to plants. Manganese availability may be further influenced by the activity of

Mn-oxidizing and Mn-reducing bacteria that colonize plant roots (Rengel et al., 1996).

Since differential Mn efficiency can only be demonstrated for plants growing in soil, but not for those growing in the nutrient solution (Huang, Webb, and Graham, 1994), it appears obvious that a change in the biology and/or chemistry of the rhizosphere precedes an increase in Mn availability to plants. However, the nature and activity of root exudate components that might be involved in mobilization of Mn is still unclear. The importance of organic acids (e.g., malic and citric) (Godo and Reisenauer, 1980) remains somewhat blurred because their effectiveness in forming stable complexes with micronutrients is low at high pH (Jones and Darrah, 1994), where Mn deficiency usually occurs. Further research on root exudates effective in mobilizing Mn from the high-pH substrates for uptake by plant roots is warranted.

MICROBIAL INTERACTIONS INFLUENCING PLANT NUTRITION

Microorganisms have numerous direct and indirect effects on plant nutrition, including the direct solubilization of nutrients required for plant growth, enhanced mineralization of organic matter in the rhizosphere, and indirect effects of microorganisms on plant root growth. It has been shown long ago that plant roots inoculated with the rhizosphere microorganisms take up more P from the sparingly soluble Ca phosphates (Gerretsen, 1948). Microorganisms found on the rhizoplane are especially effective, and some strains of *Pseudomonas fluorescens* dissolve a wide range of sparingly soluble inorganic P complexes (Loneragan, 1995, and references therein). Rhizosphere microorganisms can also increase the plant availability of P from the rock phosphate (Förster and Freier, 1988). Other key plant-microbial interactions influencing plant nutrition include symbioses with N_2-fixing bacteria and mycorrhizal fungi that contribute to N nutrition and P uptake, respectively, by plants. Symbiotic N_2-fixation has been the subject of many reviews, and a detailed discussion on this topic is beyond the scope of the present chapter.

Microorganisms supported by rhizodeposition are estimated to have a biomass that is one-third that of the root biomass. During nutrient cycling in the rhizosphere, nutrients immobilized in the microbial biomass are released as waste products from soil animals that feed on roots, bacteria, and fungi. Carbon flow from plant photosynthesis thus sets up a demand for mineral nutrients by the entire soil biological community that requires mineralization of nutrients contained in soil organic matter and soil minerals. It can be speculated that when an ecosystem functions optimally, nutrients are mobilized in relation to plant demand, thus preventing losses by leaching or volatilization.

Growth-Promoting Bacteria

Phosphorus-solubilizing bacteria, and associative N_2-fixing bacteria, have been particularly well investigated for their potential role in plant nutrition. In both cases, it appears that the ability of these bacteria to enhance plant nutrition may be confounded in part by growth-promoting effects related to the release of plant hormones and growth factors. For example, in a recent study with canola *(Brassica napus)*, seven strains of bacteria, selected as the best P-solubilizing bacteria among 120 rhizosphere isolates, were tested for their effects on plant growth and P uptake (De Freitas, Banerjee, and Germida, 1997). Several bacterial isolates increased both plant growth and pod weight, but had no significant effect on P uptake from rock phosphate.

Other data support a more direct role in the contribution of these microorganisms to plant nutrition. Mixed-culture inoculations of barley with the associative N_2 fixer *Azospirillum lipoferum* and the P solubilizer *Agrobacterium radiobacter* have been shown to significantly increase grain yield and N nutrition of plants as compared to inoculation with single cultures (Belimov, Kojemiakov, and Chuvarliyeva, 1995). This appeared to involve a synergistic relationship in which both N_2-fixing activity and rates of P solubilization were increased by the coinoculation. In yet another recent example of the synergistic relationships that may arise in microbial communities, it was shown that *Bacillus polymyxa* stimulated increased *Rhizobium etli* populations and nodulation when coresident in the rhizosphere of bean *(Phaseolus vulgaris)* (Petersen, Srinivasan, and Chanway, 1996). These sorts of data are particularly intriguing in that they suggest that there may be systems-level effects of microbial communities on plant nutrition that are not evident when the rhizosphere community is disassembled and studied in simplified systems with pure cultures.

Associative Nitrogen Fixation

The importance of associative N_2-fixing symbioses has received new attention with the discovery of highly effective symbioses with sugarcane and rice (Boddey et al., 1995). In Brazilian sugar cane cultivars, which have been inadvertently selected for growth in nutrient-limiting soils with low rates of N fertilization, up to 60 percent of the N may be provided by biological N_2 fixation. In sugar cane, the symbionts include endophytic N_2-fixing bacteria that live in the roots, shoots, and leaves of the plant. Both *Acetobacter diazotrophicus* and *Herbaspirillum* spp. are found within roots and aerial tissues and are able to survive and pass from crop to crop in the seeds. However, unlike *Azospirillum* spp. and other rhizosphere diazotrophs, these species survive poorly in soil. *Herbaspirillum* spp. has also been found in several other graminaceous crops, including rice.

Phosphorus-Solubilizing Microorganisms

One of the key mechanisms by which microorganisms may enhance plant nutrition is by causing increased rates of organic matter turnover in the rhizosphere as compared to rates that occur in the bulk soil (Nardi, Reniero, and Concheri, 1997). This is particularly important for the mobilization of P from organic P, which typically constitutes from 30 to 70 percent of the total P in agricultural soils. In terms of increasing availability of organic P to plants, fungi that produce acid phosphatase and bacteria that produce alkaline phosphatase (Tarafdar and Claassen, 1988) may also contribute to P nutrition of plants. However, plant roots themselves may exude acid phosphatase (McLachlan, 1980), as well as a range of other compounds that solubilize inorganic and organic P complexes (for the review, see Pearson and Rengel, 1997), leading to a suggestion that activity of plant roots in solubilizing P complexes may be greater than the effect of the rhizosphere microorganisms (Tinker, 1984).

A number of studies showed that colonization of plant roots with bacteria may increase P uptake (*Azospirillum brasilense* and sorghum or tomato; *Pseudomonas putida* and canola; for references, see Loneragan, 1995). Generally, increased uptake was brought about by increasing root surface area, either through improving root growth or promoting root hair development. In addition, bacterial species from the genera *Bacillus*, *Enterobacter*, *Agrobacter*, *Azospirillum*, and others (e.g., see Banik and Dey, 1985; Bashan et al., 1995) and fungi (such as phosphatase-producing *Aspergillus fumigatus* [Tarafdar and Marschner, 1995] or organic acid-exuding *Penicillium bilaii*, previously reported as *P. bilaji* [Leggett et al., 1993, and references therein]) are known to solubilize P in the rhizosphere. Similarly, the endophyte *Cladorrhinum foecundissimum* increased availability of P in the rhizosphere of cotton plants growing in P-deficient soil, thus increasing uptake of P (Gasoni and De Gurfinkel, 1997). However, despite the claims made that commercial P-solubilizing biofertilizers are effective in increasing P availability in the rhizosphere of crop plants (e.g., Leggett et al., 1993), comprehensive field evaluation of potential contribution of microorganisms to P nutrition of crops has yet to be accomplished (Kloepper, 1993).

One of the problems of using bacterial or fungal biofertilizers is that these microorganisms may not establish readily in the rhizosphere (e.g., Badr El Din, Khalafallah, and Moawad, 1986), despite the increased supply of carbohydrates into the rhizosphere of P-deficient plants (Turner, Newman, and Campbell, 1985). However, *Agrobacterium radiobacter* colonized the rhizoplane of barley effectively, increasing P uptake and grain yield in both pot and field experiments (Belimov, Kojemiakov, and Chuvarliyeva, 1995). Even when good root colonization occurred, upon harvest and removal of colonized plants, populations of biofertilizer microorganisms declined rapidly, reaching undetectable levels after 90 days (Bashan et al., 1995).

Upland rice maintained greater P availability in the rhizosphere than a number of other crops, even though these crops acidified the rhizosphere and rice did not (Ahad, 1993). Even when small drop in pH occurred in the rhizosphere of upland rice cultivars, this could not account for the amount of P solubilized (Hedley, Kirk, and Santos, 1994). Increased activity of phosphatase (the enzyme that converts plant-unavailable organic P forms into plant-available inorganic ones) was relatively small (Tadano and Sakai, 1991) and also could not account for the extent of P solubilization in the rhizosphere. In contrast, there was a positive correlation between the number of P-solubilizing bacteria in the rhizosphere and P uptake, as well as grain yield in rice (Banik and Dey, 1985).

In a recent study examining the potential benefit of plant inoculation with acid and alkaline phosphatase-producing fungi, it was shown that uptake of N, P, Ca, Mg, Fe, and Zn as enhanced for cluster bean (*Cyamopsis tetragonoloba*) when grown in arid zone soil (Tarafdar, Rao, and Praveen-Kumar, 1995). The improvement in N uptake was due to enhanced nodulation and nitrogenase activity after correction of P deficiency. A significant improvement in K concentration, due to inoculation of *Aspergillus fumigatus,* and Cu concentration, with the inoculation of *Aspergillus rugulosus* and *Aspergillus terreus,* was also noted. Although phosphatase-producing fungi are undoubtedly present in most soils already, that manipulation of their population density can be used to enhance plant nutrition is encouraging, in that it points to a possible direction for improved management of soil microbial communities.

Mycorrhizae

The role of mycorrhizal symbioses in plant nutrition has been the subject of several books and recent reviews (Marschner and Dell, 1994; Marschner,

1995a). Some of the most definitive results for studies examining the direct contribution of mycorrhizae to plant nutrition have employed root boxes with net partitions that allow mycorrhizal hyphal access, but not root penetration, to soil compartments containing labeled nutrients. Using these methods, it has been shown that the external hyphae of VAM can deliver up to 80 percent of plant P, 25 percent of plant N, 10 percent of plant K, 25 percent of plant Zn, and 60 percent of plant Cu. Ectomycorrhizal fungi, which colonize conifers, oaks, and a few other tree species, also have significant effects on the host plant nutrition. Most research has focused on the improved P nutrition of mycorrhizal plants. In the forest floor, ectomycorrhizal mats can also release organic acids that provide a local weathering environment and thereby increase the availability of P, S (sulfur), and trace nutrients (Griffiths, Baham, and Caldwell, 1994).

Role of Microorganisms in Iron Nutrition

The interaction of microorganisms with plants in Fe nutrition has been studied for some time, particularly with regard to the possible production of microbial siderophores in the plant rhizosphere (for a review, see Crowley and Gries, 1994). Almost all microorganisms produce siderophores in response to Fe deficiency; over 100 different siderophores have been chemically characterized. Many studies have shown that virtually all tested microbial siderophores can serve as Fe sources for plants, but that excess unchelated siderophores can strongly compete with plants for available Fe and inhibit its uptake. Since siderophores have been speculated to sequester Fe to the detriment of plant pathogens, this has led to the question of whether siderophores might also impede plant Fe uptake. However, current models suggest that siderophores are produced only at microsites of high microbial growth in the rhizosphere and do not accumulate at substantial concentrations in the rhizosphere. Since many siderophores are not readily reduced by the inducible chelate reductase of strategy I plants, it does not appear that they are directly utilized by plants. Instead, uptake of Fe from siderophores may involve chelate degradation and release of Fe (Bar-Ness et al., 1991), uptake of the siderophore-Fe complex (Wang et al., 1993), or ligand exchange of Fe with phytosiderophores (PS) (Yehuda et al., 1996).

Despite much research on the plant use of siderophores added to hydroponic or soil media, experiments examining the actual role of microbial siderophores in Fe uptake by soil-grown plants have cast serious doubt on their ecological relevance for plant nutrition. One of the most important questions has been whether siderophores are produced in high enough quantities to supply ecologically relevant concentrations of Fe to plants. Attempts to quantify siderophore concentrations in rhizosphere soil suggest that they may be present only at nanomolar concentrations when averaged over the entire rhizosphere (Buyer, Kratzke, and Sikora, 1993), which would be too low for providing Fe to plants. Recent studies using an Fe-regulated reporter gene suggested that siderophore production by pseudomonads colonizing the rhizosphere may be restricted to sites of active growth of the bacteria and that the concentration of siderophore produced was too low to be of direct significance for plant nutrition (Marschner and Crowley, 1997). Moreover, when the plant shoots were treated with foliar Fe, siderophore production in the rhizosphere was increased, suggesting that at least some rhizosphere bacteria colonizing the roots of grasses may rely on plant-produced PS as their primary Fe source and produce their own siderophores only to augment Fe that is not provided by the plant.

Microorganisms deleteriously influence Fe nutrition of grasses by the degradation of PS. These microorganisms include bacteria occurring in the rhizosphere soil (Shi et al., 1988), as well as ubiquitous microorganisms found in laboratory nutrient solutions (Crowley et al., 1992). This may be overcome in part by the plant's ability to produce high quantities of PS, which at high concentrations can mobilize apoplasmic Fe (Von Wiren et al., 1995), and by diurnal pulsing of PS concentrations at the root apexes to temporarily saturate the rhizosphere with PS before they can be consumed by microbial degradation (Crowley and Gries, 1994). In experiments comparing the influence of continuous wick irrigation versus flood irrigation, it was shown that plants became chlorotic due to higher rates of PS degradation in the rhizosphere of plants in the flooded treatment (Von Wiren et al., 1993). This was attributed to a higher population of soil microorganisms that were able to move and colonize the root apical zones when the soils were flooded.

It is important to maintain axenic conditions during collection of PS-containing root exudates in order not to underestimate the rate of release. However, since Fe chelates (and probably Zn chelates as well) are more resistant to microbial degradation than the free ligands (Boudot et al., 1989), short-term uptake experiments with metal-DMA were conducted in nonaxenic conditions without a significant loss of the metal-PS complex, indicating negligible microbial decomposition (Von Wiren et al., 1994; Rengel, Römheld, and Marschner, 1998).

Role of Microorganisms in Manganese Nutrition

The significance of the soil microflora in Mn uptake is far from clear. Barber and Lee (1974) found greater uptake of Mn by barley in nonaxenic compared to axenic conditions, while Timonin (1946) reported the opposite findings for oats.

The availability of Mn to plants depends on its oxidation state: the oxidized form (Mn^{4+}) is not available to plants, whereas the reduced form Mn^{2+} is. Oxidation reactions are biological, while reduction can be either biological or chemical in nature (Ghiorse, 1988). Therefore, microorganisms that oxidize Mn will decrease its availability to plants. It is expected that an increased number of Mn reducers and/or a decreased number of Mn oxidizers will contribute to the increased tolerance to Mn deficiency (for a review, see Rengel, 1999).

BIOAUGMENTATION OF THE RHIZOSPHERE

During the past few decades, much research has been done on the use of microbial inoculants to improve plant nutrition, including the use of P-solubilizing bacteria, *Rhizobium* inoculants, and mycorrhizal fungi. The possibility of using genetically modified microorganisms that are responsive to specific root exudate components (Van Overbeek, Van Veen, and Van Elsas, 1997) is another novel area of research with great potential for tailoring plant-microbial systems with improved nutrient uptake efficiency. One of the major challenges for using either natural or genetically engineered bacteria has been the development of inoculation methods that can ensure the uniform distribution of active bacteria in soil able to compete with indigenous microorganisms. This is also a problem for the application of microorganisms for biocontrol of plant disease or for the addition of beneficial microorganisms producing plant growth factors that can enhance root growth and mineral nutrient uptake.

Recently, new technology has been developed that allows the in situ culture of bacteria in automated fermentors in the field, which can then be injected into the irrigation water as continuous batch applications. This precludes many of the problems associated with cost-effective inoculum production and problems with transportation and storage of soil inoculants. By repeatedly inundating the soil with the selected bacteria, it may also aid with breaking competitive exclusion by the indigenous bacteria that already occupy the niche to be utilized by the introduced bacteria. Presently, this technology is being explored for application of biocontrol organisms, but also holds promise for delivery of other soil microorganisms that might be used to bioaugment plant beneficial bacteria to enhance plant nutrition or root growth in problem soils.

Microbially produced plant hormones can alter plant root growth and development, thus increasing plant yield (Arshad and Frankenberger,

1993). Rather than applying expensive purified compounds to soils, a lower-cost approach has been to add precursors such as tryptophan, which can then be converted to auxins by rhizosphere microorganisms. The possibility of manipulating the size and species composition of the rhizosphere microbial community through the addition of vitamins or vitamin precursors to soils or as seed amendments is another recent research area. In one such study, imbibing corn seeds with thiamine resulted in fourfold increases in the population number of associative N_2-fixing bacteria and a twofold increase in total rhizosphere bacteria (Azaizeh, Neumann, and Marschner, 1996).

Effects of Microbial Communities on Plant Nutrition

Roots exert selectivity on microbial populations by changing root exudate composition, by production of growth factors and chemoattractants, and by inhibition of microbes through the production of antagonistic substances. These antagonistic effects may be equally important to the selective enrichment of bacteria and to shaping the composition of the rhizosphere community. For example, it has recently been shown that phenylpropane derivatives such as p-coumaric acid, ferulic acid, and caffeic acid, are taken up by wheat roots and converted to hydroxystyrenes, which are then released in root exudates as potent antimicrobial compounds (Kobayashi, Kim, and Kawazu, 1996). Other examples of compounds that can influence rhizosphere microbial populations are the volatile organic substances that serve as chemoattractants (Linderman and Gilbert, 1975) and S-containing compounds such as isothiocyanate, which have been studied for their ability to suppress certain root pathogens (Lewis and Papavizas, 1970).

In many studies on plant-microbial interactions, the emphasis has been on the effects of individual microorganisms on plant growth or, more recently, on the synergistic interactions that occur in tripartite symbioses involving plants, N_2-fixing bacteria, and mycorrhizae. Along this same line, the possible importance of helper bacteria to improve the survival and activity of root symbionts has been recognized. For example, plant inoculation with so-called mycorrhizal helper bacteria has been shown to be useful for enhancing the development of mycorrhizal roots (Jarstfer and Sylvia, 1993). Interactions between mycorrhizal fungi and phosphatase-producing bacteria also have been shown for certain bacteria-mycorrhizal fungus combinations (Toro, Azcon, and Herrera, 1996), but appear to involve species-specific interactions that would be impossible to predict as a general phenomenon. Together, the previous examples reveal the complex interactions of microorganisms in communities that have system-level, emergent properties that may never be dissected using a reductionist experimental approach. One starting point for future studies would be to compare microbial communities associated with native plants that are adapted to problem soils to determine to what extent they differ from those which develop on nonadapted plants.

To date, most studies on the diversity and community structure of microorganisms associated with plant roots have relied on plating of microorganisms on agar media. Early research suggested that only a few genera of bacteria were dominant on plant roots, including *Pseudomonas*, *Arthrobacter*, *Enterobacter*, and *Bacillus*. In more recent studies using a specifically designed rhizosphere bacteria isolation medium (Buyer, 1995), twenty-four genera and forty-one species were identified as common rhizosphere isolates. Genetic methods that do not require culture of microorganisms on agar plates can also be used to characterize microbial communities, but are still being developed for soil microbial ecology. Using 16sRNA sequences that allow identification of broad taxonomic groups or specific genera, the community structure can be characterized in relation to specific groups of bacteria. Other available methods employ DNA banding patterns that are generated using selected PCR primers (e.g., see Borneman and Triplett, 1997). Numerous difficulties can be anticipated in taking this approach, such as determination of which root zones should be sampled or at what stage of plant development the rhizosphere should be sampled. It will also be necessary to use microsite sampling techniques to characterize root exudate composition in these different zones to establish a correlation between root exudation and microbial community composition.

CONCLUSIONS

The importance of root exudates and soil microorganisms in plant nutrition is well established and may explain, at least in part, some of the differences in nutrient uptake efficiency by various plant species and cultivars. Many relevant traits appear to involve quantitative changes in plant responses to nutrient stress, involving, for example, upregulation of acidification, release of organic acids and PS, or increased secretion of phosphatases. The increasing availability of chromosomal maps for many plants now makes it possible to construct gene maps for plant traits that influence the availability of nutrients in the rhizosphere and also to screen for these traits using genetic techniques.

In contrast, factors that influence plant-beneficial, microbial interactions, such as regulation of mycorrhizal symbioses, endophytic N_2-fixing bacteria, and production of plant growth hormones by microorganisms, are barely understood. The future application of knowledge linking rhizosphere ecology to plant nutrition will require coordinated interdisciplinary research involving plant physiologists, microbial ecologists, and plant breeders to identify and select traits that are associated with specific beneficial interactions. Because of the importance of microbial interactions within communities, this will almost certainly involve a systems-level approach toward the study of microbial communities that are associated with nutrient-efficient plants.

In sustainable agriculture, it will always be necessary to replenish nutrients that are removed by cropping. Thus, the goal of developing plant-microbial systems that are efficient at taking up nutrients is not to eliminate the need for fertilizers, but to permit the most efficient use of fertilizer materials and to adapt plants to soils where nutrient availability is decreased as a result of adverse soil chemical conditions (Buol and Eswaran, 1994). Microorganisms that colonize plant roots are adapted to the soils in which they are found and have devised mechanisms for acquiring nutrients from soil minerals and from organic matter. It may also be possible to bioaugment soils with plant-beneficial microorganisms, or even to tailor plants that secrete exudates to support beneficial microorganisms that are introduced into soils. However, the least-intensive, most easily managed approach will be to manipulate indigenous soil microbial communities.

Analytical methods now exist that are capable of complete analysis of root exudate components (Fan et al., 1997) and of probing the details of rhizosphere chemistry. New genetic methods also are available to study plant-microbial interactions in detail at specific locations in the rhizosphere. The challenge will be to combine these methods to study these rhizosphere processes and identify important traits of plant-microbial systems that function for nutrient acquisition. In addition to this traditional reductionist approach that dissects individual processes is the need to consider overall system properties that function for enhancing root growth and plant nutrition. Eventually, a better knowledge of root exudates and rhizosphere ecology may be used to select or design more nutrient-efficient plant cultivars, as well as for the development of cultural practices that promote beneficial plant-microbial interactions.

REFERENCES

Ae, N., J. Arihara, K. Okada, T. Yoshihara, and C. Johansen (1990). Phosphorus uptake by pigeon pea and its role in cropping systems of the Indian Subcontinent. *Science* 248: 477-480.

Ahad, M.A. (1993). Phosphorus availability and pH changes in the rhizosphere of rice, maize, soybean and jute. *Indian Journal of Agricultural Research* 27: 51-56.

Arshad, M. and W.T. Frankenberger (1993). Microbial production of plant growth regulators. In *Soil Microbial Ecology: Applications in Agricultural and Environmental Management*, Ed. F.B. Metting. New York: Marcel Dekker, Inc., pp. 307-348.

Awad, F., V. Römheld, and H. Marschner (1994). Effect of root exudates on mobilization in the rhizosphere and uptake of iron by wheat plants. *Plant and Soil* 165: 213-218.

Azaizeh, H.A., G. Neumann, and H. Marschner (1996). Effects of thiamine and nitrogen fertilizer form on the number of N_2-fixing and total bacteria in the rhizosphere of maize plants. *Zeitschrift für Pflanzenernährung und Bodenkunde* 159: 183-188.

Azarabadi, S. and H. Marschner (1979). Role of the rhizosphere in utilization of inorganic iron III compounds by corn plants. *Zeitschrift für Pflanzenernährung und Bodenkunde* 142: 751-764.

Badr El Din, S.M.S., M.A. Khalafallah, and H. Moawad (1986). Response of soybean to dual inoculation with *Rhizobium japonicum* and phosphate dissolving bacteria. *Zeitschrift für Pflanzenernährung und Bodenkunde* 149: 130-135.

Banik, S. and B.K. Dey (1985). Effect of inoculation with native phosphate solubilizing microorganisms on the available phosphorus content in the rhizosphere and uptake of phosphorus by rice *Oryza sativa* cultivar IR-8 plants grown in an Indian alluvial soil. *Zentralblatt für Mikrobiologie* 140: 455-464.

Barber, D.A. and R.B. Lee (1974). The effect of micro-organisms on the absorption of manganese by plants. *New Phytologist* 73: 97-106.

Bar-Ness, E., Y. Chen, Y. Hadar, and H. Marschner (1991). Siderophores of *Pseudomonas putida* as an iron source for dicot and monocot plants. *Plant and Soil* 130: 231-241.

Bashan, Y., M.E. Puente, M.N. Rodriguez Mendoza, G. Toledo, G. Holguin, R. Ferrera Cerrato, and S. Pedrin (1995). Survival of *Azospirillum brasilense* in the bulk soil and rhizosphere of 23 soil types. *Applied and Environmental Microbiology* 61: 1938-1945.

Belimov, A.A., A.P. Kojemiakov, and C.V. Chuvarliyeva (1995). Interaction between barley and mixed cultures of nitrogen-fixing and phosphate-solubilizing bacteria. *Plant and Soil* 173: 29-37.

Bienfait, H.F. (1988). Prevention of stress in iron metabolism of plants. *Acta Botanica Neerlandica* 38: 105-129.

Bishop, M.L., A.C. Chang, and R.W.K. Lee (1994). Enzymatic mineralization of organic phosphorus in a volcanic soil in Chile. *Soil Science* 157: 238-243.

Boddey, R.M., O.C. De Oliveira, S. Urquiaga, V.M. Reis, F.L. De Olivares, V.L.D. Baldani, and J. Döbereiner (1995). Biological nitrogen fixation associated with sugar cane and rice: Contributions and prospects for improvement. *Plant and Soil* 174: 195-209.

Bolan, N.D., J. Elliott, P.E.H. Gregg, and S. Weil (1997). Enhanced dissolution of phosphate rocks in the rhizosphere. *Biology and Fertility of Soils* 24: 169-174.

Borneman, J. and E.W. Triplett (1997). Molecular microbial diversity in soils from eastern Amazonia: Evidence for unusual microorganisms and microbial population shifts associated with deforestation. *Applied and Environmental Microbiology* 63: 2647-2653.

Boudot, J.P., A.B.H. Brahim, R. Steiman, and F. Seigle-Murandi (1989). Biodegradation of synthetic organo-metallic complexes of iron and aluminium with selected metal to carbon ratios. *Soil Biology and Biochemistry* 21: 961-966.

Brancadoro, L., G. Rabotti, A. Scienza, and G. Zocchi (1995). Mechanisms of Fe-efficiency in roots of *Vitis* spp. in response to iron deficiency stress. *Plant and Soil* 171: 229-234.

Buol, S.W. and H. Eswaran (1994). Assessment and conquest of poor soils. In *Proceedings of the Workshop on Adaptation of Plants to Soil Stresses.* INTSORMIL Pub. No. 94-2, pp. 17-27.

Buyer, J.S. (1995). A soil and rhizosphere microorganism isolation and enumeration medium that inhibits *Bacillus mycoides. Applied and Environmental Microbiology* 61: 1839-1842.

Buyer, J.S., M.G. Kratzke, and L.J. Sikora (1993). A method for detection of pseudobactin, the siderophore produced by a plant growth promoting *Pseudomonas* strain, in the barley rhizosphere. *Applied and Environmental Microbiology* 59: 677-681.

Cakmak, I., K.Y. Gülüt, H. Marschner, and R.D. Graham (1994). Effect of zinc and iron deficiency on phytosiderophore release in wheat genotypes differing in zinc efficiency. *Journal of Plant Nutrition* 17: 1-17.

Cakmak, I., L. Özturk, S. Karanlik, H. Marschner, and H. Ekiz (1996). Zinc-efficient wild grasses enhance release of phytosiderophores under zinc deficiency. *Journal of Plant Nutrition* 19: 551-563.

Chaney, R.L. and P.F. Bell (1987). Complexity of iron nutrition: Lessons from plant-soil interaction research. *Journal of Plant Nutrition* 10: 963-994.

Chapin, F.S. (1987). Adaptation and physiological responses of wild plants to nutrient stress. In *Genetic Aspects of Plant Mineral Nutrition*, Eds. W.H. Gabelman and B.C. Loughman. Dordrecht, Netherlands: Martinus Nijhoff Publishers, pp. 15-25.

Cheng, W., D.C. Coleman, R. Carroll, and C.A. Hoffman (1994). Investigating short-term carbon flows in the rhizospheres of different plant species using isotopic trapping. *Agronomy Journal* 86: 782-788.

Christensen, H. and I. Jakobsen (1993). Reduction of bacterial growth by a vesicular-arbuscular mycorrhizal fungus in the rhizosphere of cucumber *Cucumis sativus* L. *Biology and Fertility of Soils* 15: 253-258.

Cieslinski, G., K.C.J. Van Rees, A.M. Szmigielska, and P.M. Huang (1997). Low molecular weight organic acids released from roots of durum wheat and flax into sterile nutrient solutions. *Journal of Plant Nutrition* 20: 753-764.

Cohen, C.K., W.A. Norvell, and L.V. Kochian (1997). Induction of the root cell plasma membrane ferric reductase: An exclusive role for Fe and Cu. *Plant Physiology* 114: 1061-1069.

Crowley, D.E., and D. Gries (1994). Modeling of iron availability in the plant rhizosphere. In *Biochemistry of Metal Micronutrients in the Rhizosphere*, Eds.

J.A. Manthey, D.E. Crowley, and D.G. Luster. Boca Raton, FL: Lewis Publishers/CRC Press, pp. 199-223.

Crowley, D.E., V. Römheld, H. Marschner, and P.J. Szaniszlo (1992). Root-microbial effects on plant iron uptake from siderophores and phytosiderophores. *Plant and Soil* 142: 1-7.

Curl, E.A. and B. Truelove (1986). *The Rhizosphere*. Berlin, Germany: Springer Verlag.

Darrah, P.R. (1993). The rhizosphere and plant nutrition: A quantitative approach. *Plant and Soil* 155/156: 1-20.

De Freitas, J.R., M.R. Banerjee, and J.J. Germida (1997). Phosphate-solubilizing rhizobacteria enhance the growth and yield but not phosphorus uptake of canola (*Brassica napus* L.). *Biology and Fertility of Soils* 24: 358-364.

Delhaize, E. (1995). Genetic control and manipulation of root exudates. In *Genetic Manipulation of Crop Plants to Enhance Integrated Nutrient Management in Cropping Systems. 1. Phosphorus*. Proceedings of an FAO/ICRISAT Expert Consultancy Workshop, March 15-18, 1994, Eds. C. Johansen, K.K. Lee, K.K. Sharma, G.V. Subbarao, and E.A. Kueneman. Patancheru, Andhra Pradesh, India: International Crops Research Institute for the Semi-Arid Tropics, pp. 145-152.

Dinkelaker, B.E., C. Hengeler, and H. Marschner (1995). Distribution and function of proteoid roots and other root clusters. *Botanica Acta* 108: 183-200.

Dinkelaker, B.E., V. Römheld, and H. Marschner (1989). Citric acid excretion and precipitation of calcium citrate in the rhizosphere of white lupin (*Lupinus albus* L.). *Plant, Cell and Environment* 12: 285-292.

Duncan, R.R. and V.C. Baligar (1990). Genetics, breeding, and physiological mechanisms of nutrient uptake and use efficiency: An overview. In *Crops As Enhancers of Nutrient Use*, Eds. V.C. Baligar and R.R. Duncan. San Diego, CA: Academic Press, pp. 3-36.

Ernst, W.H.O. (1996). Bioavailability of heavy metals and decontamination of soils by plants. *Applied Geochemistry* 11: 163-167.

Fan, T.W-M., A.N. Lane, J. Pedler, D.E. Crowley, and R.M. Higashi (1997). Comprehensive analysis of organic ligands in whole root exudates using nuclear magnetic resonance and gas chromatography-mass spectrometry. *Analytical Biochemistry* 251: 57-68.

Förster, I. and K. Freier (1988). Contributions to the mobilization of phosphorus by soil microorganisms: Investigation of the efficiency of phosphorus-mobilizing microrganisms *in vitro* and in the rhizosphere of sunflower and winter wheat. *Zentralblatt für Mikrobiologie* 143: 125-138.

Gahoonia, T.S., N. Claassen, and A. Jungk (1992). Mobilization of residual phosphate of different phosphate fertilizers in relation to pH in the rhizosphere of ryegrass. *Fertilizer Research* 33: 229-237.

Gardner, W.K., D.A. Barber, and D.G. Parbery (1983). Non-infecting rhizosphere micro-organisms and the mineral nutrition of temperate cereals. *Journal of Plant Nutrition* 6: 185-199.

Gasoni, L. and B.S. De Gurfinkel (1997). The endophyte *Cladorrhinum foecundissimum* in cotton roots: Phosphorus uptake and host growth. *Mycological Research* 101: 867-870.

Gerke, J. (1992). Phosphate, aluminum and iron in the soil solution of three different soils in relation to varying concentrations of citric acid. *Zeitschrift für Pflanzenernährung und Bodenkunde* 155: 339-343.

Gerretsen, F.C. (1948). The influence of microorganisms on the phosphate intake by plants. *Plant and Soil* 1: 51-81.

Ghiorse, W.C. (1988). The biology of manganese transforming microorganisms in soils. In *Manganese in Soils and Plants*, Eds. R.D. Graham, R.J. Hannam, and N.C. Uren. Dordrecht, Netherlands: Kluwer Academic Publishers, pp. 75-85.

Gillespie, A.R. and P.E. Pope (1990). Rhizosphere acidification increases phosphorus recovery of black locust. II. Model predictions and measured recovery. *Soil Science Society of America Journal* 54: 538-541.

Godo, G.H. and H.M. Reisenauer (1980). Plant effects on soil manganese availability. *Soil Science Society of America Journal* 44: 993-995.

Gollany, H.T. and T.E. Schumacher (1993). Combined use of colorimetric and microelectrode methods for evaluating rhizosphere pH. *Plant and Soil* 154: 151-159.

Grierson, P.F. (1992). Organic acids in the rhizosphere of *Banksia integrifolia* L. *Plant and Soil* 144: 259-265.

Gries, D., S. Brunn, D.E. Crowley, and D.R. Parker (1995). Phytosiderophore release in relation to micronutrient metal deficiencies in barley. *Plant and Soil* 172: 299-308.

Gries, D. and M. Runge (1992). The ecological significance of iron mobilization in wildgrasses. *Journal of Plant Nutrition* 15: 1727-1737.

Griffiths, R.P., J.E. Baham, and B.A. Caldwell (1994). Soil solution chemistry of ectomycorrhizal mats in forest soil. *Soil Biology and Biochemistry* 26: 331-337.

Hedley, M.J., G.J.D. Kirk, and M.B. Santos (1994). Phosphorus efficiency and the forms of soil phosphorus utilized by upland rice cultivars. *Plant and Soil* 158: 53-62.

Helal, H.M. and A. Dressler (1989). Mobilization and turnover of soil phosphorus in the rhizosphere. *Zeitschrift für Pflanzenernährung und Bodenkunde* 152: 175-180.

Hodge, A., S.J. Grayston, and B.G. Ord (1996). A novel method for characterization and quantification of plant root exudates. *Plant and Soil* 184: 97-104.

Hoffland, E., G.R. Findenegg, and J.A. Nelemans (1989). Solubilization of rock phosphate by rape. II. Local root exudation of organic acids as a response to phosphorus starvation. *Plant and Soil* 113: 161-166.

Hoffland, E., R. Van Den Boogaard, J. Nelemans, and G. Findenegg (1992). Biosynthesis and root exudation of citric and malic acids in phosphate-starved rape plants. *New Phytologist* 122: 675-680.

Hoffmann, C., E. Ladewig, N. Claassen, and A. Jungk (1994). Phosphorus uptake of maize as affected by ammonium and nitrate nitrogen—Measurements and model calculations. *Zeitschrift für Pflanzenernährung und Bodenkunde* 157: 225-232.

Huang, C., M.J. Webb, and R.D. Graham (1994). Manganese efficiency is expressed in barley growing in soil system but not in a solution culture. *Journal of Plant Nutrition* 17: 83-95.

Jarstfer, A.G. and D.M. Sylvia (1993). Inoculum production and inoculation strategies for vesicular-arbuscular mycorrhizal fungi. In *Soil Microbial Ecology: Applications in Agricultural and Environmental Management*, Ed. F.B. Metting. New York: Marcel Dekker, Inc., pp. 349-378.

Johnson, J.F., D.L. Allan, and C.P. Vance (1994). Phosphorus stress-induced proteoid roots show altered metabolism in *Lupinus albus*. *Plant Physiology* 104: 657-665.

Johnson, J.F., D.L. Allan, C.P. Vance, and G. Weiblen (1996). Root carbon dioxide fixation by phosphorus-deficient *Lupinus albus*. *Plant Physiology* 112: 19-30.

Jolley, V.D., K.A. Cook, N.C. Hansen, and W.B. Stevens (1996). Plant physiological responses for genotypic evaluation of iron efficiency in strategy I and strategy II plants. A review. *Journal of Plant Nutrition* 19: 1241-1255.

Jones, D.L. and P.R. Darrah (1993). Influx and efflux of amino acids from *Zea mays* L. roots and their implications for N nutrition and the rhizosphere. *Plant and Soil* 155: 87-90.

Jones, D.L. and P.R. Darrah (1994). Amino-acid influx at the soil-root interface of *Zea mays* L. and its implications in the rhizosphere. *Plant and Soil* 163: 1-12.

Jones, D.L. and P.R. Darrah (1995). Influx and efflux of organic acids across the soil-root interface of *Zea mays* L. and its implications in rhizosphere C flow. *Plant and Soil* 173: 103-109.

Jones, D.L. and P.R. Darrah (1996). Re-sorption of organic compounds by roots of *Zea mays* L. and its consequences in the rhizosphere. III. Characteristics of sugar influx and efflux. *Plant and Soil* 178: 153-160.

Jones, D.L., P.R. Darrah, and L.V. Kochian (1996). Critical evaluation of organic acid mediated iron dissolution in the rhizosphere and its potential role in root iron uptake. *Plant and Soil* 180: 57-66.

Jones, D.L., A.C. Edwards, K. Donachie, and P.R. Darrah (1994). Role of proteinaceous amino acids released in root exudates in nutrient acquisition from the rhizosphere. *Plant and Soil* 158: 183-192.

Jones, D.L, A.M. Prabowo, and L.V. Kochian (1996). Kinetics of malate transport and decomposition in acid soils and isolated bacterial populations: The effect of microorganisms on root exudation of malate under Al stress. *Plant and Soil* 182: 239-247.

Jungk, A., B. Seeling, and J. Gerke (1993). Mobilization of different phosphate fractions in the rhizosphere. *Plant and Soil* 155/156: 91-94.

Kirk, J.G.D. and L.V. Du (1997). Changes in rice root architecture, porosity, and oxygen and proton release under phosphorus deficiency. *New Phytologist* 135: 191-200.

Kloepper, J.W. (1993). Plant growth-promoting rhizobacteria as biological control agents. In *Soil Microbial Ecology: Applications in Agricultural and Environ-*

mental Management. Ed. F.B. Metting. New York: Marcel Dekker, Inc., pp. 255-274.

Kobayashi, A., M.J. Kim, and K. Kawazu (1996). Uptake and exudation of phenolic compounds by wheat and antimicrobial components of the root exudate. *Zeitschrift für Naturforschung, Section C, Journal of Biosciences* 51: 527-533.

Kraffczyk, I., G. Trolldenier, and H. Beringer (1984). Soluble root exudates of maize *Zea mays:* Influence of potassium supply and rhizosphere microorganisms. *Soil Biology and Biochemistry* 16: 315-322.

Landsberg, E.C. (1981). Organic acid synthesis and release of hydrogen ions in response to Fe-deficiency stress of mono- and dicotyledonous plant species. *Journal of Plant Nutrition* 3: 579-591.

Leggett, M.E., S.C. Gleddie, E. Prusinkiewicz, and J. Xie (1993). Development of guidelines for the use of PROVIDE™ on wheat and canola. In *Plant Nutrition—From Genetic Engineering to Field Practice*, Ed. N.J. Barrow. Dordrecht, Netherlands: Kluwer Academic Publishers, pp. 375-378.

Lewis, J.A. and G.C. Papavizas (1970). Effect of sulfur-containing volatile compounds and vapors from decomposition of crucifers in soil. *Soil Biology and Biochemistry* 2: 239-246.

Li, M., T. Shinano, and T. Tadano (1997). Distribution of exudates of lupin roots in the rhizosphere under phosphorus deficient conditions. *Soil Science and Plant Nutrition* 43: 237-245.

Li, M. and T. Tadano (1996). Comparison of characteristics of acid phosphatases secreted from roots of lupin and tomato. *Soil Science and Plant Nutrition* 42: 753-763.

Linderman, R.G. and R.G. Gilbert (1975). Influence of volatile substances of plant origin on soil-borne plant pathogens. In *Biology and Control of Soil-Borne Plant Pathogens*, Ed. G.W. Bruehl. St. Paul, MN: American Society of Phytopathology, pp. 90-99.

Loneragan, J.F. (1995). The role of rhizosphere microorganisms in influencing phosphorus uptake, and prospects for favorable manipulation. In *Genetic Manipulation of Crop Plants to Enhance Integrated Nutrient Management in Cropping Systems. 1. Phosphorus.* Proceedings of an FAO/ICRISAT Expert Consultancy Workshop, March 15-18, 1994. Eds. C. Johansen, K.K. Lee, K.K. Sharma, G.V. Subbarao, and E.A. Kueneman. Patancheru, Andhra Pradesh, India: International Crops Research Institute for the Semi-Arid Tropics, pp. 75-78.

Lynch, J.M. and J.M. Whipps (1990). Substrate flow in the rhizosphere. *Plant and Soil* 129: 1-10.

Ma, J.F., G. Kusano, S. Kimura, and K. Nomoto (1993). Specific recognition of mugineic acid-ferric complex by barley roots. *Phytochemistry* 34: 599-603.

Ma, J.F. and K. Nomoto (1993). Inhibition of mugineic acid-ferric complex uptake in barley by copper, zinc and cobalt. *Physiologia Plantarum* 89: 331-334.

Ma, J.F. and K. Nomoto (1996). Effective regulation of iron acquisition in graminaceous plants. The role of mugineic acids as phytosiderophores. *Physiologia Plantarum* 97: 609-617.

Ma, J.F., T. Shinada, C. Matsuda, and K. Nomoto (1995). Biosynthesis of phytosiderophores, mugineic acids, associated with methionine cycling. *Journal of Biological Chemistry* 270: 16549-16554.

Marschner, H. (1995a). *Mineral Nutrition of Higher Plants*, Second Edition. London, UK: Academic Press.

Marschner, H. (1995b). Rhizosphere pH effects on phosphorus nutrition. In *Genetic Manipulation of Crop Plants to Enhance Integrated Nutrient Management in Cropping Systems. 1. Phosphorus.* Proceedings of an FAO/ICRISAT Expert Consultancy Workshop, March 15-18, 1994, Eds. C. Johansen, K.K. Lee, K.K. Sharma, G.V. Subbarao, and E.A. Kueneman. Patancheru, Andhra Pradesh, India: International Crops Research Institute for the Semi-Arid Tropics, pp. 107-115.

Marschner, P. and D.E. Crowley (1997). Iron stress and pyoverdin production by a fluorescent pseudomonad in the rhizosphere of white lupin (*Lupinus albus*) and barley (*Hordeum vulgare* L.). *Applied and Environmental Microbiology* 63: 277-281.

Marschner, P., D.E. Crowley, and R.M. Higashi (1997). Root exudation and physiological status of a root-colonizing pseudomonad in mycorrhizal and nonmycorrhizal pepper (*Capsicum annum* L.). *Plant and Soil* 189: 11-20.

Marschner, H. and B. Dell (1994). Nutrient uptake in mycorrhizal symbiosis. *Plant and Soil* 159: 89-102.

Marschner, H. and V. Römheld (1983). *In vivo* measurement of root induced pH changes at the soil root interface: Effect of plant species and nitrogen source. *Zeitschrift für Pflanzenphysiologie* 111: 241-251.

Marschner, H. and V. Römheld (1994). Strategies of plants for acquisition of iron. *Plant and Soil* 165: 261-274.

Marschner, H., V. Römheld, and M. Kissel (1986). Different strategies in higher plants in mobilization and uptake of iron. *Journal of Plant Nutrition* 9: 695-713.

Masaoka, Y., M. Kojima, S. Sugihara, T. Yoshikara, M. Koshino, and A. Ichikara (1993). Dissolution of ferric phosphate by alfalfa (*Medicago sativa* L.) root exudates. *Plant and Soil* 155/156: 75-78.

Matar, A.E., J.L. Paul, and H. Jenny (1967). Two phase experiments with plants growing in phosphate-treated soil. *Soil Science Society of America Proceedings* 31: 235-237.

McLachlan, K.D. (1980). Acid phosphatase activity of intact roots and phosphorus nutrition in plants. I. Assay conditions and phosphatase activity. *Australian Journal of Agricultural Research* 31: 429-440.

Meharg, A.A. (1994). A critical review of labelling techniques used to quantify rhizosphere carbon-flow. *Plant and Soil* 166: 55-62.

Meharg, A.A. and K. Killham (1995). Loss of exudates from the roots of perennial ryegrass inoculated with a range of microorganisms. *Plant and Soil* 170: 345-349.

Mench, M. and E. Martin (1991). Mobilization of cadmium and other metals from two soils by root exudates of *Zea mays* L., *Nicotiana tabacum* L. and *Nicotiana rustica* L. *Plant and Soil* 132: 187-196.

Morel, J.L., M. Mench, and A. Guckert (1986). Measurement of Pb^{2+}, Cu^{2+}, and Cd^{2+} binding with mucilage exudates from maize (*Zea mays* L.) roots. *Biology and Fertility of Soils* 2: 886-893.

Mori, S. (1994). Mechanisms of iron acquisition by Graminaceous (strategy II) plants. In *Biochemistry of Metal Micronutrients in the Rhizosphere*, Eds. J.A. Manthey, D.E. Crowley, and D.G. Luster. Boca Raton, FL: Lewis Publishers/ CRC Press, pp. 225-250.

Nagarajah, S., A.M. Posner, and J.P. Quirk (1970). Competitive adsorption of phosphate with polygalacturonate and other organic anions on kaolinite and oxide surfaces. *Nature* 228: 83-85.

Nambiar, E.K.S. (1976). Uptake of 65Zn from dry soil by plants. *Plant and Soil* 44: 267-271.

Nardi, S., F. Reniero, and G. Concheri (1997). Soil organic matter mobilization by root exudates of three maize hybrids. *Chemosphere* 362: 2237-2244.

O'Connell, K.P., R.M. Goodman, and J. Handelsman (1996). Engineering the rhizosphere: Expressing a bias. *Trends in Biotechnology* 14: 83-88.

Oida, F., N. Ota, Y. Mino, K. Nomoto, and Y. Sugiura (1989). Stereospecific iron uptake mediated by phytosiderophore in gramineous plants. *Journal of the American Chemical Society* 111. 3436-3437.

Olsen, R.A., J.H. Bennett, D. Blume, and J.C. Brown (1981). Chemical aspects of the Fe stress response mechanism in tomatoes. *Journal of Plant Nutrition* 3: 905-921.

Otani, T., N. Ae, and H. Tanaka (1996). Phosphorus (P) uptake mechanisms of crops grown in soils with low P status. II. Significance of organic acids in root exudates of pigeonpea. *Soil Science and Plant Nutrition* 42: 553-560.

Pearson, N.J. and Z. Rengel (1997). Mechanisms of plant resistance to nutrient deficiency stresses. In *Mechanisms of Environmental Stress Resistance in Plants*, Eds. A.S. Basra and R.K. Basra. Amsterdam, Netherlands: Harwood Academic Publishers, pp. 213-240.

Pepper, I.L. and D.F. Bezdicek (1990). Root microbial interactions and rhizosphere nutrient dynamics. In *Crops As Enhancers of Nutrient Use*, Eds. V.C. Baligar and R.R. Duncan. San Diego, CA: Academic Press, pp. 375-410.

Petersen, D.J., M. Srinivasan, and C.P. Chanway (1996). *Bacillus polymyxa* stimulates increased *Rhizobium etli* populations and nodulation when co-resident in the rhizosphere of *Phaseolus vulgaris*. *FEMS Microbiology Letters* 142: 271-276.

Rabotti, G. and G. Zocchi (1994). Plasma membrane-bound H^+-ATPase and reductase activities in Fe-deficient cucumber roots. *Physiologia Plantarum* 90: 779-785.

Rengel, Z. (1997). Root exudation and microflora populations in rhizosphere of crop genotypes differing in tolerance to micronutrient deficiency. *Plant and Soil* 196: 255-260.

Rengel, Z. (1999). Physiological mechanisms underlying differential nutrient efficiency of crop genotypes. In *Mineral Nutrition of Crops: Fundamental Mecha-*

nisms and Implications, Ed. Z. Rengel. Binghamton, NY: The Haworth Press, Inc., pp. 227-265.

Rengel, Z. and R.D. Graham (1996). Uptake of zinc from chelate-buffered nutrient solutions by wheat genotypes differing in Zn efficiency. *Journal of Experimental Botany* 47: 217-226.

Rengel, Z., R. Gutteridge, P. Hirsch, and D. Hornby (1996). Plant genotype, micronutrient fertilization and take-all infection influence bacterial populations in the rhizosphere of wheat. *Plant and Soil* 183: 269-277.

Rengel, Z., V. Römheld, and H. Marschner (1998). Uptake of zinc and iron by wheat genotypes differing in zinc efficiency. *Journal of Plant Physiology* 152: 433-438.

Romera, F.J., E. Alcantara, and M.D. De la Guardia (1992). Effects of bicarbonate, phosphate and high pH on the reducing capacity of Fe-deficient sunflower and cucumber plants. *Journal of Plant Nutrition* 15: 1519-1530.

Römheld, V. (1991). The role of phytosiderophores in acquisition of iron and other micronutrients in graminaceous species: An ecological approach. *Plant and Soil* 130: 127-134.

Römheld, V. and H. Marschner (1986). Mobilization of iron in the rhizosphere of different plant species. *Advances in Plant Nutrition* 2: 155-204.

Rovira, A.D. (1979). Biology of the soil-root interface. In *The Soil-Root Interface*, Eds. J.L. Harley and R.S. Russell. London, UK: Academic Press, pp. 145-160.

Sakai, H. and T. Tadano (1993). Characteristics of response of acid phosphatase secreted by the roots of several crops to various conditions in the growth media. *Soil Science and Plant Nutrition* 39: 437-444.

Schaffert, R.E. (1994). Discipline interactions in the quest to adapt plants to soil stresses through nutrient improvement. In *Proceedings of the Workshop on Adaptation of Plants to Soil Stresses*. INTSORMIL Pub. No. 94-2, pp. 1-16.

Schwertmann, U. (1991). Solubility and dissolution of iron oxides. In *Iron Nutrition and Interactions in Plants*, Eds. Y. Chen and Y. Hadar. Dordrecht, Netherlands: Kluwer Academic Publishers, pp. 3-28.

Shepherd, T. and H.V. Davies (1993). Carbon loss from the roots of forage rape *Brassica napus* L. seedlings following pulse-labelling with carbon-14 labelled carbon dioxide. *Annals of Botany* 72: 155-163.

Shi, W.M., M. Chino, R.A. Youssef, S. Mori, and S. Takagi (1988). The occurrence of mugineic acid in the rhizosphere soil for barley plant. *Soil Science and Plant Nutrition* 34: 585-592.

Simons, M., H.P. Permentier, L.A. De Weger, C.A. Wijffelman, and B.J.J. Lugtenberg (1997). Amino acid synthesis is necessary for tomato root colonization by *Pseudomonas fluorescens* strain WCS365. *Molecular Plant-Microbe Interactions* 10: 102-106.

Sundin, P., A. Valeur, S. Olsson, and G. Odham (1990). Interactions between bacteria-feeding nematodes and bacteria in the rape rhizosphere: Effects on root exudation and distribution of bacteria. *FEMS Microbiology and Ecology* 73: 13-22.

Tadano, T. and H. Sakai (1991). Secretion of acid phosphatase by roots of several crop species under phosphorus deficient conditions. *Soil Science and Plant Nutrition* 37: 129-140.

Tagliavini, M., A. Masia, and M. Quartieri (1995). Bulk soil pH and rhizosphere pH of peach trees in calcareous and alkaline soils as affected by the form of nitrogen fertilizers. *Plant and Soil* 176: 263-271.

Tarafdar, J.C. and N. Claassen (1988). Organic phosphorus compounds as a phosphorus source of higher plants through the activity of phosphatase produced by plant roots and microorganisms. *Biology and Fertility of Soils* 5: 308-312.

Tarafdar, J.C. and A. Jungk (1987). Phosphatase activity in the rhizosphere and its relation to the depletion of soil organic phosphorus. *Biology and Fertility of Soils* 3: 199-204.

Tarafdar, J.C. and H. Marschner (1994). Phosphatase activity in the rhizosphere and hyposphere of VA mycorrhizal wheat supplied with inorganic and organic phosphorus. *Soil Biology and Biochemistry* 26: 387-395.

Tarafdar, J.C. and H. Marschner (1995). Dual inoculation with *Aspergillus fumigatus* and *Glomus mosseae* enhances biomass production and nutrient uptake in wheat (*Triticum aestivum* L.) supplied with organic phosphorus as Na-phytate. *Plant and Soil* 173: 97-102.

Tarafdar, J.C., A.V. Rao, and Praveen-Kumar (1995). Role of phosphatase-producing fungi on the growth and nutrition of clusterbean (*Cyamopsis tetragonoloba* (L.) Taub.). *Journal of Arid Environment* 29: 331-337.

Thibaud, J-B. (1994). The role of root apoplast acidification by the H^+ pump in mineral nutrition of terrestrial plants. In *Biochemistry of Metal Micronutrients in the Rhizosphere*, Eds. J.A. Manthey, D.E. Crowley, and D.G. Luster. Boca Raton, FL: Lewis Publishers/CRC Press, pp. 309-323.

Timonin, M.I. (1946). Microflora of the rhizosphere in relation to the manganese-deficiency disease of oats. *Soil Science Society of America Proceedings* 11: 284-292.

Tinker, P.B. (1984). The role of microorganisms in mediating and facilitating the uptake of plant nutrients from soil. *Plant and Soil* 76: 77-91.

Toro, M., R. Azcon, and R. Herrera (1996). Effects on yield and nutrition of mycorrhizal and nodulated *Pueraria phaseoloides* exerted by P-solubilizing rhizobacteria. *Biology and Fertility of Soils* 21: 23-29.

Trofymow, J.A., D.C. Coleman, and C. Cambardella (1987). Rates of rhizodeposition and ammonium depletion in the rhizosphere of axenic oats roots. *Plant and Soil* 97: 333-344.

Turner, S.M., E.I. Newman, and R. Campbell (1985). Microbial population of ryegrass *Lolium perenne* root surfaces. Influence of nitrogen and phosphorus supply. *Soil Biology and Biochemistry* 17: 711-716.

Uren, N.C. (1993). Mucilage secretion and its interaction with soil, and contact reduction. *Plant and Soil* 155/156: 79-82.

Uren, N.C. and H.M. Reisenauer (1988). The role of root exudates in nutrient acquisition. *Advances in Plant Nutrition* 3: 79-114.

Van Overbeek, L.S., J.A. Van Veen, and J.D. Van Elsas (1997). Induced reporter gene activity, enhanced stress resistance, and competitive ability of a genetically modified *Pseudomonas fluorescens* strain released into a field plot planted with wheat. *Applied and Environmental Microbiology* 63: 1965-1973.

Vetterlein, D. and H. Marschner (1993). Use of a microtensiometer technique to study hydraulic lift in a sandy soil planted with pearl millet *Pennisetum americanum* L. Leeke. *Plant and Soil* 149: 275-282.

Von Wiren, N., H. Marschner, and V. Römheld (1995). Uptake kinetics of iron-phytosiderophores in two maize genotypes differing in iron efficiency. *Physiologia Plantarum* 93: 611-616.

Von Wiren, N., H. Marschner, and V. Römheld (1996). Roots of iron-efficient maize also absorb phytosiderophore-chelated zinc. *Plant Physiology* 111: 1119-1125.

Von Wiren, N., S. Mori, H. Marschner, and V. Römheld (1994). Iron inefficiency in maize mutant *ys1* (*Zea mays* L. cv yellow-stripe) is caused by a defect in uptake of iron phytosiderophores. *Plant Physiology* 106: 71-77.

Von Wiren, N., V. Römheld, J.L. Morel, A. Guckert, and H. Marschner (1993). Influence of microorganisms on iron acquisition in maize. *Soil Biology and Biochemistry* 25: 371-376.

Von Wiren, N., V. Römheld, T. Shioiri, and H. Marschner (1995). Competition between microorganisms and roots of barley and sorghum for iron accumulated in the root apoplasm. *New Phytologist* 130: 511-521.

Wang, Y., H.N. Brown, D.E. Crowley, and P.J. Szaniszlo (1993). Evidence for direct utilization of a siderophore, ferrioxamine B, in axenically grown cucumber. *Plant, Cell and Environment* 16: 579-585.

Walter, A., A. Pich, G. Scholz, H. Marschner, and V. Römheld (1995). Effects of iron nutritional status and time of day on concentrations of phytosiderophores and nicotianamine in different root and shoot zones of barley. *Journal of Plant Nutrition* 18: 1577-1593.

Walter, A., V. Römheld, H. Marschner, and S. Mori (1994). Is the release of phytosiderophores in zinc-deficient wheat plants a response to impaired iron utilization? *Physiologia Plantarum* 92: 493-500.

Welch, R.M. (1995). Micronutrient nutrition of plants. *CRC Critical Reviews in Plant Science* 14: 49-82.

Yehuda, Z., M. Shenker, V. Römheld, H. Marschner, Y. Hadar, and Y. Chen (1996). The role of ligand exchange in the uptake of iron from microbial siderophores by gramineous plants. *Plant Physiology* 112: 1273-1280.

Zhang, F.S. (1993). Effect of zinc nutritional status on the zinc uptake in wheat. *Acta Phytophysiologica Sinica* 19: 143-148.

Zhang, F., V. Römheld, and H. Marschner (1989). Effect of zinc deficiency in wheat on the release of zinc and iron mobilizing root exudates. *Zeitschrift für Pflanzenernährung und Bodenkunde* 152: 205-210.

Zhang, F. S., V. Römheld, and H. Marschner (1991). Diurnal rhythm of release of phytosiderophores and uptake rate of zinc in iron-deficient wheat. *Soil Science and Plant Nutrition* 37: 671-678.

Chapter 2

Kinetics of Nutrient Uptake by Plant Cells

Robert J. Reid

Key Words: carriers, enzyme kinetics, multiphasic, plant growth, plant membranes, surface potential.

INTRODUCTION

The cell membrane must rank as one of the most remarkable structures on this planet. With a basic thickness of only 7 nm and being composed of little more than greasy protein, it represents a formidable boundary between the living cell and its hostile environment. In its simplest guise, it is a selective filter, identifying those elements which are needed and those which are not, in measured proportions. The complexity that underlies this basic function derives from the interaction of a multitude of different polypeptides with membrane lipids and the aqueous environments that the membrane separates. The importance of membrane transport in the proper functioning of the cell is evidenced by the investment of approximately 12 percent of the total genome encoding for perhaps 1,000 transport proteins (Tanner and Caspari, 1996). At the moment, we can only guess at how many individual transporters this represents, but if it is accepted that plants require only about sixteen different nutrients (Marschner, 1995), it is obvious, in theory at least, that some overcapacity in transport capability exists for each element. The reason for this lies undoubtedly in the need for flexibility, to be able to adapt to variable or even rapidly fluctuating conditions in the external environment.

The author gratefully acknowledges research support from the Australian Research Council.

Our understanding of membrane transport is rudimentary, and we are still struggling to draw up a framework within which an integrated transport system can be described. We are at the crossroads of historical (and continuing) attempts to analyze transport in terms of a formalism based on enzyme reaction kinetics and a new technology that is becoming increasingly capable of identifying discrete molecular components of the process. This chapter provides a view of the current state of the debate over the meaning and value of kinetic descriptions of nutrient transport in plants.

TRANSPORT MECHANISMS

Plants require the twenty or so essential elements in widely varying amounts, ranging from the main structural elements C (carbon), H (hydrogen), O (oxygen), and N (nitrogen) to vanishingly small amounts of such elements as Ni (nickel) and Mo (molybdenum), which are needed for the functioning of only one, or a few, enzymes. The availability of nutrients in the soil also varies widely and bears no relationship to plant requirement. The elemental composition of plant tissues broadly reflects the relative amounts required for adequate plant growth. Thus, the principle function of the plant plasma membrane is, when presented with a soil soup of variable composition, to extract in the correct amounts those elements listed in the recipes for a defined intracellular menu. The processes that contribute to this reordering of ingredients are many and varied, but, unfortunately, in only a few cases do we have any real understanding of the basic molecular mechanisms.

Nearly all plant nutrients are obtained in ionic form. The direction of their passive movement will therefore be determined by concentration and charge differences. From comparison of cytoplasmic concentrations with those usually available in soil solution, and taking into account electrical potential differences across the plasma membrane, it appears that all anions, as well as most macronutrient cations, are taken up against electrochemical gradients. The known exceptions are Ca^{2+} (calcium) and K^+ (potassium), whose uptake is mediated by ion channels (Tester, 1990). Even for K^+, the soil concentration is often low enough that an active uptake mechanism needs to be invoked (Maathius and Sanders, 1996). Efflux of Ca^{2+}, H^+, and probably Na^+ (sodium) is directly energized by ATPases, while efflux of K^+ and Cl^- (chlorine) occurs via ion channels (Tester, 1990; Tyerman, 1992). Little is known about uptake or efflux mechanisms for divalent micronutrient cations such as Zn^{2+} (zinc), Mn^{2+} (manganese), Ni^{2+}, or Cu^{2+} (copper) (Clarkson and Lüttge, 1989; Kochian, 1991), although the negative membrane potential difference (PD) could support passive accumulation against a concentration gradient of 10^3 to 10^6.

It now seems likely that for NO_3^-, $H_2PO_4^-$, Cl^-, and K^+ (Sanders and Hansen, 1981; Ullrich and Novacky, 1990), active uptake occurs via carriers driven by symport with one or more protons. The inward electrical gradient for the driver ion H^+ is generated by outward pumping of plasma membrane H^+-ATPases. In animal systems, cotransport is most commonly driven by Na^+, and there is evidence for its limited use for cotransport in fungi, blue-green algae, green algae, and aquatic angiosperms (Skulachev, 1985; Scott, 1987; Maathius et al., 1996), but not as yet for terrestrial higher plants.

CONTROL OF UPTAKE

In simple systems such as unicellular algae or yeast, growth rates may be limited by the availability of one or more specific transporters. Extrapolation of this phenomenon to complex tissues such as roots is more difficult, since uptake must satisfy individual cell, tissue, and whole plant requirements. It is clear, though, that uptake systems are responsive to environmental conditions and that they do not operate independently. Lee (1982) showed that plants which had been made deficient in a particular nutrient tended to absorb the remaining nondeficient nutrients at a slower rate.

How then is control of net uptake of nutrients achieved? Around twenty-five years ago, W. J. Cram (currently Professor of Biological and Nutritional Sciences, University of Newcastle on Tyne, United Kingdom) explained to me that transport rates could be adjusted according to an inflow of energy or of information—a simple statement, but one that was too abstract for me, as a junior scientist, to understand fully. What sort of information? The idea that the rate of transport depended directly on the energy status of the cell was widely held in the 1970s, supported by the rampant use of metabolic inhibitors. It now seems that in only a very few cases is the rate of transport of a nutrient directly determined by electrochemical gradients or the supply of metabolic energy. Most reactions operate sufficiently far from thermodynamic equilibrium that moderate changes in energy status are not important (Sanders, 1990). That most active transport processes do respond to metabolic inhibition indicates that inflow of energy is ultimately necessary, but it seems likely that this is a coarse control reserved for extreme circumstances. The fine control must come from the inflow of information. This can take many guises and must involve signal-response pathways. An increase in net uptake may occur as a result of either increased influx or decreased efflux. Increased influx may be achieved by synthesis of more molecules of a transporter or synthesis of a different transporter with characteristics that confer a greater efficiency under the prevailing conditions (e.g., low nutrient concentrations). Conversely, reduced efflux may occur due to repression of the synthesis of the efflux carrier.

The alternative to repression/induction is the activation/deactivation of existing transporters by feedback signals. It is obvious that plants are able to adjust transport rapidly in response to sudden environmental changes (e.g., osmotic, light/dark, nutrient flushes), but identification of levels or signals that regulate the rate has proved extremely difficult.

ENZYME KINETICS AND MEMBRANE TRANSPORT

The observation that rates of nutrient uptake could be saturated by increasing the external concentration prompted Epstein and his co-workers (Epstein and Hagen, 1952; Elzam, Rains, and Epstein, 1964; Epstein, 1966) to consider membrane carriers as enzymes with selective binding sites for particular transported substrates. They noted that when ion uptake was plotted against concentration, the relationship for most ions conformed to a rectangular hyperbola, remarkably similar to that of classical Michaelis-Menten enzyme kinetics. Epstein and Hagen (1952) proposed that the characteristics of individual membrane carriers could be described by two kinetic constants: K_m (Michaelis-Menten constant or the concentration at half saturation) and V_{max} (maximum velocity). The rate (V) at any given ion concentration (C) can be estimated from the Michaelis-Menten equation:

$$V = \frac{V_{max} \times C}{K_m + C}$$

The use of enzyme kinetic principles to describe nutrient absorption has been widely embraced, and, despite some criticism of the validity of the approach, it remains the only commonly used method of analyzing plant membrane transport characteristics. Figure 2.1 outlines the main features of enzyme kinetics and methods used to derive the kinetic constants. In true Michaelis-Menten kinetics, the relationship between substrate concentration and rate is a rectangular hyperbola (Figure 2.1a). It is more convenient when estimating K_m and V_{max} from experimental data to use an algebraic transformation to give a linear representation. The most commonly used rearrangement is the Lineweaver-Burk equation, which simply uses the reciprocal of both sides of the Michaelis-Menten equation to obtain a straight line plot of $^1/v$ versus $^1/c$ (Figure 2.1b). Whereas with a rectangular hyperbola V_{max} is approached asymptotically, in a Lineweaver-Burk plot, the intercept on the ordinate equals $^1/V_{max}$ and the intercept on the abscissa equals $^1/K_m$. Another useful transformation of the Michaelis-Menten equation is the Eadie-Hofstee plot (Figure 2.1c), which magnifies departures of the data from linearity. Such deviations can occur

FIGURE 2.1. (a) Rate of an Enzymic Reaction As a Function of Concentration According to Michaelis-Menten kinetics, (b) Lineweaver-Burk Plot, (c) Eadie-Hofstee Plot, and (d) Modified Representation of Michaelis-Menten Kinetics As Applied to Net Uptake, to Incorporate C_{min}

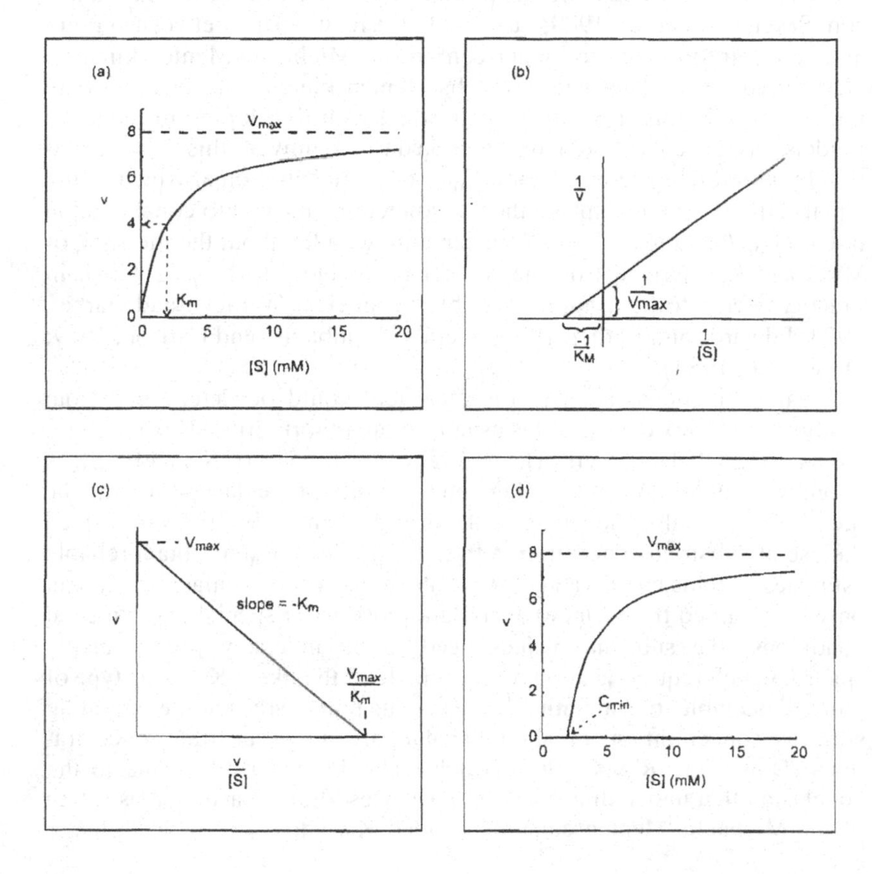

if the relationship between rate and substrate concentration is not truly a rectangular hyperbola. This is most commonly the case with allosteric or regulatory enzymes, which show a sigmoidal response. These enzymes contain, in addition to the catalytic site, an allosteric site that may bind the substrate and increase the catalytic activity (homotropic) or bind some other substance that increases or decreases the catalytic activity (heterotropic). Allosteric effects provide a theoretical basis for a mechanism for regulation of the rate of carrier proteins.

Deviation from linearity is also commonly observed when dealing with net uptake, which is really the composite of two competing processes, influx and efflux. Obviously, if there is a concentration at which influx equals efflux, the net flux will be zero. If efflux becomes significant compared to influx, as appears to be the case for NO_3^- uptake in barley; for which efflux accounts for approximately 60 percent of influx (Clarkson, Saker, and Purves, 1989; Lee, 1993), the relationship between net flux and concentration will no longer conform to Michaelis-Menten kinetics. Undeterred by the fact that Michaelis-Menten kinetics are inappropriate under such circumstances, those concerned with developing mechanistic models for nutrient uptake have attempted to circumvent this minor irritation by introducing an extra term, C_{min}, the concentration at which influx equals efflux. Zero net uptake then no longer occurs at zero concentration but at C_{min} (see Figure 2.1d). One can only wonder about the meaning of V_{max} and K_m derived from net flux data, or indeed of C_{min}, which in practice seems to be a highly variable parameter (Warncke and Barber, 1974; Edwards and Barber, 1976; Therios, Weinbaum, and Carlson, 1979; Drew et al., 1984).

Ideally, kinetic parameters for a carrier should be determined from unidirectional flux data, and this usually entails short-term influx measurements using labeled tracers. The accuracy of estimates of K_m and V_{max} is strongly influenced by the number and reliability of the data points and the method used to plot and calculate the data. Cleland (1967) recommended the use of substrate concentration from 0.2 K_m to 5 K_m to obtain reliable estimates. Ritchie and Prvan (1996) contend that valid estimates of K_m can only be obtained from Lineweaver-Burk plots under special experimental conditions; the substrate values need to be inversely proportionally spaced, which requires a good prior estimate of the likely K_m. This type of plot is unsuitable for estimating K_m if no data points are near the saturating substrate concentration or if aberrant values are at low substrate concentrations (Dowd and Riggs, 1965). Ritchie and Prvan (1996) came to the conclusion that under almost all circumstances, direct least-squares fitting of the Michaelis-Menten equation yielded K_m and V_{max} with a greater accuracy than double reciprocal plotting methods such as Lineweaver-Burk. They found that least-squares fitting to logarithmically spaced substrate concentrations gave reasonable estimates of K_m over a wide concentration range, although more accurate values were obtained if substrate values were arithmetically spaced around the expected K_m.

A review of published values for NH_4^+ uptake in plants gives values of K_m ranging from 1 to 166 μM and for NO_3^- uptake from 7 to 224 μM (see Table 2.1), with similar variability in V_{max} (Garnett, 1996). Such a wide spread of values implies either that there are many different carriers for N or that the value obtained is strongly dependent on how it was measured. The latter would certainly be cause for concern and call into question the usefulness of the carrier kinetic approach for describing plant nutrient uptake.

DUAL ISOTHERMS AND MULTIPHASIC KINETICS

It was obvious even in the early studies of Epstein, Rains, and Elzam (1963); Elzam, Rains, and Epstein (1964); and Epstein and Rains (1965) that when influx was examined over a wide concentration range, the isotherm (influx versus concentration) could not be described by a single rectangular hyperbola, but appeared to be two or more sequential hyperbolas (Epstein and Rains, 1965) or one or more hyperbolas with a linear phase in the higher concentration range (Kochian and Lucas, 1982; Pace and McClure, 1986; Siddiqi et al., 1989). These observations have given rise to lively debate in the literature about the underlying molecular mechanisms. Epstein, Rains, and Elzam (1963) considered the biphasic saturation in the concentration dependence of K^+ uptake ("dual isotherms") to represent two separate carriers operating over different concentration ranges. The alternative view championed by Nissen (Nissen, 1971; 1980; 1991) was that the "bumpiness," or multiphasic nature, of uptake isotherms (see Figure 2.2) was due to concentration-dependent phase changes in a single carrier that possesses two binding sites, one a transport site and the other a transition site. Interactions of ions with the transition site would cause conformational changes of the carrier protein and therefore affect the affinity (K_m) of the transport site and change V_{max}. In support of the multiphasic carrier concept, Nissen (1971; 1991) showed that sulfate transport exhibited a concentration dependence with sharp transitions and discontinuities (see Figure 2.3), which were considered inconsistent with the simultaneous operation of more than one carrier. The number of phases varies with the particular ion; in the case of SO_4^{2-}, Nissen (1971) considered that there were eight distinct Michaelis-Menten phases.

TABLE 2.1. Published Kinetic Parameters for Uptake of NH_4^+ and NO_3^-

Plant	General method	K_m µM	V_{max} µmol g^{-1} h^{-1}	C_{min} µM	Reference
			NH_4^+		
tomato	steady state	6-12	9-20*		Smart and Bloom (1988)
barley	steady state	15-28	9-10*		Bloom (1985)
carob	steady state	10	2.3		Cruz, Lips, and Martinsloucao (1993)
onions	6-h depletion	14	32-55*	5	Abbes et al. (1995)
Spartina	10 to 20-h depletion	1	1.3*		Bradley and Morris (1990)
Norway spruce	unidirectional flux of ^{13}N	20-40	1.9-2.4		Kronzucker, Siddiqi, and Glass (1996)
rice	unidirectional flux of ^{13}N	30-190	3-13		Wang et al. (1993)
spruce	5-h depletion			1.5	Marschner, Häussling, and George (1991)
barley	steady state	153-166	3.7-8.3		Mäck and Tischner (1994)
wheat	8-h depletion	50	3.5		Goyal and Huffaker (1986)
Douglas fir	steady state	8-14	0.7-0.9*		Kamminga-Van Wijk and Prins (1993)
			NO_3^-		
tomato	steady state	34-54	19-31*		Smart and Bloom (1988)
barley	steady state	7-187	3-15*		Bloom (1985)
corn	steady state	224	10.7*		Pace and McClure (1986)
various catch crop species	9-h depletion	5-36	9.8-35	10	Laine et al. (1993)
carob	steady state	19	2.03		Cruz, Lips, and Martinsloucao (1993)
onions	6-h depletion	12	14-30*	4.54	Abbes et al. (1995)
barley	long-term depletion	34-36	8		Aslam, Travis, and Huffaker (1992)
spruce	unidirectional flux of ^{13}N	112	0.7		Kronzucker, Siddiqi, and Glass (1996)

barley	unidirectional flux of ^{13}N	30-79	1-9		Siddiqi et al. (1990)
wheat	8-h depletion	27	2.6		Goyal and Huffaker (1986)
barley	unidirectional flux of ^{13}N	14	2.6*		Lee and Drew (1986)
lettuce	7-h depletion	9.3	9.6*	0.6	Swiader and Freiji (1996)
spruce	5-h depletion			22	Marschner, Häussling, and George (1991)
corn, soybean, sorghum, bromegrass	long-term depletion			1.4-2.7	Warncke and Barber (1974)
spruce	not specified	200	18		Peuke and Tischner (1991)

Source: Adapted from Garnett (1996).

The V_{max} values were not always in units of $\mu mol\ g^{-1}\ h^{-1}$, and where indicated (*), values were estimated. The V_{max} values are in terms of root fresh weight for the high-affinity system.

Three other slightly differing interpretations of transport phenomena broadly support a multiphasic nature for a single carrier. Hodges (1973) suggested that positive and negative cooperation between various binding sites of the one carrier could explain the apparent inflections in uptake isotherms. In practice, however, it would be difficult kinetically to distinguish a two-site carrier with negative cooperativity from two different carriers (Borst-Pauwels, 1973).

Two other schemes acknowledge that for most ions in the low concentration range, transport is active and is likely to involve symport with H^+ or some other driver ion. Hence, two or more binding sites must exist. Komor and Tanner (1974; 1975) explained the existence of two Michaelis-Menten phases for sugar uptake by *Chlorella* in terms of a carrier that could cross the membrane in two loaded forms: with the sugar molecule alone (representing the low-affinity phase—passive uptake), or fully loaded with the sugar molecule and a proton (higher affinity—active uptake). Dual schemes such as this have been termed "slip" models (Eddy, 1980) and have been used to account for uptake of various solutes, including high-affinity K^+ uptake driven by Na^+ in *Chara* (Walker, Reid, and Smith, 1993).

FIGURE 2.2. "Bumpy" Isotherm for the Low-Affinity Uptake of K^+ in Barley Roots

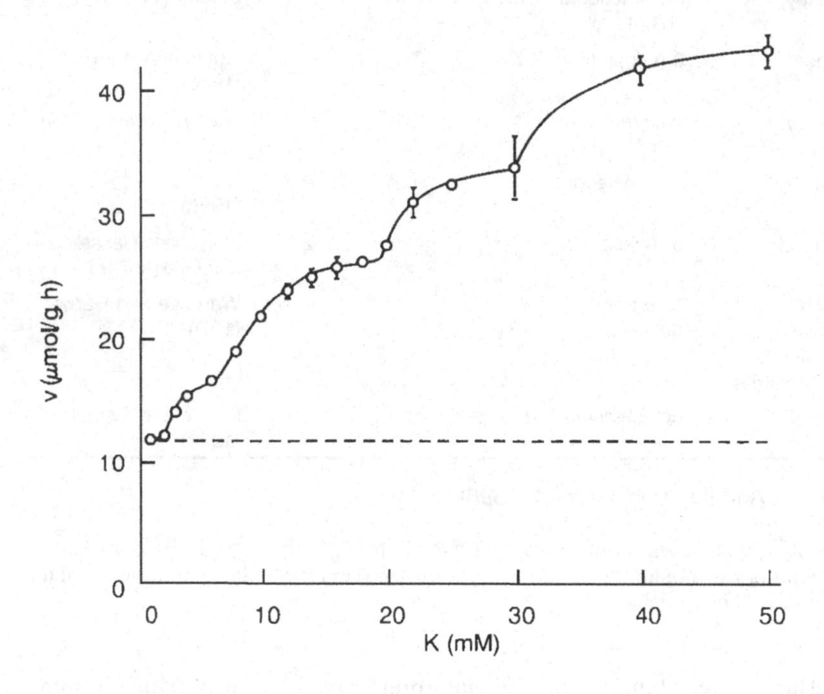

Source: Adapted from Epstein and Rains (1965).

Sanders (1986) pointed out that the slip model does not account for those cases which show both high- and low-affinity systems operating against electrochemical gradients (e.g., Van Bel et al., 1982; Glass, Shaff, and Kochian, 1992), and, therefore, a driver ion would be required in both phases. He proposed that biphasic kinetics would result from the operation of a carrier in which binding of solute or driver ion could occur in random order, but only the fully loaded carrier could traverse the membrane. Mathematical modeling showed that through selection of the appropriate binding and rate constants, either a monophasic (high driver ion concentration) or biphasic response could be obtained (Sanders, 1986).

The validity of the multiphasic interpretation of experimental data has been challenged by Borstlap (1983), who performed statistical analyses on published data from a variety of sources. In all cases, it was considered that the data could be satisfactorily explained by the sum of independent

FIGURE 2.3. Eadie-Hofstee Plot of SO_4^{2-} Uptake in Barley Showing Inflections and Discontinuities Between Phases

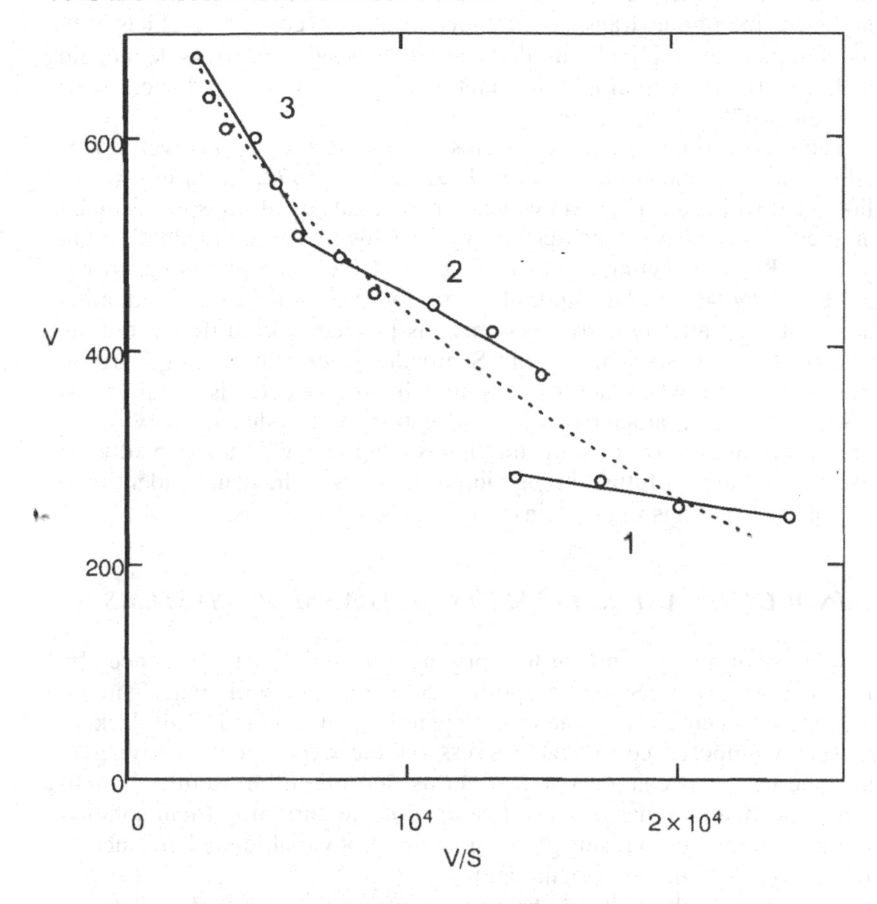

Note: The dashed line was fitted to the data assuming a continuous isotherm. Adapted from Nissen (1991).

Michaelis-Menten or linear phases. The data on K^+ and Na^+ (Epstein and Rains, 1965) could be best explained by the sum of two Michaelis-Menten terms, while Cl^- data (Elzam, Rains, and Epstein, 1964), 3-*o*-methyl glucose (Linask and Laties, 1973), SO_4^{2-} (Nissen, 1971), and lysine (Soldal and Nissen, 1978) could all be fitted to the sum of two Michaelis-

Menten terms and a linear term. Arguments about how the empirical flux data should be interpreted have unfolded ever since enzyme kinetics were applied to membrane transport, and there is still no consensus. It has been claimed that data originally used to support a "dual" uptake system could be better fitted by multiphasic kinetics (Nissen, 1991) and vice versa (Epstein, 1973).

The use of multiple uptake systems for a single nutrient is well established in fungi and bacteria (Ferro-Luzzi Ames, 1972; Iaccarino, Guardiola, and DeFelice, 1978), in which genetic analysis of transport is much simpler than in plants. Patch-clamp studies suggest that up to ten different types of K^+-conducting channels in roots of the same plant (Vogelzang and Prins, 1994) may be responsible for low-affinity K^+ uptake. Identification of high-affinity carrier systems has proved more difficult, but the precedent set by Schachtman and Schroeder (1994) in cloning a cation transporter from wheat and expressing it in frog oocytes is certain to be followed by the characterization of other transport systems. Many of the kinetic arguments concerning multiphasic uptake will undoubtedly be resolved in the near future through improvements in the genetic identification of specific transport proteins.

INDUCTION AND REPRESSION OF TRANSPORT SYSTEMS

It is important to note that transport activity is strongly influenced by plant nutrient status. Some transporters come and go, while others simply increase or decrease their activity depending on demand. This lack of constancy imposes considerable stress on the experimenter striving to describe transport characteristics in terms that might be useful in conditions other than the narrow specifications of the current nutrient solution (extrapolation to crop plants growing in a highly variable medium such as soil is too horrifying to contemplate).

Despite the multiplicity of phases encountered in some studies, it seems convenient these days to divide transport of a nutrient into a high-affinity transport system (HATS) and a low-affinity transport system (LATS), equivalent to the older System I and System II (Epstein, Rains, and Elzam, 1963). The low-affinity system is invariably constitutive, while the high-affinity system might only appear after induction (e.g., by starvation), in which case the acronym is elaborated to IHATS to distinguish it from CHATS.

Evidence suggests that inducible/repressible high-affinity systems operate for all plant macronutrients except Ca^{2+} and Mg^{2+} (magnesium) (Clarkson and Lüttge, 1991), but little evidence indicates that uptake of divalent micronutrient cations is regulated in the same way (Clarkson and Lüttge, 1989). The following sections give some insight into the variable nature of plant membrane transport.

Nitrate Uptake

High-affinity systems most commonly respond to nutrient deprivation by increasing activity. The exception is NO_3^-, for which induction is effected by the presence of NO_3^- in the external solution rather than by its absence. This phenomenon is well documented (Lee and Drew, 1986; Siddiqi et al., 1990; Clarkson and Lüttge, 1991), and some consistency exists among species. A recent study with spruce exemplifies the main features of NO_3^- uptake. Kronzucker, Siddiqi, and Glass (1995) identified three kinetically distinct transport systems. In seedlings not previously exposed to NO_3^- (uninduced), two systems were observed, a high-affinity system with a K_m of 15 μM termed CHATS and a low-affinity system that operated at concentrations greater than 1 mM. Following exposure to NO_3^-, the CHATS activity increased threefold and another system with a K_m of 100 μM (IHATS) appeared.

Phosphate and Sulfate Uptake

Deprivation of $H_2PO_4^-$ or SO_4^{2-} invariably results in an increase in activity of the respective high-affinity systems, without major changes in K_m. For SO_4^{2-} (sulfur), the activity can be very sensitive to S (sulfur) status; Datko and Mudd (1984) reported a 500-fold increase in SO_4^{2-} transport capacity in *Lemna paucicostata* following starvation. Phosphate uptake does not respond so spectacularly to starvation, increasing by only two- to fourfold in a range of species (Clarkson and Scattergood, 1982; Lee, 1982; Lefebvre and Glass, 1982; Cogliatti and Clarkson, 1983; McPharlin and Bieleski, 1987). A curious feature about P (phosphorus) deprivation is the lack of significant changes in cytoplasmic phosphate (Lee and Ratcliffe, 1983; Rebeille et al., 1983; Mimura et al., 1990) that might act as the signal for altering transport of $H_2PO_4^-$ across the plasma membrane. The decrease in vacuolar phosphate concentration during starvation has therefore been considered to be the controlling factor (Lee, Ratcliffe, and Southon, 1990), although it is not understood precisely how signals from the vacuole might be sensed at the plasma membrane. Mimura, Reid, and Smith (1998) argued that the controlling factor which is sensed by the cell is the concentration of phosphate in the external solution and not the vacuolar concentration. In starvation experiments with plant roots, two factors must be considered: the reduction in external phosphate concentration and the decrease in vacuolar phosphate concentration due to dilution by plant growth and transport to the shoot. The cell could therefore respond to the change in either level. Using mature (i.e., nongrowing) giant algal cells, Mimura, Reid, and Smith (1998) were able to show that the increase in influx capacity following starvation occurred even in the absence of changes in internal phosphate status. In this species, at least, it appeared that transport was regulated by external rather than internal phosphate concentrations.

Potassium and Ammonium

In contrast to the anions described earlier, for which changes in V_{max} occur without large changes in K_m, internal NH_4^+ status affects both parameters. Wang and colleagues (1993) showed that the K_m for NH_4^+ uptake in rice roots was highly dependent on the concentration of NH_4^+ in the growth solution, ranging from 32 μM to 188 μM for pretreatment concentrations varying from 2 μM to 1000 μM. Similarly, the apparent K_m for K^+ influx in barley roots has been shown to be closely correlated with tissue K^+ content (Siddiqi and Glass, 1987). In both cases, as the K_m increased, the V_{max} decreased. It is easy to accept that V_{max} could be determined by controlling the synthesis of the transport protein or by synthesizing a different transport protein, but the prospect of a single transporter with a variable K_m presents greater, though not insurmountable, difficulties. Allosteric regulation, by internal or external concentrations, mediated by changes in the affinity of the transport sites, is entirely feasible, as is the expression of a suite of transporters differing only slightly in kinetic characteristics (e.g., multiple K^+ channels in roots of *Plantago*; Vogelzang and Prins, 1994), but we need to come to terms with the implications this would have for the way in which we view the carrier kinetic approach to membrane transport. The V_{max} and K_m would then only be seen as "constants" in the context of a narrowly defined set of conditions.

ROOT STRUCTURE AND OBSERVED KINETICS

Michaelis-Menten Kinetics and Membranes

Although Michaelis-Menten kinetics are widely employed in descriptions of membrane transport processes, their use has also attracted a degree of criticism, especially in relation to whole plant uptake. Even at the membrane level, the validity of applying Michaelis-Menten principles has been questioned, since it is implicit in traditional enzyme kinetics that the reaction occurs in dilute solutions that are spatially homogenous. Obviously, this will rarely, if ever, be the case at the surface of a root where diffusional limitation and electrostatic effects apply. It has been demonstrated by theory, computer simulation, and experimentation that elementary chemical kinetics are quite different when reactions are diffusionally limited or dimensionally restricted, such as might occur when reactions are confined to a two-dimensional membrane or a one-dimensional ion channel (Savageau, 1995).

Diffusion Limitation

Plant roots are not homogeneous structures. They contain cells in a number of layers and at various stages of development. Spatial variation in nutrient uptake is known to occur along roots (Eshel and Waizel, 1972; 1973; Robards et al., 1973; Clarkson and Sanderson, 1978; Dalton, 1984). In soil and in unstirred solutions, demand for nutrients may exceed the ability of diffusion to maintain supply, leading to depletion at the root surface. For K^+, the depletion may be as high as 50 percent (Drew and Nye, 1969; Newman et al., 1987). Although vigorous stirring can reduce depletion around the root, it has little effect on cell layers below the surface, where diffusion is greatly reduced by tortuosity, viscosity, and electrostatic interactions with ionic groups on cell wall polymers and the membrane surface. It could reasonably be expected, therefore, that in a whole root, there will be considerable spatial variation in the nutrient concentration to which membrane carriers in different parts of the root are exposed, and this might depend on the demand as influenced by plant nutrient status. In experimental terms, this will appear as noise in the system and lead to imprecision in the determination of kinetic constants. Nevertheless, despite the existence of several root zones, various cell types and cell compartments, cell walls and multiple cell layers, there is occasionally good correspondence between kinetic parameters measured in roots and in protoplasts prepared from the roots (Kochian and Lucas, 1982) and between excised roots and whole plants (Epstein, 1973).

Electrostatic Effects

In this final section, I would expand upon an aspect of plant membranes that has hitherto received very little attention in kinetic studies of transport, but which could go a long way to explaining why observed kinetics tend to be so variable. It concerns the electrostatic nature of membranes and the implications that this has for ion activities around transport sites.

The passive uptake of charged nutrients through ion channels is a simple manifestation of electrophoretic movement caused by differences in charge density on either side of the membrane. It is readily accepted that there is normally a voltage difference across cell membranes, but it is apparently conceptually difficult to accept that voltage differences could exist between bathing solution and the cell wall or the external surface of the plasma membrane. This observation is based on the almost total absence of any reference to surface charge in kinetic studies of plant membrane transport. Arguments about the relevance of cell wall electrostatic properties, although still unresolved, have diverted attention from the potentially more important question of the charge on the membrane itself. These charges are closer to the membrane transporters than are the cell wall charges. The significance of these charges in fungal cells (yeast) has been considered by Cerbon and Calderon (1990; 1992). Interpretation of ionic effects on animal cell membrane transport in terms of effects on surface potential is common, especially in relation to ion channels (e.g., anions, Hayman and Ashley, 1993; Ca^{2+}, Thibault, Porter, and Landfield, 1993; Na^{+}, Hanck and Sheets, 1992; K^{+}, Bielefeldt, Rotter, and Jackson, 1992; mechanosensitive, Lane, McBride, and Hamill, 1993) or to the potency of charged drugs such as amiloride (Lane, McBride, and Hamill, 1993). With few exceptions (e.g., Jacoby and Hanson, 1985; Blatt, 1992; Carandang et al., 1992), plant membrane physiologists have largely ignored the effect of surface potential on the observed characteristics of membrane transporters.

The surface potential arises mainly from the charges on phospholipids, with a lesser contribution from exposed membrane proteins. Membrane composition, and the ratio of anionic/zwitterionic phospholipids varies from species to species, between root and shoot cells (Sandstrom and Cleland, 1989), and between cortex and stele (Cowan et al., 1993). The surface potential is a function of the concentration of anionic phospholipids, as demonstrated by Cerbon and Calderon (1990) in yeast, for which modification of the culture conditions caused a twofold enrichment of anionic phospholipids; this led to a doubling of the surface potential and considerable changes in the kinetics of some membrane transporters (Calderon and Cerbon, 1992). The surface potential estimated under low ionic conditions for protoplasts from barley leaves was −48 mV (Abe and Takeda, 1988) and −35 mV in tobacco leaves (Nagata and Melchers, 1978), while in artificial bilayers containing a high proportion of anionic phospholipids, the surface potential can be as high as −120 mV (McLaughlin, Szabo, and Eisenman, 1971; McLaughlin et al., 1981). The surface potential will have a variety of consequences for membrane transport: surface pH will be lower than in the bulk solution (1 pH unit/59 mV); the membrane voltage measured between the external solution and the cytoplasm will be overestimated; the activities of cations, particularly multivalent cations, will be enhanced (tenfold/59 mV for monovalent cations, 100-fold for divalent, and 1000-fold for trivalent cations), while anions will show an inverse relation.

What effect will surface charges have on transport kinetics? The surface potential will not affect V_{max} because it only modifies the substrate concentration and does not directly alter the characteristics of the transporter (ignoring, for the moment, the fact that pH is changed and arguments about charge effects on transport protein conformation and binding sites). For transport processes controlled by thermodynamic gradients, the surface potential will not alter the rate (except near saturation) because the surface potential does not alter the overall electrochemical potential across the membrane. The main consequence will be for K_m, which, for a negatively charged membrane, will be underestimated for cations and overestimated for anions. Furthermore, the K_m determined from the bulk solution concentration will vary with the ionic composition of the medium, particularly the presence of divalent and trivalent cations. The K_m measured in this way should therefore be considered an "apparent" K_m, a variable that is a function of the experimental conditions rather than a true constant (Körner et al., 1985). A good illustration of this is given by Gage and colleagues (1985), who showed that the apparent K_m for Rb^+ (rubidium) transport in yeast increased by a factor of five as the Ca^{2+} concentration was varied from 5 to 35 mM (see Figure 2.4). They showed that a precise correlation existed between the effects of Ca^{2+} on surface charge and on the K_m and therefore interpreted the response to Ca^{2+} as being via an indirect effect on surface charge rather than a direct effect of Ca^{2+} on the transporter. They also found no difference in K_m between intact cells and protoplasts, which is perhaps relevant to plant cells because of showing that the membrane surface is independent of the charged cell-wall phase.

The impact of surface charge will be greater for divalent than for monovalent cations. The potential for error in estimating kinetic constants is increased markedly. Reid, Tester, and Smith (1997) suggested that the fiftyfold difference in estimated K_m for plant Ca^{2+} channels from various species was more to do with the effect on surface charge of the measurement conditions than to any real differences in the channel protein itself. It is also interesting to speculate whether differences between species and

FIGURE 2.4. The Effect of Ca^{2+} Concentration on the K_m for Uptake of Rb^+ (Circles) and on the Surface Potential Determined by the Fluorescence Ratio (Free/Bound) of 9-Aminoacridine in Yeast (Squares)

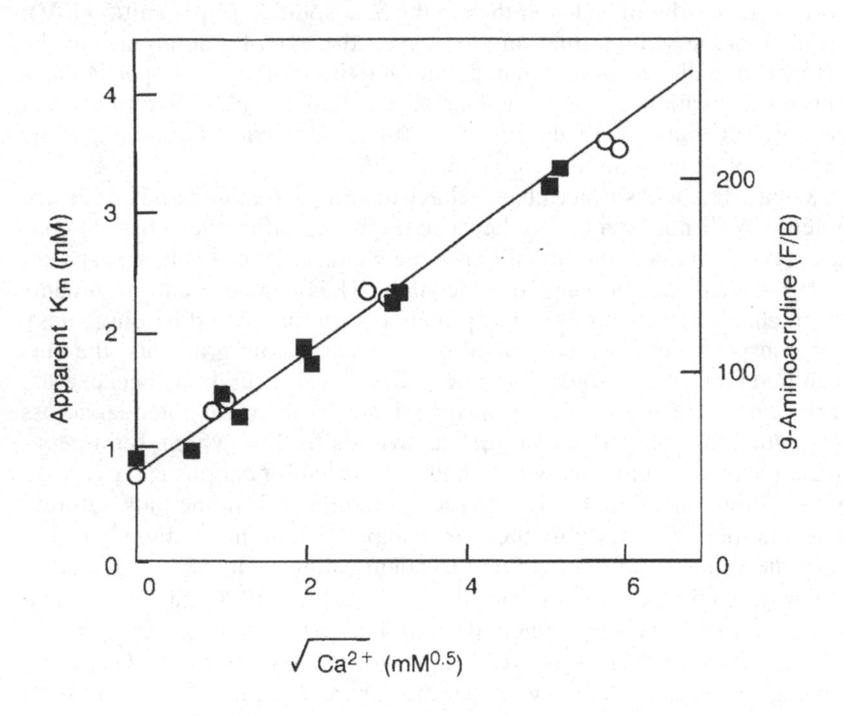

Source: Adapted from Gage et al. (1985).

cultivars, with respect to efficiency of uptake of divalent micronutrient cations such as Zn^{2+} and Mn^{2+}, might be due to variations in membrane composition. For example, a higher proportion of anionic phospholipids would generate a more negative surface potential that would concentrate cations at the uptake sites.

It is obvious that surface potentials exist and that there are strong theoretical arguments for surface charge effects on transport kinetics (Theuvenet and Borst-Pauwels, 1976), but at present, there are few measurements of surface charge in plant cells, and none that would allow useful prediction of how the composition of experimental solutions would affect solute concentrations at the membrane surface. There is clearly a need to examine more closely the relationship between surface charge and membrane transport, a view supported by Kinraide (1994), one of the few plant physiologists to interpret ion toxicities and transport phenomena in terms of surface charge.

CONCLUSIONS

The carrier kinetic approach provides a useful framework within which to begin to understand how membrane transport operates at the cellular level. In the whole plant, however, clear definition of individual transport characteristics is compromised by the complexity of regulatory responses and interactions between nutrients. The interplay of physicochemical factors has likewise confounded attempts to understand how transporters operate at the molecular level.

Major advances are likely to come from the identification by genetic methods of transport proteins that can be matched to observed kinetics and from a better understanding of control of induction and repression of carrier systems. Isolation, followed by characterization of transporters in simplified systems, will reveal much about the basic mechanism, but it is uncertain how much benefit this will bring to our understanding of plant growth in complex media such as soil, for which improvements in nutrient availability and nutrient mobilization strategies may outweigh the benefit of subtle differences in the properties of membrane transporters.

REFERENCES

Abbes, C., L.E. Parent, A. Karam, and D. Isfan (1995). Effect of NH_4^+/NO_3^- ratios on growth and nitrogen uptake by onions. *Plant and Soil* 171: 289-296.

Abe, S. and J. Takeda (1988). Effects of La^{3+} on surface charges, dielectrophoresis, and electrofusion of barley protoplasts. *Plant Physiology* 87: 389-394.

Aslam, M., R.L. Travis, and R.C. Huffaker (1992). Comparative kinetics and reciprocal inhibition of nitrate and nitrite uptake in the roots of uninduced and induced barley (*Hordeum vulgare* L.) seedlings. *Plant Physiology* 99: 1124-1133.

Bielefeldt, K., J. Rotter, and M. Jackson (1992). Three potassium channels in rat posterior pituitary nerve terminals. *Journal of Physiology* 458: 41-67.

Blatt, M.R. (1992). Potassium channels of stomatal guard cells: Characteristics of the inward rectifier and its control by pH. *Journal of General Physiology* 99: 615-644.

Bloom, A.J. (1985). Wild and cultivated barleys show similar affinities for mineral nitrogen. *Oecologia* 65: 555-557.

Borstlap, A.C. (1983). The use of model-fitting in the interpretation of "dual" uptake isotherms. *Plant, Cell and Environment* 6: 407-416.

Borst-Pauwels, G.W.F.H. (1973). Two site-single carrier transport kinetics. *Journal of Theoretical Biology* 40: 19-31.

Bradley, P.M. and J.T. Morris (1990). Influence of oxygen and sulphide concentration on nitrogen uptake kinetics in *Spartina alterniflora*. *Ecology* 71: 282-287.

Calderon, V. and J. Cerbon (1992). Interfacial pH modulation of membrane protein function *in vivo*. Effects of anionic phospholipids. *Biochimica et Biophysica Acta* 1106: 251-256.

Carandang, J., U. Pick, I. Sekler, and H. Gimmler (1992). Potassium fluxes and potassium content in *Dunaliella acidophila*, an alga with positive electrical potentials: I. Low affinity uptake system. *Journal of Plant Physiology* 139: 413-421.

Cerbon, J. and V. Calderon (1990). Proton-linked transport systems as sensors of changes in the membrane surface potential. *Biochimica et Biophysica Acta* 1028: 261-267.

Cerbon, J. and V. Calderon (1992). Surface potential regulation of phospholipid composition and in-out translocation in yeast. *European Journal of Biochemistry* 219: 195-200.

Clarkson, D.T. and U. Lüttge (1989). Divalent cations, transport and compartmentation. *Progress in Botany* 51: 91-112.

Clarkson, D.T. and U. Lüttge (1991). Mineral nutrition: inducible and repressible nutrient transport systems. *Progress in Botany* 52: 61-83.

Clarkson, D.T., L.R. Saker, and J.V. Purves (1989). Depression of nitrate and ammonium transport in barley plants with diminished sulphate status. Evidence of co-regulation of nitrate and sulphate intake. *Journal of Experimental Botany* 40: 953-963.

Clarkson, D.T. and J. Sanderson (1978). Sites of absorption and translocation of iron in barley roots. Tracer and microautoradiographic studies. *Plant Physiology* 61: 731-736.

Clarkson, D.T. and C.B. Scattergood (1982). Growth and phosphate transport in barley and tomato plants during the development of, and recovery from, phosphate stress. *Journal of Experimental Botany* 33: 865-875.

Cleland, W.W. (1967). The statistical analysis of enzyme kinetic data. *Advances in Enzymology* 29: 1-32.

Cogliatti, D.H. and D.T. Clarkson (1983). Physiological changes in the phosphate uptake by potato plants during development of, and recovery from, phosphate deficiency. *Physiologia Plantarum* 58: 287-294.

Cowan, D.S.C., D.T. Cooke, D.T. Clarkson, and J.L. Hall (1993). Lipid and sterol composition of plasma membranes isolated from stele and cortex of maize roots. *Journal of Experimental Botany* 44: 991-994.

Cruz, C., S.H. Lips, and M.A. Martinsloucao (1993). Uptake of ammonium and nitrate by carob (*Ceratonia siliqua*) as affected by root temperature and inhibitors. *Physiologia Plantarum* 89: 532-543.

Dalton, F.N. (1984). Dual pattern of potassium transport in plant cells: A physical artifact of a single uptake mechanism. *Journal of Experimental Botany* 35: 1723-1732.

Datko, A.H. and S.H. Mudd (1984). Sulfate uptake and its regulation in *Lemna paucicosta* Hegelm. 6746. *Plant Physiology* 75: 466-473.

Dowd, J.E. and D.S. Riggs (1965). A comparison of estimates of Michaelis-Menten kinetic constants from various linear transformations. *Journal of Biological Chemistry* 240: 863-869.

Drew, M.C. and P.H. Nye (1969). The supply of nutrient ions by diffusion to plant roots in soil. *Plant and Soil* 31: 407-424.

Drew, M.C., L.R. Saker, S.A. Barber, and W. Jenkins (1984). Changes in the kinetics of phosphate absorption in nutrient-deficient barley roots measured by a solution depletion technique. *Planta* 160: 490-499.

Eddy, A.A. (1980). Slip and leak models of gradient-coupled transport. *Transactions of the Biochemical Society of London* 8: 271-273.

Edwards, J.H. and S.A. Barber (1976). Nitrate flux into corn roots as influenced by shoot requirements. *Agronomy Journal* 68: 471-473.

Elzam, O.E., D.W. Rains, and E. Epstein (1964). Ion transport kinetics in plant tissue: Complexity of the chloride absorption isotherm. *Biochemical and Biophysical Research Communications* 15: 273-276.

Epstein. E. (1966). Dual pattern of ion absorption by plant cells and by plants. *Nature* 212: 1324-1327.

Epstein, E. (1973). Mechanisms of ion transport through plant cell membranes. *International Review of Cytology* 34: 123-168.

Epstein, E. and C.E. Hagen (1952). A kinetic study of the absorption of alkali cations by barley roots. *Plant Physiology* 27: 457-474.

Epstein, E. and D.W. Rains (1965). Carrier-mediated cation transport in barley roots: Kinetic evidence for a spectrum of active sites. *Proceedings of the National Academy of Sciences USA* 53: 1320-1324.

Epstein, E., D.W. Rains, and O.E. Elzam (1963). Resolution of dual mechanisms of potassium absorption by barley roots. *Proceedings of the National Academy of Sciences USA* 49: 684-692.

Eshel, A. and Y. Waisel (1972). Variations in sodium uptake primary roots of corn seedlings. *Plant Physiology* 49: 585-589.

Eshel, A. and Y. Waisel (1973). Variations in uptake of sodium and rubidium along barley roots. *Physiologia Plantarum* 28: 557-560.

Ferro-Luzzi Ames, G. (1972). Components of histidine transport. In *Membrane Research*, Ed. C.D. Fox. New York: Academic Press, pp. 409-426.

Gage, R.A., W. Van Wijngaarden, A.P.R. Theuvenet, G.W.F.H. Borst-Pauwels, and A.J. Verkleij (1985). Inhibition of Rb^+ uptake in yeast by Ca^{2+} is caused by a reduction in the surface potential and not in the Donnan potential of the cell wall. *Biochimica et Biophysica Acta* 812: 1-8.

Garnett, T.P. (1996). Ammonium and Nitrate Uptake by *Eucalyptus nitens*. PhD thesis, University of Tasmania, Hobart, Australia.

Glass, A.D.M., J.E. Shaff, and L.V. Kochian (1992). Studies of nitrate uptake in barley. *Plant Physiology* 99: 456-463.

Goyal, S.S. and R.C. Huffaker (1986). The uptake of NO_3^-, NO_2^- and NH_4^+ by intact wheat (*Triticum aestivum*) seedlings. I. Induction and kinetics of transport systems. *Plant Physiology* 82: 1051-1056.

Hanck, D. and M. Sheets (1992). Extracellular divalent and trivalent cation effects on sodium current kinetics in single canine cardiac Purkinje cells. *Journal of Physiology* 454: 267-298.

Hayman, K. and R. Ashley (1993). Structural features of a multisubstate cardiac mitoplast anion channel: Inferences from single channel recording. *Journal of Membrane Biology* 136: 191-197.

Hodges, T.K. (1973). Ion absorption by plant roots. *Advances in Agronomy* 25: 163-207.

Iaccarino, M., J. Guardiola, and M. DeFelice (1978). On the permeability of biological membranes. *Journal of Membrane Science* 3: 287-302.

Jacoby, B. and J.B. Hanson (1985). Control on sodium-22 influx in corn (*Zea mays*) roots. *Plant Physiology* 77: 930-934.

Kamminga-Van Wijk, C. and H.B.A. Prins (1993). The kinetics of uptake of NH_4^+ and NO_3^- by Douglas fir from single N solutions and from solutions containing both NH_4^+ and NO_3^-. *Plant and Soil* 151: 91-96.

Kinraide, T. (1994). Use of a Gouy-Chapman-Stern model for membrane-surface electrical potential to interpret some features of mineral rhizotoxicity. *Plant Physiology* 106: 1583-1592.

Kochian, L.V. (1991). Mechanisms of micronutrient uptake and translocation in plants. In *Micronutrients in Agriculture*, Second Edition, Eds. J.J. Mortvedt, F.R. Cox, L.M. Shuman, and R.M. Welch. Madison, WI: Soil Science Society of America, pp. 229-295.

Kochian, L.V. and W.J. Lucas (1982). Potassium transport in corn roots. I. Resolution of kinetics into a saturable and linear component. *Plant Physiology* 70: 1723-1731.

Komor, E. and W. Tanner (1974). The hexose-proton cotransport system of *Chlorella*, pH-dependent change in K_m values and the translocation constants of the uptake system. *Journal of General Physiology* 64: 568-581.

Komor, E. and W. Tanner (1975). Simulation of a high- and low-affinity sugar uptake system in *Chlorella* by a pH-dependent change in the K_m of the uptake system. *Planta* 123: 195-198.

Körner, L.E., P. Kjellbom, C. Larsson, and I.M. Moller (1985). Surface properties of right side-out plasma membrane vesicles isolated from barley roots and leaves. *Plant Physiology* 79: 72-79.

Kronzucker, H.J., M.Y. Siddiqi, and A.D.M. Glass (1995). Kinetics of NO_3^- influx in spruce. *Plant Physiology* 109: 319-326.

Kronzucker, H.J., M.Y. Siddiqi, and A.D.M. Glass (1996). Kinetics of NH_4^+ influx in spruce. *Plant Physiology* 110: 773-779.

Laine, P., A. Ourry, J. Macduff, J. Boucard, and J. Salette (1993). Kinetic parameters of nitrate uptake by different catch crop species: Effects of low temperatures or previous nitrate starvation. *Physiologia Plantarum* 88: 85-92.

Lane, J., D.W. McBride, and O. Hamill (1993). Ionic effects on amiloride block of the mechanosensitive channel in Xenopus oocytes. *British Journal of Pharmacology* 108: 116-119.

Lee, R.B. (1982). Selectivity and kinetics of ion uptake by barley plants following nutrient deficiency. *Annals of Botany* 50: 429-449.

Lee, R.B. (1993). Control of net uptake of nutrients by regulation of influx in barley plants recovering from nutrient deficiency. *Annals of Botany* 72: 223-230.

Lee, R.B. and M.C. Drew (1986). Nitrogen-13 studies of nitrate fluxes in barley roots. II. Effect of plant N-status on the kinetic parameters of nitrate influx. *Journal of Experimental Botany* 37: 1768-1779.

Lee, R.B. and R.G. Ratcliffe (1983). Phosphorus nutrition and the intracellular distribution of inorganic phosphate in pea root tips: A quantitative study using ^{31}P-NMR. *Journal of Experimental Botany* 34: 1222-1244.

Lee, R.B., R.G. Ratcliffe, and T.E. Southon (1990). ^{31}P-NMR measurements of the cytoplasmic and vacuolar Pi content of mature maize roots: Relationships with phosphorus status and phosphate fluxes. *Journal of Experimental Botany* 41: 1063-1078.

Lefebvre, D.D. and A.D.M. Glass (1982). Regulation of phosphate influx in barley roots: Effects of phosphate deprivation and reduction of influx with provision of orthophosphate. *Physiologia Plantarum* 54: 199-206.

Linask, J. and G.G. Laties (1973). Multiphasic absorption of glucose and 3-O-methylglucose by aged potato slices. *Plant Physiology* 51: 289-294.

Maathius, F.J.M. and D. Sanders (1996). Mechanisms of potassium absorption by higher plant roots. *Physiologia Plantarum* 96: 158-168.

Maathius, F.J.M., D. Verlin, F.A. Smith, D. Sanders, J.A. Fernandez, and N.A. Walker (1996). The physiological relevance of Na^+-coupled K^+-transport. *Plant Physiology* 112: 1609-1616.

Mäck, G. and R. Tischner (1994). Constitutive and inducible net NH_4^+ uptake of barley (*Hordeum vulgare* L.) seedlings. *Journal of Plant Physiology* 144: 351-357.

Marschner, H. (1995). *Mineral Nutrition of Higher Plants*, Second Edition. London, UK: Academic Press.

Marschner, H., M. Häussling, and E. George (1991). Ammonium and nitrate uptake rates and rhizosphere pH in non-mycorrhizal roots of Norway spruce (*Picea abies* (L.) Karst). *Trees* 5: 14-21.

McLaughlin, S., N. Mulrine, T. Gresalfi, G. Vaio, and A. McLaughlin (1981). Adsorption of divalent cations to bilayer membranes containing phosphatidylserine. *Journal of General Physiology* 77: 445-473.

McLaughlin, S., G. Szabo, and G. Eisenman (1971). Divalent ions and the surface potential of charged phospholipid membranes. *Journal of General Physiology* 58: 667-687.

McPharlin, I.R. and R.L. Bieleski (1987). Phosphate uptake by *Spirodela* and *Lemna* during early phosphorus deficiency. *Australian Journal of Plant Physiology* 14: 561-572.

Mimura, T., K.-J. Dietz, W. Kaiser, M.J. Schramm, G. Kaiser, and U. Heber (1990). Phosphate transport across biomembranes and cytosolic phosphate homeostasis in barley leaves. *Planta* 180: 139-146.

Mimura, T., R.J. Reid, and F.A. Smith (1998). Control of phosphate transport across the plasma membrane of *Chara*. *Journal of Experimental Botany* 49: 13-19.

Nagata, T. and G. Melchers (1978). Surface charge of protoplasts and their significance in cell-cell interaction. *Planta* 142: 235-238.

Newman, I.A., L.V. Kochian, M.A. Grusak, and W.J. Lucas (1987). Fluxes of H^+ and K^+ in corn roots: Characterization and stoichiometries using ion-selective microelectrodes. *Plant Physiology* 84: 1177-1184.

Nissen, P. (1971). Uptake of sulfate by roots and leaf slices of barley: Mediated by single, multiphasic mechanisms. *Physiologia Plantarum* 24: 315-324.

Nissen, P. (1980). Multiphasic uptake of potassium by barley roots of high and low potassium content: Separate sites for uptake and transitions. *Physiologia Plantarum* 48: 193-200.

Nissen, P. (1991). Multiphasic uptake mechanisms in plants. *International Review of Cytology* 126: 89-134.

Pace, G.M. and P.R. McClure (1986). Comparison of nitrate uptake kinetic parameters across maize inbred lines. *Journal of Plant Nutrition* 9: 1095-1111.

Peuke, A.D. and R. Tischner (1991). Nitrate uptake and utilization of aseptically cultivated Spruce seedlings, *Picea abies* (L.) Karst. *Journal of Experimental Botany* 42: 723-728.

Rebeille, R., R. Bligny, J.-B. Martin, and R. Douce (1983). Relationship between the cytoplasm and vacuole phosphate pool in *Acer pseudoplatanus* cells. *Archives of Biochemistry and Biophysics* 225: 143-148.

Reid, R.J., M.A. Tester, and F.A. Smith (1997). Voltage control of calcium influx in intact cells. *Australian Journal of Plant Physiology* 24: 805-810.

Ritchie, R.J. and T. Prvan (1996). A simulation study on designing experiments to measure the K_m of Michaelis-Menten kinetics curves. *Journal of Theoretical Biology* 178: 239-254.

Robards, A.W., S.M. Jackson, D.T. Clarkson, and J. Sanderson (1973). The structure of barley roots in relation to the transport of ions into the stele. *Protoplasma* 77: 291-311.

Sanders, D. (1986). Generalized kinetic analysis of ion-driven cotransport systems: II. Random ligand binding as a simple explanation for non-Michaelian kinetics. *Journal of Membrane Biology* 90: 67-87.

Sanders, D. (1990). Kinetic modeling of plant and fungal membrane transport systems. *Annual Review of Plant Physiology and Plant Molecular Biology* 41: 77-107.

Sanders, D. and U.-P. Hansen (1981). Mechanism of Cl^- transport at the plasma membrane of *Chara corallina:* II. Transinhibition and determination of H^+/

Cl^- binding order from a reaction kinetic model. *Journal of Membrane Biology* 58: 139-153.

Sandstrom, R.P. and R.E. Cleland (1989). Comparison of the lipid composition of oat root and coleoptile plasma membranes. Lack of short term change in response to auxin. *Plant Physiology* 90: 1207-1213.

Savageau, M.A. (1995). Michaelis-Menten mechanism reconsidered: Implications of fractal kinetics. *Journal of Theoretical Biology* 176: 115-124.

Schachtman, D.P. and J.L. Schroeder (1994). Structure and transport mechanism of a high-affinity potassium uptake transporter from higher plants. *Nature* 370: 655-658.

Scott, D.M. (1987). Sodium cotransport systems: Cellular, molecular and regulatory aspects. *Bioessays* 7: 71-78.

Siddiqi, M.Y. and A.D.M. Glass (1987). Regulation of K^+ influx in barley: Evidence for a direct control of influx by K^+ concentration of root cells. *Journal of Experimental Botany* 38: 935-947.

Siddiqi, M.Y., A.D.M. Glass, T.J. Ruth, and M. Fernando (1989). Studies of the regulation of nitrate influx by barley seedlings using $^{13}NO_3^-$. *Plant Physiology* 90: 806-813.

Siddiqi, M.Y., A.D.M. Glass, T.J. Ruth, and T.W. Rufty (1990). Studies on the uptake of nitrate in barley. 1. Kinetics of $^{13}NO_3$ influx. *Plant Physiology* 93: 1426-1432.

Skulachev, V.P. (1985). Membrane-linked energy transductions. Bioenergetic functions of sodium: H^+ is not unique as a coupling ion. *European Journal of Biochemistry* 151: 199-208.

Smart, D.R. and A.J. Bloom (1988). Kinetics of ammonium and nitrate uptake among wild and cultivated tomatoes. *Oecologia* 73: 336-340.

Soldal, T. and P. Nissen (1978). Multiphasic uptake of amino acids by barley roots. *Physiologia Plantarum* 43: 181-188.

Swiader, J.M. and F.G. Freiji (1996). Characterizing nitrate uptake in lettuce using very sensitive ion chromatography. *Journal of Plant Nutrition* 19: 15-27.

Tanner, W. and T. Caspari. (1996). Membrane transport carriers. *Annual Review of Plant Physiology and Plant Molecular Biology* 47: 595-626.

Tester, M. (1990). Plant ion channels: Whole cell and single channel studies. *New Phytologist* 114: 305-340.

Therios, I.N., S.A. Weinbaum, and R.M. Carlson (1979). Nitrate compensation points of several plum clones and relationship to nitrate uptake effectiveness. *Journal of the American Society for Horticultural Science* 104: 768-770.

Theuvenet, A.P.R. and G.W.F.H. Borst-Pauwels (1976). The influence of surface charge on the kinetics of ion translocation across biological membranes. *Journal of Theoretical Biology* 57: 313-329.

Thibault, O., N. Porter, and P. Landfield (1993). Low Ba^{2+} and Ca^{2+} induce a sustained high probability of repolarisation openings of L-type Ca^{2+} channels in hippocampal neurons: Physiological implications. *Proceedings of the National Academy of Sciences USA* 90: 11792-11796.

Tyerman, S.D. (1992). Anion channels in plants. *Annual Review of Plant Physiology and Plant Molecular Biology* 43: 351-373.

Ullrich, C.I. and A.J. Novacky (1990). Extra- and intracellular pH and membrane potential changes induced by K^+, Cl^-, $H_2PO_4^-$, and NO_3^- uptake and fusicoccin in root hairs of *Limnobium stoloniferum*. *Plant Physiology* 94: 1561-1567.

Van Bel, A.J.E., A.C. Borstlap, A. Pinxteren-Bazuine, and A. Van Ammerlaan (1982). Analysis of valine uptake by *Commelina* mesophyll cells in a biphasic active and a diffusional component. *Planta* 155: 335-341.

Vogelzang, S.A. and H.B.A. Prins (1994). Patch clamp analysis of the dominant plasma membrane K^+ channel in root cell protoplasts of *Plantago media* L. Its significance for the P and K state. *Journal of Membrane Biology* 141: 113-122.

Walker, N.A., R.J. Reid, and F.A. Smith (1993). The uptake and metabolism of urea by *Chara australis*: IV. Symport with sodium—a slip model for the high and low affinity systems. *Journal of Membrane Biology* 136: 263-271.

Wang, M.Y., M.Y. Siddiqi, T.J. Ruth, and A.D.M. Glass (1993). Ammonium uptake by rice roots. 2. Kinetics of $^{13}NH_4^+$ influx across the plasmalemma. *Plant Physiology* 103: 1259-1267.

Warncke, D.D. and S.A. Barber (1974). Nitrate uptake effectiveness of four plant species. *Journal of Environmental Quality* 3: 28-30.

Chapter 3

Molecular Biology of Nutrient Transporters in Plant Membranes

Frank W. Smith

Key Words: cotransporters, ion transport, membrane proteins, nitrogen transport, nutrient uptake, phosphate transport, potassium transport, sulfate transport.

INTRODUCTION

The uptake of nutrient solutes by plants, and their subsequent redistribution throughout the plant, necessitates passage through lipid membranes. Transfer of solutes through membranes is mediated and controlled by membrane proteins. The physiological characteristics of many of the processes that transport nutrients through plant membranes are well-known. The kinetics of nutrient uptake processes were elegantly described by experiments of Epstein and colleagues more than forty years ago (Epstein, 1953). The manner in which the processes are energized have also been established by electrophysiological studies. The molecular nature of the processes has been inferred from these studies and through analogy with transport processes in microbial and mammalian systems. However, the actual membrane proteins through which nutrients are transported into plant cells have been identified only recently. Application of molecular techniques has enabled the genes that encode some of these proteins to be cloned and the proteins to be identified. These developments offer exciting new insights into both the theory and practice of crop nutrition.

Transport of nutrients across membranes occurs at many places throughout a plant. Nutritionists have concentrated their attention on the initial uptake process by the root. However, nutrient ions also traverse membranes when they are unloaded from the symplast into xylem vessels, unloaded again following translocation, and loaded into leaves and storage organs, such as fruits, grains, tubers, and roots, or loaded into phloem for remobilization. It must also be noted that plants possess membranes other than the plasma membrane around the exterior of the cell's cytoplasm. An important example is the tonoplast membrane that surrounds cell vacuoles and through which many nutrients are moved for storage. The transport of ions in and out of these cell types and organelles requires involvement of membrane proteins. We should therefore expect plant genomes to encode a large number of membrane transport proteins.

This chapter will concentrate on molecular aspects of transporters involved in transport of nutrients from relatively low external concentrations. In contrast, the molecular biology of specific ion channels, commonly involved in transfer of solutes from higher external concentrations, is covered in some of the excellent recent reviews on the biochemistry and physiology of biological membranes (Tyerman, 1992; Kramer, 1994; Sussman, 1994; Maniol and Traxler, 1995; Tanner and Caspari, 1996).

ISOLATION OF ION TRANSPORTERS

Transporter proteins lose their tertiary structure, and hence their function, when separated from the lipid environment of the membrane in which they normally reside. This has made it difficult to detect the polypeptides involved in the transport of ions across plant membranes and has been an obstacle to isolation and purification of these proteins. It has therefore been necessary to employ indirect means to identify specific ion transporter proteins from plants and to ascertain their amino acid sequences, structure, and function. The methods that have led to success have involved isolation from plants of the genetic sequences that encode the transporter proteins rather than isolation of the proteins themselves.

Heterologous Expression of Genes Encoding Plant Ion Transporters in Yeast

The yeast *Saccharomyces cerevisiae* has played a major role in isolation of genes encoding plant transporter proteins. Similar to plants, *Saccharomyces cerevisiae* is a eukaryote. It is therefore possible to functionally express many genes from other eukaryotic organisms in yeast. This technology has been reviewed recently by Frommer and Ninnemann (1995) and Rentsch and Frommer (1996). Where it can be used, this cloning strategy has an advantage of not requiring prior knowledge of any part of the sequence of the gene or protein being sought. As such, it is particularly valuable in initial identification of members of a gene family.

Gene cloning by this strategy relies upon the capacity of an introduced gene to functionally complement a mutation in the haploid yeast, thereby restoring the metabolic capacity and/or growth of the mutant (see Figure 3.1). The articles of Smith and colleagues (1993) and Hawkesford and Smith (1997a) provide an illustration of the successful application of this approach to isolating specific plant ion transporters. A key requirement for isolating genes encoding ion transporters by this strategy is the availability

FIGURE 3.1. Isolation of Genes Encoding Plant Nutrient Transporters by Heterologous Expression in Yeast

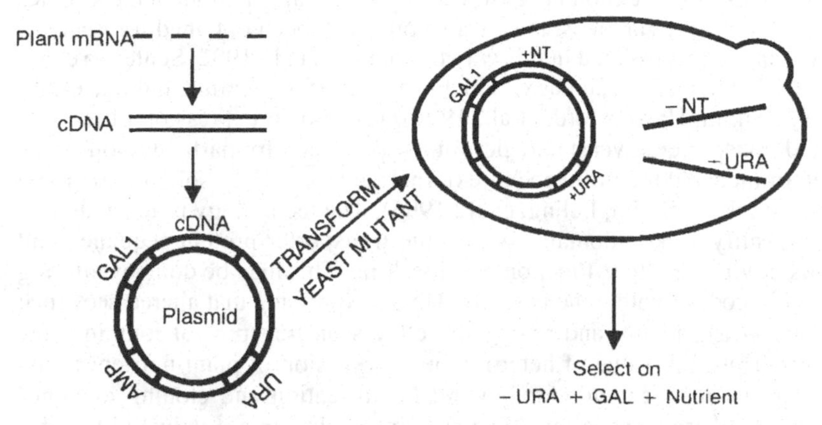

Note: cDNA is prepared by reverse transcription of plant mRNA and inserted in a "shuttle vector" that can replicate in both yeast and *Escherichia coli*. This plasmid vector contains markers that enable cells harboring the plasmid to be selected in yeast (URA) and *E. coli* (AMP). It also contains a strong promoter (GAL1) that drives transcription of the cDNA in yeast. The plasmid is introduced into a yeast strain that has a mutation in the gene corresponding to the auxotrophic marker (− URA) and a mutation (− NT) in the gene that normally encodes the transporter protein responsible for uptake of that nutrient in wild-type yeast. Transformed yeasts are grown out on medium that selects for the yeast auxotrophic marker (− URA), enables the promoter (GAL1) in the plasmid, and contains the nutrient of interest. Under these conditions, the only yeast cells that can grow are those transformed with plasmids that contain cDNAs which encode functional transporters (+NT) capable of restoring uptake of the nutrient. Shuttle vectors that contain other selectable markers and promoters are also available.

of a yeast mutant that lacks the capacity to take up the ion of interest. Such mutants can be naturally occurring or can be generated by either chemical or irradiation mutagenesis, as was done to create the specific mutant YSD1 used to clone plant SO_4^{2-} transporters (Smith, Hawkesford, et al., 1995). This mutant is then used to screen cDNA libraries constructed in a "shuttle vector" that replicates in yeast. This vector contains an auxotrophic marker (URA in Figure 3.1) for selection of yeast cells that contain the vector and a strong promoter (GAL1 in Figure 3.1) that drives transcription of the lengths of cDNA inserted in the vector. For isolation of plant ion transporters, cDNA is best prepared from mRNA derived from plant tissue known to have a high capacity to transport the ion of interest.

Successful isolation of cDNAs encoding plant nutrient transporters by this strategy results in identification of full-length, functional cDNAs. This often enables further characterization of these ion transporters through heterologous expression in yeast. For example, information on the kinetics of ion transport via the isolated transporter has been obtained for many of the transporters isolated in this way (Anderson et al., 1992; Sentenac et al., 1992; Ninnemann, Jauniaux, and Frommer, 1994; Smith, Ealing, et al., 1995; Smith, Hawkesford, et al., 1997). Functional expression of the isolated transporter in yeast may permit its specificity for particular ions to be determined and the influence of external factors such as solution pH to be characterized (Smith, Ealing, et al., 1995). The technology is also valuable for identifying key domains within the transporter protein sequence and for studying structure-function relationships. This may be done by altering specific codons within the isolated cDNA, expressing that altered sequence in the yeast mutant, and noting the effects on transport of ions into the yeast. Thus, the value of heterologous expression of plant nutrient transporters in yeast extends far beyond identification and cloning of genes encoding these transporters. The system has become a valuable tool for characterizing plant transporter proteins. However, as will be noted later in this chapter, caution is required, particularly where interactions between proteins may be involved.

Hybridization Screening for Plant Ion Transporters

Another technique that has been successfully employed to isolate genes that encode plant ion transporters is hybridization screening. This technique is based upon the specific pairing between C-G and A-T base residues within DNA. It is a well-documented technique (Maniatis, Fritsch, and Sambrook, 1992) that is commonly used by molecular biologists. The technique employs specific "probes" composed of lengths of single-stranded DNA that complement the sequence of the gene of interest and therefore bind specifically to that gene. Prior information on the likely sequence of the gene of interest is therefore required to design probes for successful isolation of genes by this technique. Such sequence information is being accumulated at an increasing pace. Once the first DNA sequences of families of plant ion transporters have been obtained, regions in which the sequence has been conserved during evolution can be identified and probes to isolate similar genes from the same species or other plant species can be designed. Part of an isolated sequence itself may be used as a probe. Alternatively, probes can be constructed using the polymerase chain reaction to amplify DNA from the target species by designing two short primers specific to conserved regions within that sequence.

Sequence data contained within international databases, together with computer searching techniques, now provide the necessary information required to identify many plant nutrient transporter sequences. Expressed sequence tags (ESTs—short sequences obtained from cDNA libraries), particularly those derived from *Arabidopsis* and rice, are proving to be a valuable source of information on potential plant nutrient transporter sequences. The availability of these ESTs has led to cloning of genes encoding root phosphate (Muchhal, Pardo, and Ragothama, 1996; Smith, Ealing, et al., 1997) and SO_4^{2-} transporters (Takahashi et al., 1996, 1997). Sequence data from organisms other than plants are also proving useful in recognizing plant transporter sequences. This arises because many of these sequences have been highly conserved during evolution and bear striking similarities, even across different classes of organisms. As examples, considerable homology exists between SO_4^{2-} transporters isolated from fungi, yeasts, plants, and mammals (Smith, Ealing, et al., 1995) and between phosphate transporters isolated from fungi, yeasts, and plants (Harrison and Van Buuren, 1995; Smith, Ealing, et al., 1997). These databases play a vital role in identification and isolation of genes associated with nutrient transport. It might be expected that as the major international genome sequencing programs yield their outputs, the amount of sequence information available will make isolation of genes that encode nutrient transport proteins in most plant species a relatively trivial exercise.

COMMON MOLECULAR FEATURES OF PLANT NUTRIENT TRANSPORTERS

The Topology of Plant Nutrient Transporters

The plant ion transporters isolated to date are relatively large polypeptides ranging from approximately 53 kDa to 75 kDa in size. This corresponds to polypeptides approximately 500 to 660 amino acids in length.

Being integral membrane proteins, all possess a number of highly hydrophobic regions that traverse the membrane. These hydrophobic regions, known as membrane-spanning domains (MSDs), are predicted from hydropathy analyses of the polypeptides. Such analyses, which examine the charge associated with particular sequences of amino acids within the polypeptide, are aided by computer models such as Kyte's and Doolittle's (1982), MEMSAT (Jones, Taylor, and Thornton, 1994), TMAP (Persson and Argos, 1994), and PHDhtm (Rost et al., 1995). Each MSD commonly contains seventeen to twenty-five amino acids arranged in a helix that traverses the membrane and is "capped" on either side. Intracellular and extracellular hydrophilic loops join the MSDs. It should be borne in mind that these analyses are only predictive in nature—confirmation of the structure of any of the plant ion transporters that have been isolated must await physical three-dimensional mapping of the tertiary structure of the protein embedded within its lipid membrane.

Hydrophobicity analyses of most of the plant nutrient transporters isolated to date predict twelve MSDs. However, depending upon the parameters set in the computer models, predictions may range from nine to fourteen MSDs for a number of the sequences that have been reported. An even number of MSDs results in both the N and C terminal hydrophilic tails of the protein lying on the same side of the membrane. This is usually the intracellular side where the tails may serve to anchor the transporter protein to the cell's cytoskeleton. An odd number of MSDs would result in one hydrophilic terminal end being extracellular and would raise the possibility that the terminus may act as a receptor of some external signal involved in activation of the protein.

Predictive models of plant ion transporters containing twelve MSDs are supported by analogy with the topology of similar molecules in other eukaryotic organisms. Many such transporters have been extensively studied and belong to a superfamily of proteins, the major facilitator superfamily (Marger and Saier, 1993). The characteristics and evolutionary significance of the various families and subfamilies of transporters have been reviewed by Saier and Reizer (1991). The major facilitator superfamily includes proteins that transport solutes as diverse as sugars, organic acids, and inorganic ions, such as phosphate and NO_3^-. In many of these proteins, the MSDs are arranged in a "6 + 6" pattern—two sets of six MSDs separated by a long intracellular loop. This arrangement is typical of those plant phosphate transporters that have been isolated to date (see Figure 3.2). However, it contrasts with the topology of the family of plant SO_4^{2-} transporters, which also appear to contain twelve MSDs but lack the large central intracellular loop (see Figure 3.2).

FIGURE 3.2. Topology of High-Affinity SO_4^{2-} and Phosphate Transporter Membrane Proteins Isolated from Plants

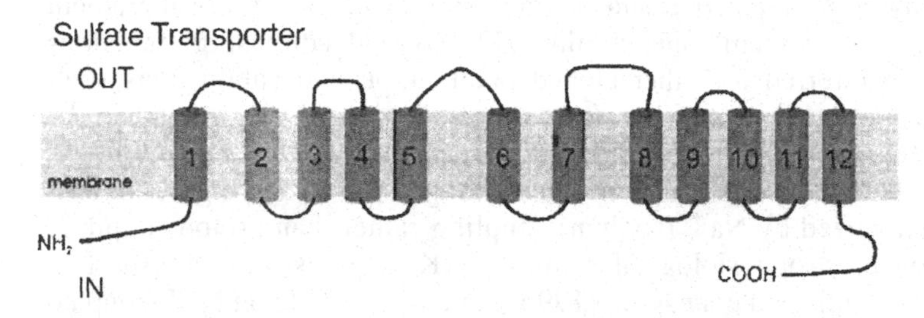

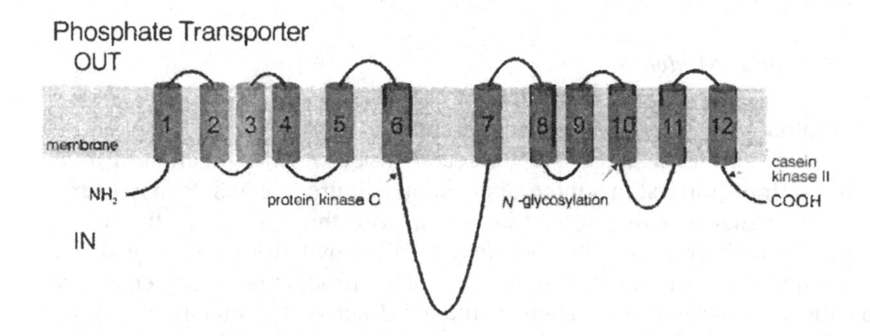

Source: Adapted from Smith, Ealing, et al. (1995, 1997).

Note: Each transporter contains twelve membrane spanning domains, but these are arranged in a "6 + 6" configuration in phosphate transporters. The location of potential sites for posttranslational modification of the phosphate transporter by phosphorylation (kinase sites) and *N*-glycosylation are shown.

Cotransporter Activity

Many nutrient uptake and transport processes are sensitive to external pH. Electrophysiological measurements indicate that this is often associated with the cotransport of protons by transporter proteins. A number of the isolated plant ion transporters exhibit this phenomenon. For example, when the SO_4^{2-} transporters isolated from *Stylosanthes* and barley were expressed in yeast, an increase in the external pH from 5.6 to 7.6 progressively reduced SO_4^{2-} uptake rates (Smith, Ealing, et al., 1995; Smith, Hawkesford, et al., 1997). It was also found that increasing the pH buffering capacity of the external medium reduced SO_4^{2-} uptake rates by these transporters. These results are consistent with SO_4^{2-} influx being coupled to H^+ (hydrogen) transport and energized by the electrochemical gradient across the plant membrane. Similar H^+-coupled cotransport functions have been inferred for other cloned plant anion and cation transporter proteins. Cotransport is a common feature of many of the proteins in the major facilitator superfamily (Marger and Saier, 1993). For many mammalian ion transporters and some aquatic angiosperms and algae, cotransport is energized by Na^+ (sodium) coupling rather than proton coupling. Based on electrophysiological studies on K^+ (potassium) transport in plants, Maathuis and colleagues (1996) have concluded that Na^+-coupled K^+ transport has no, or limited, physiological significance in terrestrial plant species.

A Molecular Model

Proton-coupled cotransport and the predicted topology of the plant ion transporter proteins suggests a molecular model for the function of these nutrient transporters. In a three-dimensional figure, the MSDs may form a doughnut-shaped structure that acts as a pore through which the solute, together with cotransported protons, could move from one side of the membrane to the other (see Figure 3.3). This process would be energized by the electrochemical gradient maintained across the membrane by the action of a H^+-ATPase, the proton pump. Critical domains, possibly on the external hydrophilic loops and/or on the interior surfaces of the pore formed by the MSDs, may determine the specificity of the transporter for particular ions. Such domains may also determine the "binding capacity," hence the affinity of the transport process and the K_m.

Regulation of Transporters

The rate of transport of an ion across a plant membrane is determined by the number and activity of transporters in the membrane. The number of transporters is ultimately determined by the rates of transcription of the genes encoding the transporters, the rates of decay of the mRNAs, and the half-lives of the transporter proteins themselves. Those studies of the regulation of plant nutrient transporters which have been done suggest that they are rapidly turned over in the membranes. Physiological studies with inhibitors of protein synthesis (Hawkesford and Belcher, 1991; Clarkson et al., 1992) and molecular studies (Smith, Hawkesford, et al., 1997) indicate that the turnover rate of the HVST1 SO_4^{2-} transporter protein in

FIGURE 3.3. Model of the Proposed Molecular Mechanism of SO_4^{2-} Transport Through a Plant Membrane by a High-Affinity SO_4^{2-}-H^+ Symporter

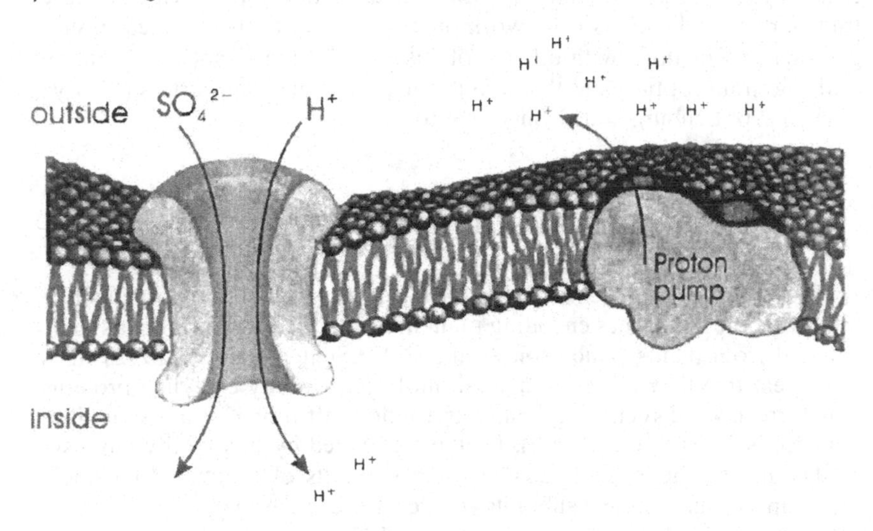

Note: Transport is energized by a large integral membrane proton-ATPase that pumps H^+ to the outside of the membrane.

barley roots is approximately 2.5 hours and the mRNA has a half-life of less than two hours. Such turnover rates would enable rapid adjustment of transport rates by direct transcriptional control. Transcription is switched on through the promoter region that lies upstream of the coding region of a gene. High levels of expression of some genes encoding nutrient transporters (Smith, Ealing, et al., 1995; Muchhal, Pardo, and Ragothama, 1996) indicate that plants can rapidly increase the number of transporters in the plasma membranes when necessary.

The short-term fine-tuning of nutrient transport rates is probably achieved by regulating the activity of the transporter proteins. Although isolated plant ion transporters are now available, very little work has been done on regulation of their activity. Many of the transporter proteins that have been isolated have sites that could potentially interact with other molecules. Notably, many transporters contain sites that could be phosphorylated or dephosphorylated, thereby altering the activity of the protein. Similarly, sites that might be glycosylated can be found on many plant transporter sequences. Conservation of such sites on most of the phosphate transporter sequences reported to date (see Figure 3.2) suggests that they may play an important role in regulating the activity of these transporters. Indications from work on phosphate transport in yeasts suggest that interactions with other proteins may be important in regulating both the transcription and the activity of phosphate transporters (Bun-ya et al., 1996; Lenburg and O'Shea, 1996).

POTASSIUM TRANSPORT

The *Arabidopsis* genes *AKT1* and *KAT1* are historically important since they were the first genes encoding plant membrane transport proteins to be isolated from plants (Anderson et al., 1992; Sentenac et al., 1992). It is now clear that they represent a small multigene family encoding proteins that form inward rectifying channels for low-affinity K^+ transport into plant cells (Smart et al., 1996). Proteins encoded by this family have six MSDs and are believed to be the basic subunits of functional channels made up of four identical subunits (Chilcott et al., 1995).

High affinity K^+ transport is mediated by a number of transporters. The first of these to be isolated was *HKT1*, cloned from wheat roots (Schachtman and Schroeder, 1994). *HKT1* encodes a protein of 58.9 kDa predicted to contain twelve MSDs. The K_m for K^+ when *HKT1* is expressed in yeast is 29 μM. *HKT1* is primarily expressed in root cortical cells. However, controversy has arisen regarding the physiological significance of HKT1 after it was found that HKT1 acted as a Na^+-coupled K^+ transporter (Rubio, Gassmann, and Schroeder, 1995). Studies of Maathuis and colleagues (1996) and Walker, Sanders, and Maathuis (1996) have shown that high-affinity K^+ uptake could proceed in the absence of Na^+, suggesting that additional transporters may be involved in high-affinity K^+ uptake into plant cells. This has recently been confirmed by isolation of another distinct family of K^+ transporters from *Arabidopsis* (Quintero and Blatt, 1997; Fu and Luan, 1998; Kim et al., 1998). Several members of this *AtKUP* family of K^+ transporters have been identified and sequenced. They have over 700 amino acids and are predicted to contain twelve MSDs. Members of this family have distinct patterns of expression, and some members are induced by K^+ starvation. An important characteristic of one member of the family, the *AtKUP1* transporter expressed in *Arabidopsis* roots (Fu and Luan, 1998), is that it exhibits both high- and low-affinity uptake characteristics when expressed in yeast. This single protein therefore displays "biphasic" enzyme kinetics. The high-affinity phase has a K_m of 44 μM and the low-affinity phase a K_m of 11 mM, with a transition between the two phases occurring at about 200 μM. This single transporter therefore operates over a very wide range of external K^+ concentrations. It is now obvious that many genes encode high-affinity K^+ transporters, and this may account for the apparent inconsistencies that have been observed in the characteristics of high-affinity K^+ uptake by plant cells.

SULFATE TRANSPORT

The first high-affinity anion transporters to be cloned from plants were SO_4^{2-} transporters (Smith, Ealing, et al., 1995). Using the yeast complementation strategy, three genes encoding SO_4^{2-} transporters (*SHST1, SHST2,* and *SHST3*) were isolated from the tropical forage legume species *Stylosanthes* and one (*HVST1*) from barley (Smith, Hawkesford, et al., 1997). These transporters are predicted to contain twelve MSDs and appear to represent two subfamilies of SO_4^{2-} transporters (Hawkesford and Smith, 1997b). *SHST1, SHST2,* and *HVST1* were shown to be expressed in roots, where they encode high-affinity transporters with a K_m for SO_4^{2-} of 10 μM. These proteins are likely to be the primary transporters involved in uptake of SO_4^{2-} by root cells. *SHST3* belongs to a different subfamily. It is expressed in both roots and leaves and encodes a transporter with a higher K_m for SO_4^{2-} of 100 μM. It has been proposed that SHST3 is involved in the internal intercellular transport of SO_4^{2-}. More recently, EST clones have been used to isolate a number of homologues of these genes from *Arabidopsis* (Takahashi et al., 1996; 1997), increasing the members of this family of plant transporters.

Expression of the high-affinity SO_4^{2-} transporters *SHST1, SHST2,* and *HVST1* is regulated by the S (sulfur) status of the plant. Transcription of these genes is down-regulated in plants adequately supplied with SO_4^{2-}, and low levels of the corresponding mRNAs are found in roots. If the plants are deprived of an external SO_4^{2-} supply, however, there are marked increases in the rates of transcription of these genes, which result in accumulation of high levels of the corresponding mRNAs, particularly near the root tips. Upon resupplying SO_4^{2-} to these S-starved plants, transcription is again down-regulated, and both the levels of the corresponding mRNAs and the capacity of the roots for SO_4^{2-} uptake decline to low levels within a few hours. This negative feedback regulation of genes encoding high-affinity SO_4^{2-} transporters in plant roots is consistent with physiological data on the effects of SO_4^{2-} status of the plant on the capacity of the roots to take up SO_4^{2-} (Clarkson, Smith, and Vanden Berg, 1983; Hawkesford and Belcher, 1991).

Another form of regulation of the transcription of these plant root SO_4^{2-} transporters has been demonstrated in barley (Smith, Hawkesford, et al., 1997). *O*-acetylserine, a precursor of the important S amino acid cysteine, was found to up-regulate transcription of the gene encoding the HVST1 transporter in barley roots, even when those roots had a high external supply of SO_4^{2-}. This represents a positive regulatory control by a key metabolite involved in the assimilation of SO_4^{2-} within the cell.

These data serve to illustrate that transcriptional regulation of genes encoding nutrient transporters plays a major role in controlling the nutrient content of plant cells. They also demonstrate that this control can be very responsive to changes in both the supply of the nutrient and the internal demand for that nutrient for synthesis of key metabolites.

PHOSPHATE TRANSPORT

The importance of P (phosphate) in agricultural systems ensured that genes encoding plant phosphate transporters were early targets for isolation and molecular characterization. Following the cloning of the first of these genes from *Arabidopsis* using an EST-derived probe (Muchhal, Pardo, and Ragothama, 1996; Smith, Ealing, et al., 1997), sequences have been published from potato (Leggewie, Willmitzer, and Riesmeier, 1997), tomato (Liu, Muchhal, et al., 1998), and *Medicago truncatula* (Liu, Trieu, et al., 1998). Additional genes encoding phosphate transporters have also been isolated from *Arabidopsis* (Mitsukawa, Okumura, and Shibata, 1997), barley (F. W. Smith, D. H. Cybinski, and A. L. Rae, unpublished results), and rice (R. M. Godwin and F. W. Smith, unpublished results). As might be expected, a number of phosphate transporters appear to be associated with different functions within any plant species. All of the genes that have been isolated exhibit a high level of homology with one another and with phosphate transporters isolated from yeast (Bun-ya et al., 1991), mycorrhizal fungi (Harrison and Van Buuren, 1995), and filamentous fungi (Versaw, 1995). These key transporters have been highly conserved during evolution. The proteins are approximately 520 to 540 amino acids in length and have twelve MSDs arranged in the "6 + 6" configuration (see Figure 3.2). As noted earlier, they contain conserved sites that may permit posttranslational modification by phosphorylation and/or glycosylation.

Regrettably, plant P transporters are not expressed well in the heterologous yeast expression system. This has slowed down full characterization of these transporters. Although some kinetic data have been reported from heterologous expression in a yeast mutant that lacks a functional PHO84 phosphate transporter (Leggewie, Willmitzer, and Riesmeier,

1997; Liu, Trieu, et al., 1998), K_m values in the range 130 to 280 μM appear to be too high for effective transport of phosphate from the low concentrations that occur in the vicinity of plant root cells. These concerns may be well-founded, since a K_m for phosphate of 3.1 μM was measured when one of the P transporters isolated from *Arabidopsis* was overexpressed in cultured tobacco cells (Mitsukawa et al., 1997).

Possible reasons for poor expression of plant phosphate transporters in yeast are worth exploring, since they may point to additional requirements for fully functional phosphate transporters in plants. Genetic and molecular studies with yeast have identified genes that encode a number of proteins which interact with the yeast phosphate transporter PHO84 (Bun-ya, Harashima, and Oshima, 1992; Bun-ya et al., 1996; Yompakdee, Bun-ya, et al., 1996; Yompakdee, Ogawa, et al., 1996). It seems likely that these yeast proteins may not interact effectively with phosphate transporters derived from plants, resulting in poor functional expression of the plant transporters in yeast. This raises the important issue of whether a fully effective phosphate transport system in plants requires interaction between the phosphate transporter proteins that have been isolated and unidentified plant proteins similar to those in yeast.

Most genes isolated to date that encode phosphate transporters are solely or primarily expressed in plant roots. Their expression is regulated by the P status of the plant. Depriving plants of an external P supply induces expression of these genes and results in marked increases in the levels of phosphate transporter transcript in roots (Muchhal, Pardo, and Ragothama, 1996; Leggewie, Willmitzer, and Riesmeier, 1997; Smith, Ealing, et al., 1997; Liu, Muchhal, et al., 1998; Liu, Trieu, et al., 1998). Feedback regulation of the expression of these genes is consistent with physiological measurements showing increased capacity for phosphate uptake by roots of phosphate-deprived plants (Clarkson and Scattergood, 1982; Lefebvre and Glass, 1982; Cogliatti and Clarkson, 1983; Clarkson and Lüttge, 1991). Studies with genes encoding the tomato phosphate transporters have indicated that their expression is localized in the epidermal cells in roots (Liu, Muchhal, et al., 1998), making them well placed to function in high-affinity phosphate uptake from the soil solution. One of these phosphate transporter genes, *LePT1*, was also expressed in the palisade parenchyma and phloem cells of the leaves of phosphate-starved plants, suggesting it may also be involved in internal transport of phosphate.

Mycorrhizal fungi play an important role in the P nutrition of most plants. The molecular mechanisms involved in the uptake and transfer of phosphate by the symbiotic association between plants and mycorrhizal fungi are therefore of keen interest. A gene encoding a phosphate transporter, *GvPT*, has been isolated from the VA mycorrhizal fungus *Glomus versiforme* by Harrison and Van Buuren (1995). This transporter is 521 amino acids in length and very similar in structure to those phosphate transporters from yeast, fungi, and plants already described. *GvPT* has a K_m for phosphate of 18 μM when expressed in yeast, but this value may be subject to similar reservations as those which apply to K_m measurements of plant phosphate transporters expressed in this system. *GvPT* is primarily expressed in external hyphae and is likely to be involved in the initial uptake of phosphate into the hyphae from the soil solution.

It is probable that at least two other phosphate transporters are involved in this symbiotic relationship: one in the efflux of phosphate from the fungus at the arbuscule interface and the other in the uptake of phosphate by plant cortical cells at this interface. Efforts are under way to isolate these two transporters. Further work at Harrison's laboratory has resulted in the cloning of two genes encoding phosphate transporters in the host plant *Medicago truncatula* (Liu, Trieu, et al., 1998). These genes, *MtPT1* and *MtPT2* are expressed in roots, and their expression is regulated by the phosphate status of the plant in the manner already described. Initial evidence suggests that it is unlikely that either MtPT1 or MtPT2 are involved in phosphate transport at the interface between mycorrhizal arbuscules and root cortical cells. However, it was found that expression of both *MtPT* genes was down-regulated following colonization of *M. truncatula* with a mycorrhizal fungus. Interestingly, this down-regulation appears to result from the mycorrhizal infection itself rather than from the improved phosphate transport to the plant arising from mycorrhizal infection. If this can be verified on a wider scale, it is likely to have an important bearing on what role plant root phosphate transporters play in plants infected with mycorrhizal fungi.

NITROGEN TRANSPORT

Uptake of N (nitrogen) by plant roots and translocation of primary nitrogenous compounds throughout plants involve transport of NO_3^-, NH_4^+, and amino acids across a number of membrane barriers. These membranes lie in many different organs and cell types so it is expected that plants possess a large number of membrane proteins involved in the transport of N compounds.

Nitrate is the major N compound absorbed from the soil by most plant species. It is also a form of N readily translocated through xylem and stored in shoot tissues for later remobilization and reduction. *Arabidopsis* mutants generated by T-DNA tagging and screened with chlorate, a toxic analogue of NO_3^-, have led to isolation of the *CHL1* gene (Tsay et al., 1993). This gene encodes a transporter protein of 590 amino acids containing twelve MSDs arranged in a "6 + 6" pattern. Electrophysiological studies have indicated that the *CHL1* gene encodes a low-affinity NO_3^--H^+ symporter with a K_m of 8.5 mM (Huang et al., 1996). *CHL1* is mainly expressed in roots and is regulated by both positive and negative feedback, transcription being induced by NO_3^- but attenuated after NO_3^- accumulates in the tissues. Subsequently, additional genes in the same family have been identified (Crawford, 1995). Interestingly, some of these genes appear to be involved in peptide transport.

Two high-affinity NO_3^- transporter genes, *BCH1* and *BCH2*, have been isolated from barley roots (Trueman, Richardson, and Forde, 1996). These genes encode 507 and 509 amino acid proteins, respectively, that are predicted to contain twelve MSDs. Their topology indicates that they belong to the major facilitator superfamily, but the central loop is shorter than in many other "6 + 6" members of the family, and they have a long hydrophilic tail. These important transporters are closely related to the *crn* NO_3^- transporters from the fungus *Aspergillus* and the *nar* NO_3^- transporters from the alga *Chlamydomonas* (Quesada, Galvan, and Fernandez, 1994). The *BCH* genes are expressed in barley roots and are rapidly induced by NO_3^-, but not by NH_4^+. This is consistent with physiological observations on the induction of the NO_3^- transport system (Jackson, Flesher, and Hageman, 1973; Touraine, Clarkson, and Muller, 1994). Analyses of the barley genome suggest that it may contain seven to ten related genes.

Ammonium is an important source of N for many plants. There has been debate for many years over the mechanism of NH_4^+ transport, since NH_3 appears to diffuse across many membranes. However, NH_4^+ uptake has been shown to be affected by genetic mutations, suggesting that proteins may be involved in NH_4^+ transport. An NH_4^+ transporter has been isolated from *Arabidopsis* using yeast complementation (Ninnemann, Jauniaux, and Frommer, 1994). This gene encodes a protein of 501 amino acids that contains 9 to 12 MSDs. This protein mediates high-affinity NH_4^+ transport. It is expressed in roots, stems, and leaves.

Amino acids, amines, and ureides are major forms of reduced N that are transported throughout plants. Yeast complementation technology has recently enabled identification of genes that encode the membrane proteins involved in the transport of these important compounds. Members of several gene families have been isolated from *Arabidopsis* and characterized (Frommer, Hummel, and Riesmeier, 1993; Hsu et al., 1993; Frommer, Hummel, and Rentsch, 1994). These transporters seem to recognize a variety of amino acids with different affinity, a property which may be an important adaptive mechanism that enables cells to scavenge for a range of amino acids. A detailed examination of the complex molecular and physiological aspects of this rapidly developing field of amino acid transport is beyond the scope of this chapter. The topic has been covered in the excellent reviews of Frommer and colleagues (1994), Frommer and Ninnemann (1995), and Von Wiren, Gazzarrini, and Frommer (1997); interested readers are referred to those articles.

TOOLS FOR PHYSIOLOGICAL STUDIES

Isolation of genes encoding membrane proteins involved in nutrient uptake and transport in plants is opening up new horizons for physiologists and plant nutritionists. These breakthroughs have placed powerful new tools in their hands. The isolated genetic sequences and their protein products are being used to help delineate the molecular mechanisms underlying ion transport processes. Capabilities now exist for unraveling the complexity of ion transport through membranes by combining modern electrophysiological methods with heterologous expression of isolated nutrient transporters in *Xenopus* oocytes, yeast, *E. coli*, and cultured plant cells (Kim et al., 1998). These methodologies will also assist in defining structure/function relationships associated with these transporters.

Isolated clones of nutrient transporters provide an opportunity to manipulate the expression of that specific gene in a transgenic plant. These transgenic systems enable overexpression of the gene and antisense down-regulation of the gene. They therefore provide opportunities for delineating the function of particular transporters and for studying the effects of nutrient and environmental stresses on the functioning of specific transport processes.

Of particular interest to physiologists is identifying the cells in which various transporters operate and how these transporters respond to nutrient stress and environmental variables. Isolated clones permit specific probes to be constructed for in situ hybridization studies that identify the tissues and cell types in which the transcripts are being expressed, thus enabling the effects of environmental variables to be determined. The protein sequences themselves can also be used in a similar way to generate specific antibodies for immunocytology studies.

Among the most exciting uses to which these molecular tools are being applied are studies on the coordinate regulation of nutrient transport. The promoter regions of these genes have domains in them that respond to stimuli which are ultimately associated with the growth and development of the plant and the environment in which it is being grown. These domains determine whether the gene will be transcribed. Orderly growth and development and the maintenance of chemical homeostasis by cells (Mimura, 1995; Mimura, Sakano, and Shimmen, 1996) require coordinated regulation of the expression of genes encoding nutrient transporters. Such regulatory circuits are bound to be complex, as has been found in simpler bacteria and fungi. However, these new molecular tools are now enabling physiologists to progress in this important area of plant science. It is likely that as the number of plant transporters available increases and their identification is simplified through the international genome sequencing programs, attention will focus more strongly upon their regulation.

IMPLICATIONS FOR AGRICULTURE AND THE ENVIRONMENT

Transformation techniques for many important agricultural species are now available. These techniques enable specific genetic traits to be introduced into these plants. They may be particularly useful where available germplasm does not include sufficient variation in some desirable trait. The availability of clones that encode nutrient transporters provides an opportunity to introduce these into agricultural species in a manner that would enable their expression to be manipulated. For instance, by putting their expression under control of different promoters, the feedback regulation that attenuates expression of many of these transporters could be overridden. Although far from certain, it is possible that intervention of this type may offer new approaches to improving the use of nutrients by crop plants. The feasibility of improving the efficiency of nutrient uptake by this technology is already being assessed with transgenic plants overexpressing high-affinity nutrient transporters. Improving internal use of nutrients by manipulating expression of transporters involved in the internal cycling of those nutrients may also warrant investigation. Of particular interest is the quality and nutrient composition of grain. Opportunities may arise to increase or reduce the accumulation of particular nutrients or assimilates in grain by altering expression of the transporters responsible for grain loading.

The environment may also benefit from the application of these technologies. Increased efficiency of nutrient utilization by crop plants could reduce problems associated with chemical fertilizers. Plants or organisms with a high capacity to take up and accumulate nutrients and heavy metals, for instance, might play a role in reclaiming contaminated sites. Or, plants with increased sensitivity to nutrient accumulation might be used to monitor potential environmental problems.

Although one can speculate on potential agricultural and environmental applications of this technology, a great deal of caution is required. Clarkson and Hawkesford (1993) have drawn attention to quantitative problems in attempting to increase the efficiency of NO_3^- uptake by plants. The toxic effects of excessive accumulation of phosphate, as is likely to occur by overexpression of a high-affinity phosphate transporter, are well documented. Plants survive and grow because they have evolved regulatory controls to maintain the nutrient content of their cells within well-defined limits that are related to the rate at which nutrients are required for growth. Manipulating the expression of genes encoding nutrient transporters so that nutrient concentrations exceed those limits is unlikely to be successful. A far better understanding of the coordinate regulation of nutrient transport processes and their linkages to growth and development is required before we can appreciate the implications of these types of molecular interventions.

REFERENCES

Anderson, J.A., S.S. Huprikar, L.V. Kochian, W.J. Lucas, and R.F. Gaber (1992). Functional expression of a probable *Arabidopsis thaliana* potassium channel in *Saccharomyces cerevisiae*. *Proceedings of National Academy of Sciences USA* 89: 3736-3740.

Bun-ya, M., S. Harashima, and Y. Oshima (1992). Putative GTP-binding protein, *Grt1*, associated with the function of the Pho84 inorganic phosphate transporter in *Saccharomyces cerevisiae*. *Molecular and Cellular Biology* 12: 2958-2966.

Bun-ya, M., M. Nishimura, S. Harashima, and Y. Oshima (1991). The *PHO84* gene of *Saccharomyces cerevisiae* encodes an inorganic phosphate transporter. *Molecular and Cellular Biology* 11: 3229-3238.

Bun-ya, M., K. Shikata, S. Nakade, C. Yompakdee, S. Harashima, and Y. Oshima (1996). Two new genes, *PHO86* and *PHO87*, involved in inorganic phosphate uptake in *Saccharomyces cerevisiae*. *Current Genetics* 29: 344-351.

Chilcott, T.C., S. Frost-Shartzer, M.W. Iverson, D.F. Garvin, L.V. Kochian, and W.J. Lucas (1995). Potassium transport kinetics of *KAT1* expressed in *Xenopus* oocytes: A proposed molecular structure and field effect mechanism for membrane transport. *C.R. Academy of Science Paris Life Sciences* 318: 761-771.

Clarkson, D.T. and M.J. Hawkesford (1993). Molecular biological approaches to plant nutrition. In *Plant Nutrition—From Genetic Engineering to Field Practice*, Ed. N.J. Barrow. Dordrecht, Netherlands: Kluwer Academic Publishers, pp. 23-33.

Clarkson, D.T., M.J. Hawkesford, J.-C. Davidian, and C. Grignon (1992). Contrasting responses of sulphate and phosphate transport in barley (*Hordeum vulgare*

L.) roots to protein-modifying reagents and inhibition of protein synthesis. *Planta* 187: 306-314.

Clarkson, D.T. and U. Lüttge (1991). Mineral nutrition: Inducible and repressible nutrient transport systems. *Progress in Botany* 52: 61-83.

Clarkson, D.T. and C.B. Scattergood (1982). Growth and phosphate transport in barley and tomato plants during development of and recovery from phosphate stress. *Journal of Experimental Botany* 33: 865-875.

Clarkson, D.T., F.W. Smith, and P.J. Vanden Berg (1983). Regulation of sulphate transport in a tropical legume, *Macroptilium atropurpureum* cv. Siratro. *Journal of Experimental Botany* 34: 1463-1483.

Cogliatti, D.H. and D.T. Clarkson (1983). Physiological changes in phosphate uptake by potato plants during development of and recovery from phosphate deficiency. *Physiologia Plantarum* 58: 287-294.

Crawford, N.M. (1995). Nitrate: Nutrient and signal for plant growth. *Plant Cell* 7: 859-868.

Epstein, E. (1953). Mechanism of ion absorption by roots. *Nature* 171: 83-84.

Frommer, W.B. and O. Ninnemann (1995). Heterologous expression of genes in bacterial, fungal, animal, and plant cells. *Annual Review of Plant Physiology and Plant Molecular Biology* 46: 1463-1483.

Frommer, W.B., S. Hummel, and D. Rentsch (1994). Cloning of an *Arabidopsis histidine* transporting protein related to nitrate and peptide transporters. *FEBS Letters* 10: 63-67.

Frommer, W.B., S. Hummel, and J.W. Riesmeier (1994). Expression cloning in yeast of a cDNA encoding a broad specificity amino acid permease from *Arabidopsis thaliana*. *Proceedings of National Academy of Sciences USA* 90: 5944-5948.

Frommer, W.B., M. Kwart, B. Hirner, W.N. Fischer, S. Hummel, and O. Ninnemann (1994). Transporters for nitrogenous compounds in plants. *Plant Molecular Biology* 26: 1651-1670.

Fu, H-H. and S. Luan (1998). *AtKUP1*: A dual-affinity K^+ transporter from *Arabidopsis*. *Plant Cell* 10: 63-67.

Harrison, M.J. and M.L. Van Buuren (1995). A phosphate transporter from the mycorrhizal fungus *Glomus versiforme*. *Nature* 378: 626-629.

Hawkesford, M.J. and A.R. Belcher (1991). Differential protein-synthesis in response to sulfate and phosphate deprivation—Identification of possible components of plasma-membrane transport-systems in cultured tomato roots. *Planta* 185: 323-329.

Hawkesford, M.J. and F.W. Smith (1997a). Molecular biology of higher plant sulphate transporters. In *Sulphur Metabolism in Higher Plants*, Eds. W.J. Cram, L.J. De Kok, I. Stulen, C. Brunold, and H. Rennenberg. Leiden, Netherlands: Backhuys Publishers, pp. 13-25.

Hawkesford, M.J. and F.W. Smith (1997b). Structure and function of two higher plant sulfate transporters. In *Plant Nutrition—For Sustainable Food Production and Environment*, Eds. T. Ando, K. Fujita, T. Mae, H. Matsumoto, S. Mori, and J. Sekiya. Dordrecht, Netherlands: Kluwer Academic Publishers, pp. 233-234.

Hsu, L-C., T-J. Chiou, L. Chen, and D.R. Bush (1993). Cloning a plant amino acid transporter by functional complementation of a yeast amino acid transport mutant. *Proceedings of National Academy of Sciences USA* 90: 7441-7445.

Huang, N.C., C.S. Chiang, N.M. Crawford, and Y.F. Tsay (1996). *CHL1* encodes a component of the low-affinity nitrate uptake system in *Arabidopsis* and shows cell type-specific expression in roots. *Plant Cell* 8: 2183-2191.

Jackson, W.A., D. Flesher, and R.H. Hageman (1973). Nitrate uptake by dark-growing corn seedlings. *Plant Physiol*ogy 51: 120-127.

Jones, D.T., W.R. Taylor, and J.M. Thornton (1994). A model recognition approach to the prediction of all-helical membrane protein structure and topology. *Biochemistry* 33: 3038-3049.

Kim, E.J., J.M. Kwak, N. Uozumi, and J.I. Schroeder (1998). *AtKUP1*: An *Arabidopsis* gene encoding high-affinity potassium transport activity. *Plant Cell* 10: 51-62.

Kramer, R. (1994). Functional principles of solute transport systems: Concepts and perspectives. *Biochimica et Biophysica Acta* 1185: 1-34.

Kyte, J. and R.F. Doolittle (1982). A simple method for displaying the hydropathic character of a protein. *Journal of Molecular Biology* 157: 105-132.

Lefebvre, D.D. and A.D.M. Glass (1982). Regulation of phosphate influx in barley roots: Effects of phosphate deprivation and reduction of influx with provision of orthophosphate. *Physiologia Plantarum* 54: 199-206.

Leggewie, G., L. Willmitzer, and J.W. Riesmeier (1997). Two cDNAs from potato are able to complement a phosphate uptake-deficient yeast mutant: Identification of phosphate transporters from higher plants. *Plant Cell* 9: 381-392.

Lenburg, M.C. and E.K. O'Shea (1996). Signalling phosphate starvation. *Trends in Biochemical Science* 21: 383-387.

Liu, C., U.S. Muchhal, M. Uthappa, A.K. Kononowicz, and K.G. Ragothama (1998). Tomato phosphate transporter genes are differentially regulated in plant tissues by phosphorus. *Plant Physiology* 116: 91-99.

Liu, H., A.T. Trieu, L.A. Blaycock, and M.J. Harrison (1998). Cloning and characterization of two phosphate transporters from *Medicago truncatula* roots: Regulation in response to phosphate and to colonization by arbuscular mycorrhizal (AM) fungi. *Molecular Plant-Microbe Interactions* 11: 14-22.

Maathuis, F.J.M., D. Verlin, F.A. Smith, D. Sanders, J.F. Fernandez, and N.A. Walker (1996). The physiological relevance of Na^+-coupled K^+-transport. *Plant Physiology* 112: 1609-1616.

Maniatis, T., E.F. Fritsch, and J. Sambrook (1992). *Molecular Cloning: A Laboratory Manual.* Cold Spring Harbor, NY: Cold Spring Harbor Laboratory Press.

Maniol, C. and B. Traxler (1995). Membrane protein assembly: Genetic, evolutionary and medical perspectives. *Annual Review of Genetics* 29: 131-150.

Marger, N.D. and M.H. Saier (1993). A major superfamily of transmembrane facilitators that catalyse uniport, symport and antiport. *Trends in Biochemical Science* 18: 13-20.

Mimura, T. (1995). Homeostasis and transport of inorganic phosphate in plants. *Plant and Cell Physiol*ogy 36: 1-7.

Mimura, T., K. Sakano, and T. Shimmen (1996). Studies on the distribution, re-translocation and homeostasis of inorganic phosphate in barley leaves. *Plant, Cell and Environment* 19: 311-320.

Mitsukawa, N., S. Okumura, and D. Shibata (1997). High-affinity phosphate transporter genes of *Arabidopsis thaliana*. In *Plant Nutrition—For Sustainable Food Production and Environment*, Eds. T. Ando, K. Fujita, T. Mae, H. Matsumoto, S. Mori, and J. Sekuja. Dordrecht, Netherlands: Kluwer Academic Publishers, pp. 187-190.

Mitsukawa, N., S. Okumura, Y. Shirano, S. Sato, T. Kato, S. Harashima, and D. Shibata (1997). Overexpression of an *Arabidopsis thaliana* high-affinity phosphate transporter gene in tobacco cultured cells enhances cell growth under phosphate-limited conditions. *Proceedings of National Academy of Sciences USA* 94: 7098-7102.

Muchhal, U.S., J.M. Pardo, and K.G. Ragothama (1996). Phosphate transporters from the higher plant *Arabidopsis thaliana*. *Proceedings of National Academy of Sciences USA* 93: 10519-10523.

Ninnemann, O., J-C. Jauniaux, and W.B. Frommer (1994). Identification of a high affinity NH_4^+ transporter from plants. *EMBO Journal* 13: 3464-3471.

Persson, B. and P. Argos (1994). Prediction of transmembrane segments in proteins utilising multiple sequence alignments. *Journal of Molecular Biology* 237: 182-192.

Quesada, A., A. Galvan, and E. Fernandez (1994). Identification of nitrate transporter genes in *Chlamydomonas reinhardtii*. *Plant Journal* 5: 407-419.

Quintero, F.J. and M.R. Blatt (1997). A new family of K^+ transporters from *Arabidopsis* that are conserved across phyla. *FEBS Letters* 415: 206-211.

Rentsch, D. and W.B. Frommer (1996). Molecular approaches towards an understanding of loading and unloading of assimilates in higher plants. *Journal of Experimental Botany* 47: 1199-1204.

Rost, B., R. Casadio, P. Fariselli, and C. Sander (1995). Transmembrane helices predicted at 95% accuracy. *Protein Science* 4: 521-533.

Rubio, F., W. Gassmann, and J.I. Schroeder (1995). Sodium-driven potassium uptake by the plant potassium transporter *HKT1* and mutations conferring salt tolerance. *Science* 270: 1660-1663.

Saier, M.H. and J. Reizer (1991). Families and superfamilies of transport proteins common to prokaryotes and eukaryotes. *Current Opinion in Structural Biology* 1: 362-368.

Schachtman, D.P. and J.I. Schroeder (1994). Structure and transport mechanism of a high-affinity potassium uptake transporter from higher plants. *Nature* 370: 655-658.

Sentenac, H., N. Bonneaud, M. Minet, F. Lacroute, J-M. Salmon, F. Gaymard, and C. Grignon (1992). Cloning and expression in yeast of a plant potassium ion transport system. *Science* 256: 663-665.

Smart, C.J., D.F. Garvin, J.P. Prince, W.J. Lucas, and L.V. Kochian (1996). The molecular basis of potassium nutrition in plants. *Plant and Soil* 187: 81-89.

Smith, F.W., P.M. Ealing, B. Dong, and E. Delhaize (1997). The cloning of two *Arabidopsis* genes belonging to a phosphate transporter family. *Plant Journal* 11: 83-92.

Smith, F.W., P.M. Ealing, M.J. Hawkesford, and D.T. Clarkson (1995). Plant members of a family of sulfate transporters reveal functional subtypes. *Proceedings of National Academy of Sciences USA* 92: 9373-9377.

Smith, F.W., M.J. Hawkesford, P.M. Ealing, D.T. Clarkson, P.J. Vanden Berg, A.R. Belcher, and A.G.S. Warrilow (1997). Regulation of expression of a cDNA from barley roots encoding a high affinity sulphate transporter. *Plant Journal* 12: 875-884.

Smith, F.W., M.J. Hawkesford, I.M. Prosser, and D.T. Clarkson (1993). Approaches to cloning genes encoding for nutrient transporters in plants. *Plant and Soil* 155/156: 139-142.

Smith, F.W., M.J. Hawkesford, I.M. Prosser, and D.T. Clarkson (1995). Isolation of a cDNA from *Saccharomyces cerevisiae* that encodes a high affinity sulphate transporter at the plasma membrane. *Molecular and General Genetics* 247: 709-715.

Sussman, M.R. (1994). Molecular analysis of proteins in the plant plasma membrane. *Annual Review of Plant Physiology and Plant Molecular Biology* 45: 211-234.

Takahashi, H., N. Sasakura, M. Noji, and K. Saito (1996). Isolation and characterization of a cDNA encoding a sulfate transporter from *Arabidopsis thaliana*. *FEBS Letters* 392: 95-99.

Takahashi, H., M. Yamazaki, N. Sasakura, A. Watanabe, T. Leustek, J.A. Engler, G. Engler, M. Van Montagu, and K. Saito (1997). Regulation of sulfur assimilation in higher plants: A sulfate transporter induced in sulfate starved roots plays a central role in *Arabidopsis thaliana*. *Proceedings of National Academy of Sciences USA* 94: 11102-11107.

Tanner, W. and T. Caspari (1996). Membrane transport carriers. *Annual Review of Plant Physiology and Plant Molecular Biology* 47: 595-626.

Touraine, B., D.T. Clarkson, and B. Muller (1994). Regulation of nitrate uptake at the whole plant level. In *A Whole Plant Perspective on Carbon-Nitrogen Interactions*, Eds. J. Roy and E. Garnier. The Hague, Netherlands: SPB Academic Publishing, pp. 11-30.

Trueman, L.J., A. Richardson, and B.G. Forde (1996). Molecular cloning of higher plant homologues of the high-affinity nitrate transporters of *Chlamydomonas reinhardtii* and *Aspergillus nidulans*. *Gene* 175: 223-231.

Tsay, Y., J.I. Schroeder, K.A. Feldmann, and N.G. Crawford (1993). The herbicide sensitivity gene *CHL1* of *Arabidopsis* encodes a nitrate-inducible nitrate transporter. *Cell* 72: 705-713.

Tyerman, S.D. (1992). Anion channels in plants. *Annual Review of Plant Physiology and Plant Molecular Biology* 43: 351-373.

Versaw, W.K. (1995). A phosphate-repressible, high-affinity phosphate permease is encoded by the *pho-5* $^{+}$ gene of *Neurospora crassa*. *Gene* 153: 135-139.

Von Wiren, N., S. Gazzarrini, and W.B. Frommer (1997). Regulation of mineral nitrogen uptake in plants. *Plant and Soil* 196: 191-199.

Walker, N.A., D. Sanders, and F.J.M. Maathuis (1996). High-affinity potassium uptake in plants. *Science* 273: 977-978.

Yompakdee, C., M. Bun-ya, K. Shikata, N. Ogawa, S. Harashima, and Y. Oshima (1996). A putative new membrane protein, Pho86p, in the inorganic phosphate uptake system of *Saccharomyces cerevisiae*. *Gene* 171: 41-47.

Yompakdee, C., N. Ogawa, S. Harashima, and Y. Oshima (1996). A putative membrane protein, Pho88p, involved in inorganic phosphate transport in *Saccharomyces cerevisiae*. *Molecular and General Genetics* 251: 580-590.

Chapter 4

Long-Distance Nutrient Transport in Plants and Movement into Developing Grains

Pieter Wolswinkel

Key Words: cell turgor, cereals, developing seeds, legumes, macronutrients, micronutrients, phloem, pressure-flow mechanism, xylem transport.

INTRODUCTION

Two main pathways exist for long-distance transport of water and solutes in vascular plants. Long-distance transport from the root to the shoot occurs predominantly in the nonliving xylem vessels. Xylem transport is driven by a water-potential gradient from the soil through the plant to the atmosphere. The gradient in water potential between root and shoot is quite steep, particularly during the day when the stomata are open and transpiration is high. Under certain conditions in the shoot, a counterflow of water in the xylem may also occur, for example, from low-transpiring fruits back to the leaves (Pate et al., 1985; Hamilton and Davies, 1988; Lang and Thorpe, 1989; Marschner, 1995). Under conditions of low transpiration, the rate of xylem volume flow from the root to parts of the shoot may be strongly dependent on root pressure (Marschner, 1983, 1995).

In contrast to the xylem, long-distance transport in the phloem takes place in the living cells of the sieve tubes. The predominant direction of flow within this system is usually from photosynthetically active structures that serve as sources of assimilates (typically the leaves) to sinks (e.g., young leaves and growing fruits). In addition, phloem transport is an important component in cycling of mineral nutrients between shoots and roots.

Mature seeds contain carbohydrates and/or lipids as a carbon source, together with proteins and minerals. Mineral nutrients and assimilates containing C (carbon) and N (nitrogen) are translocated to developing fruits and seeds through xylem and phloem. Work on the physiology and other aspects of seed development has almost entirely been restricted to a small number of commercially important annuals, particularly legumes and cereals; this chapter on nutrient transport concentrates particularly on developing grains of such species.

In 1930, Münch introduced the "apoplast-symplast concept" for understanding solute flow in xylem and phloem of the higher plant. This concept and terminology involved will be valuable in the discussion that follows on water and solute movement in vascular plants. The interconnecting cell walls (including intercellular spaces) and the water-filled xylem elements can be considered as a single system called the apoplast. This is, in a sense, the "dead" part of the plant (Salisbury and Ross, 1992) and includes all cell walls and space outside plasma membrane-bound regions of the plant. The conducting elements in the xylem are thus part of the apoplast. Despite the blockage caused by the Casparian strips in the root, the ascent of sap in a plant could take place entirely in the apoplast, particularly the xylem. The rest of the plant, the "living" part, Münch called the symplast. This includes the cytoplasm of all cells. The symplast is a multicellular unit because the protoplasts of adjoining cells are connected through the plasmodesmata. Transport in the symplast includes cell-to-cell transfer over short distances via plasmodesmata and over longer distances through the sieve elements of the phloem.

In the vascular bundles, conducting cells of phloem and xylem are separated by only a few cells. In the regulation of long-distance transport of nutrients, exchange of solutes between xylem and phloem is very important for the mineral nutrition of plants. Import and export of mineral nutrients occur simultaneously during the life of plant organs, such as leaves. As a rule, aging (senescence) is associated with higher rates of export of mineral nutrients than rates of import and, thus, a decrease in amount per organ, such as in a leaf. In this chapter, the term remobilization will be used for this decrease in net nutrient content.

Nutrient transport into developing grains is strongly dependent on phloem transport. In recent years, studies on assimilate transport into developing seeds have contributed to a better understanding of factors controlling phloem transport into developing seeds; this chapter will focus on these studies.

LONG-DISTANCE TRANSPORT IN XYLEM AND PHLOEM

In plants, xylem and phloem accompany each other into most branches of the vascular network. This implies that throughout its length, vascular tissue has an extensive potential for solute exchange between xylem and phloem. Specialized cells called transfer cells are involved in the transfer of solutes between the xylem and phloem (Pate and Gunning, 1972). For comprehensive reviews of xylem transport and phloem transport of nutrients, including exchange between xylem and phloem, the reader is referred to earlier publications (Pate, 1975; Ziegler, 1975; Marschner, 1983, 1995). With the exception of Ca (calcium), the concentration of all mineral nutrients is usually several times higher in the phloem sap than in the xylem sap (e.g., Hocking, 1980). A consequence of the very low mobility of Ca in phloem and its good mobility in the xylem is that the Ca/K (potassium) ratio of an organ is lower the more the phloem supply predominates over the xylem supply (Ziegler, 1975; Marschner, 1983, 1995).

Xylem Transport

Composition of the Xylem Sap

Scientists have used different techniques to recover fluids from the xylem. One technique, applicable principally to herbaceous species, utilizes sap exuding under root pressure from plants cut at or near the base of their shoots. Solute concentrations in bleeding sap are likely to be higher than in the xylem of an intact, actively transpiring plant. Another technique is to extract the xylem sap from shoots or woody twigs by suction (vacuum extraction) or to force xylem sap from shoot xylem vessels by applying pressure. In addition, tissues outside the functioning xylem may also release solutes.

Xylem fluids collected by these methods have acid reaction (pH in the range of about 5.5 to 6.0 or somewhat higher; Pate, 1975, and references therein; Hocking, 1980; Milburn and Baker, 1989; Welch, 1995) and contain relatively low levels of dry matter (0.05 to 0.4 percent w/v; Pate, 1975), mostly in inorganic form. Major inorganic cations present are K, Ca, Mg (magnesium), Na (sodium), often in that order of decreasing concentration, while the main inorganic anions recorded are usually P (phosphorus), Cl (chlorine), S (sulfur), and, in certain species, NO_3^-.

The form and proportion of the various N fractions in the xylem sap is influenced by several factors, such as the form of N supply, the predominant site of NO_3^- reduction (roots or shoots), and the proportion of recycled N being transported (Marschner, 1995). Except at high external NH_4^+ supply, the concentration of NH_4^+ in the xylem is low (Van Beusichem, Kirkby, and Baas, 1988; experiments with *Ricinus communis*). In corn, NH_4^+ was found in the range of 1 mM, irrespective of whether N was supplied as NH_4^+ or NO_3^-. When supplied as NH_4^+, N was incorporated into amino compounds within the corn roots, and transport of N in the xylem occurred in the form of reduced N, mainly as amides (Engels and Marschner, 1993). The presence in certain plant species of amides such as glutamine and asparagine, and of the ureides allantoin and allantoic acid in other taxonomic groups of plants, may be an important factor in minimizing the carbon costs for root-to-shoot transport of assimilated N via the xylem and transport to developing fruits via the phloem. These N-rich compounds are characterized by high (> 0.4) N/C ratios (Streeter, 1979).

The concentration of organic acids in the xylem is influenced by the cation-anion uptake ratio (Triplett, Barnett, and Blevins, 1980) and the form of N supply (Arnozis and Findenegg, 1986). Sugar concentrations in the xylem sap of herbaceous species are usually low or even zero.

Xylem transport of micronutrients has been reviewed by Welch (1995); he focused on Fe (iron), Zn (zinc), Mn (manganese), Cu (copper), Ni (nickel), B (boron), Mo (molybdenum), and Cl. Marschner (1995) has also broadly discussed transport of micronutrients, including xylem transport. The complex organic and inorganic composition, pH, and E_H (standard redox potential) of the xylem sap will affect the types and amounts of micronutrient species transported in the xylem sap (Welch, 1995). White and co-workers have described detailed studies on metal complexation in xylem fluid. Stem exudate was collected from topped soybean and tomato plants grown in control and Zn-phytotoxic nutrient solutions and characterized in terms of natural metal-complex equilibria in a biological fluid (White, Chaney, and Decker, 1981; White, Decker, and Chaney, 1981). In addition, White, Baker, and colleagues (1981) have concluded that Fe was bound by citric acid, while Cu was bound by several amino acids in the exudate from control plants (mostly asparagine and histidine in soybean, and histidine, glutamine, and asparagine in tomato). In both plant species, Zn, Mn, Ca, and Mg were bound to citric and malic acids. Micronutrient-complexing phytometallophores have been identified in xylem sap (e.g., for Fe, see Welch, 1995).

Phytohormones are usual constituents of xylem sap, particularly cytokinins. Recently, concentration of abscisic acid (ABA) in the xylem sap has attracted wide interest as a possible root signal to the shoot of the root water status (Marschner, 1995).

Mechanism of Xylem Transport

Long-distance transport through the xylem from the root to the shoot consists of a mass flow of water, ions, and certain organic solutes upward from the root (Salisbury and Ross, 1992). The driving force is the gradient in decreasing (more negative) water potentials from the soil through the plant to the atmosphere. Water moves from the soil, through root tissues into the vascular tissues of the root, up through the xylem elements in the stem, into the leaves, and finally, by transpiration, through the stomata into the atmosphere. Under certain conditions, root pressure may be an important factor for the movement of a solution, via the xylem, from root to shoot. Transport of Ca via the xylem into low-transpiring organs therefore may strongly depend on the root pressure (Marschner, 1983; Kirkby and Pilbeam, 1984; Marschner, 1995).

After radial transport in the symplast of roots, most ions are released into the xylem part of the stele. Such a release into nonliving xylem vessels represents a retransfer from the symplast into the apoplast. Amino acids and organic acids, synthesized in the root cells, are also transported into the xylem. The mechanism of the release of ions and organic solutes into xylem ("xylem loading") is not well understood (Marschner, 1995). Nevertheless, there is general agreement that xylem loading is regulated separately from the ion uptake in root cortical cells.

Along the pathway of solute transport in the nonliving xylem vessels (representing a part of the apoplast), important interactions take place between solutes and (1) the walls of the vessels and (2) the cell walls of the surrounding xylem parenchyma. The major interactions are exchange adsorption of polyvalent cations in the cell wall and resorption (uptake) and release of mineral elements and of organic solutes by surrounding living cells (xylem parenchyma and phloem).

The long-distance transport of cations in the xylem can be compared with ion movement in a cation exchanger with a corresponding decline in the translocation rate with the distance travelled (Senden and Wolterbeek, 1990; Marschner, 1995). This cation-exchange adsorption is not restricted to the xylem vessels; the cell walls of the surrounding tissue also take part in exchange reactions (Wolterbeek, Van Luipen, and De Bruin, 1984). When the cation-exchange reactions are minimized by the ion complexation, the translocation rate of heavy metal cations in the xylem is enhanced, for example, in the case of Zn (McGrath and Robson, 1984) or Cd (cadmium) (Senden and Wolterbeek, 1990).

Solutes can also be resorbed from the xylem (apoplast) into the living cells along the xylem pathway from the roots to the leaves (e.g., function of xylem parenchyma cells). The composition of the xylem sap along the transport pathway can also be changed by the release or secretion of solutes from the surrounding cells (Pate and Jeschke, 1995).

Phloem Transport

Composition of the Phloem Sap

The phloem sap has a high pH (7.5 to 8.0, or somewhat higher) and a much higher dry matter content (15 to 25 percent w/v) than the xylem sap (Pate, 1975; Ziegler, 1975; Milburn and Baker, 1989; Marschner, 1995). In most angiosperms, sucrose is the most common, and usually most abundant, organic solute in the phloem solution (up to about 90 percent of dry matter of the sap; around 300 to 900 mM) (Ziegler, 1975; Zimmermann and Ziegler, 1975; Milburn and Baker, 1989), while in several groups of plants, for example, the Fabaceae, sucrose is the only sugar detected in the sieve-tube sap. Gamalei (1989) has stressed that sucrose is the characteristic transport sugar for seasonal herbaceous plants (e.g., annual crop plants in temperate climates). Amino compounds (amino acids and amides) are usually second to sugars in order of abundance in sieve-tube exudate and are present in concentrations higher than those in the xylem sap (up to 1 percent w/v; Pate, 1975; Ziegler, 1975; Milburn and Baker, 1989). The amides glutamine and asparagine may represent up to 90 percent of this fraction, whereas concentrations of NO_3^- and NH_4^+ are usually low or practically zero. Because developing seeds and fruits transpire at a low rate, if at all, they essentially subsist on a diet made up of phloem sap (Salisbury and Ross, 1992). Although, in higher plants, uptake of NO_3^- and NH_4^+ by the root is the major source of N, incorporation into organic compounds, such as amino acids and amides, is a prerequisite for N translocation into developing seeds. Organic acid anions such as citrate and malate are also present in the phloem sap. A broad range of other organic compounds can be found in the phloem sap, for example, vitamins, hormones, proteins, and ATP.

One method to collect phloem sap is the use of sucking insects such as aphids and plant hoppers. In the process of feeding, these insects insert their stylet into sieve tubes. If the stylet is severed, for example, with a laser beam (Hayashi and Chino, 1990), it remains in the tissue, and the high internal pressure within the sieve tubes forces phloem sap out of the open end of the stylet.

With respect to mineral elements in the phloem sap, K is usually present in the highest concentration, followed by P, Mg, and S (Ziegler, 1975; Milburn and Baker, 1989). The elements Cl and Na may also be present at high concentrations, depending on the external supply and the plant species (Jeschke and Pate, 1991). In contrast, concentration of Ca in the phloem sap is low.

In contrast to macronutrients, the reliable data on micronutrient concentrations in the phloem sap are relatively rare (e.g., Hocking, 1980). However, it can be concluded that, with the exception of Ca, the concentration of all mineral elements is usually several times higher in the phloem sap than in the xylem sap.

Welch (1995) has reviewed several aspects of phloem transport of micronutrients (see also Marschner, 1983, 1995). Relevant factors that influence the loading, transport, and unloading of micronutrients in the phloem sap include pH, E_H, and organic constituents within the sap. Phloem transport of Fe has been studied by several groups, particularly with respect to the chemical form in which this micronutrient moves through the phloem (Maas et al., 1988; Stephan and Scholz, 1993). Iron is transported in the phloem sap of *Ricinus communis* as a complex of approximately 2.4 kDa size, as determined by gel exclusion chromatography (Maas et al., 1988). Scholz and co-workers have described the presence of nicotianamine in plant tissues and its role as a chelator of several divalent metals (Scholz et al., 1988; Becker et al., 1989; Stephan and Scholz, 1993). The data collected by this group with the tomato mutant *chloronerva* (lacking nicotianamine) support the view that nicotianamine is essential for phloem transport of Fe and, possibly, has an important role in phloem transport of other heavy metals as well. A further study of metal-nicotianamine complexes and other metal complexes may lead to a better understanding of transport of micronutrients in the phloem. For more details on phloem transport of micronutrients, the reader is referred to recent reviews (Marschner, 1995; Welch, 1995).

There are groups of mineral elements with a high mobility, an intermediate mobility, and a low mobility in the phloem (Pate, 1975; Ziegler, 1975; Marschner, 1995; Welch, 1995). The nutrients K, Mg, P, S (regularly in reduced form), N (amino-N), Cl, and Na have high mobility in the phloem. In contrast, Ca has low mobility in the phloem. The nutrients Fe, Zn, Cu, and Mo have been placed in the group with intermediate mobility in the phloem, and Mn has been classified as having intermediate (Ziegler, 1975) or low mobility (Marschner, 1995). Welch (1995) has commented that the mobility of Mn in the phloem may be intermediate or low, depending on the Mn status of the plant species and the source and sink organs studied (see also Hill, Robson, and Loneragan, 1979; Nable and Loneragan, 1984). Ziegler (1975) has placed B in the group with low mobility in the phloem; in contrast, in recent reviews, B has been placed in the group with intermediate mobility in the phloem (Marschner, 1995; see also Welch, 1995).

Mechanism of Phloem Transport

Phloem transport has been described in comprehensive reviews (Zimmermann and Milburn, 1975; Baker and Milburn, 1989). In the phloem sieve tubes, sucrose and other substances flow in solution from regions of production (the leaves) to regions of consumption, referred to, respectively, as "source" and "sink."

Long-distance transport of assimilates via the phloem pathway is strongly dependent on phloem loading and unloading in the source and sink regions, respectively. According to the hypothesis proposed by Münch (1930), the flow of the phloem solution is driven by a turgor pressure difference between sections of the phloem pathway (Milburn, 1975; Delrot, 1989; Milburn and Kallarackal, 1989; Salisbury and Ross, 1992). Pressure-driven mass flow, as described by Münch, is still the most widely accepted mechanism for long-distance transport via the sieve tubes. The difference in turgor pressure results from an "active" (or energy-requiring) and selective accumulation of solutes (phloem loading) at the source (Giaquinta, 1983; Delrot, 1989) and removal of solutes (phloem unloading) at the sink (Thorne, 1985; Wolswinkel, 1985, 1992; Patrick, 1990; Patrick and Offler, 1995, 1996; Fisher and Oparka, 1996). Continued entry of sugar at the source and removal at the sink, with the accompanying osmotic movement of water into the source end and out of the sink end of the sieve-tube system, maintains the flow.

The development of the empty-ovule technique in three different laboratories (Patrick, 1983; Thorne and Rainbird, 1983; Wolswinkel and Ammerlaan, 1983) (see Figure 4.1; see also Figure 8.25 in Salisbury and Ross, 1992) has permitted new lines of research on assimilate transport into developing seeds. The osmotic environment of seed tissues has a strong effect on the rate of phloem transport into empty seeds and on solute efflux into the seed apoplast (Wolswinkel, 1990, 1992; Wolswinkel, Ammerlaan, and Koerselman-Kooij, 1992). A treatment of seed tissues with a solution having an osmotic potential of approximately −1.0 MPa (e.g., a solution containing 400 mM mannitol or sucrose) can be used as a "control treatment" (imitating the situation in intact seeds; Wolswinkel and Koerselman-Kooij, 1992). This is based on several points discussed in more detail elsewhere (Wolswinkel, 1990, 1992). There is much evidence

FIGURE 4.1. Illustration of the Empty-Ovule Technique

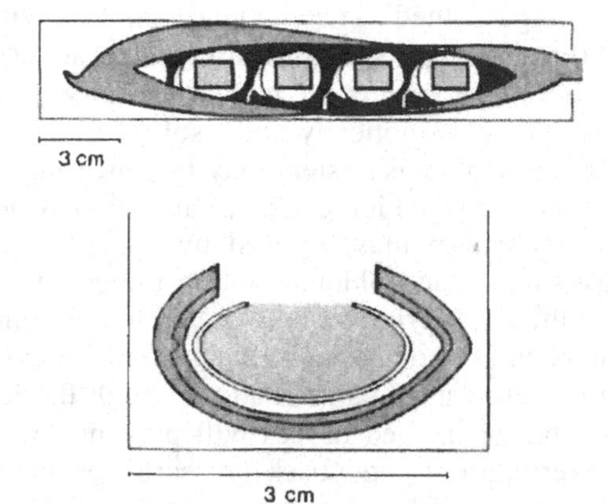

Source: Reproduced from Wolswinkel and Ammerlaan (1983), by kind permission from Springer-Verlag, Stuttgart, Germany.

Note: The diagram shows a developing pod of *Vicia faba* with a window cut in its wall and a window cut in the seed coat of each of a series of four developing seeds. After removal of the embryo, each empty ovule can be filled with a buffer solution (illustrated by a cross-section of an ovule; some authors have used an agar medium). Unloading of assimilates occurs into the seed-coat apoplast and the assimilates, which would normally be absorbed by the developing embryo, diffuse into the buffer or agar trap.

in several plant species that high solute concentration occurs in the apoplast of developing seeds, for example, an apoplast solution with an osmotic potential similar to that of at least 300 to 400 mM mannitol. High solute concentrations in the sink apoplast have also been observed in vegetative storage tissues of plants such as sugarbeet and sugarcane (Wolswinkel, 1985; Welbaum and Meinzer, 1990; Fisher and Oparka, 1996). In pulse-labeling experiments in which empty seeds of three species of dicotyledonous plants were filled with a solution of approximately 400 mM sucrose or mannitol, transport of sucrose and amino acids into attached empty ovules was almost equal to that into intact ovules within the fruit (Wolswinkel and Ammerlaan, 1984; Wolswinkel, 1992). Sucrose and amino transport into attached empty ovules was strongly reduced by low osmolality of the medium (e.g., a solution without osmoticum; see also Grusak and Minchin, 1988).

The high sink strength of developing seeds, at least in many taxonomic groups of dicotyledonous plants, is brought about by the sink end of the phloem pathway being "bathed" in an apoplast solution with a high concentration of osmotically active solutes. Unlike the basic model of Münch (1930), in which both osmometers are bathed in water in the same container, the concentration of osmotically active solutes in the apoplast of the sink end of the source-path-sink system may be much higher than in the apoplast of the source end (see Figure 4.2; see also discussion in Wolswinkel, 1992). This mechanism may be used by many plants to bias the partitioning of assimilates and phloem-mobile mineral ions in favor of reproductive and other high-priority sinks (Lang and Düring, 1991). Evidence suggests that a turgor homeostat mechanism maintains cell turgor of the seed coats at or below a set value (Patrick, 1990; Patrick and Offler, 1995). A low cell turgor in seed tissues will prevent the sink end of a mass-flow path exerting a strong "back pressure" on phloem transport into that sink (Wolswinkel and Koerselman-Kooij, 1992).

In contrast to data for dicotyledons, transport of assimilates into corn pedicel cups or perfused wheat grains was not reduced by the absence of a high solute concentration in the apoplast of maternal seed tissues (Porter, Knievel, and Shannon, 1987; Wang and Fisher, 1994a). Other factors may stimulate phloem transport into developing wheat grains. Lee and Atkey (1984) have discussed water loss from the developing caryopsis of wheat, suggesting that the rate of water loss from the developing wheat grain may be consistent with a transpiration-driven system of assimilate transport to the developing grain via the phloem. If this is a relevant mechanism, it may also have a role in the translocation of phloem-mobile mineral ions into developing grains.

Transfer Between the Xylem and the Phloem

In two classical reviews, Pate (1975, 1986) has described several aspects of the transfer of solutes between the xylem and the phloem. In a more recent review, Marschner (1995) has also paid much attention to this transfer. Organic and inorganic solutes can be transferred from the xylem to the phloem all along the pathway from roots to shoot, with the stem playing an important role. Transfer cells (Pate and Gunning, 1972) are involved in this process. In stems, the nodes represent sites of intensive xylem-to-phloem transfer (Pate and Jeschke, 1995). Information is scarce about the opposite process, phloem-to-xylem transfer (Marschner, 1995). However, an important role of phloem-to-xylem transfer of mineral ions has been demonstrated in certain situations (Martin, 1982; Jeschke, Pate, and Atkins, 1987).

FIGURE 4.2. Diagram of a Pressure-Flow Mechanism

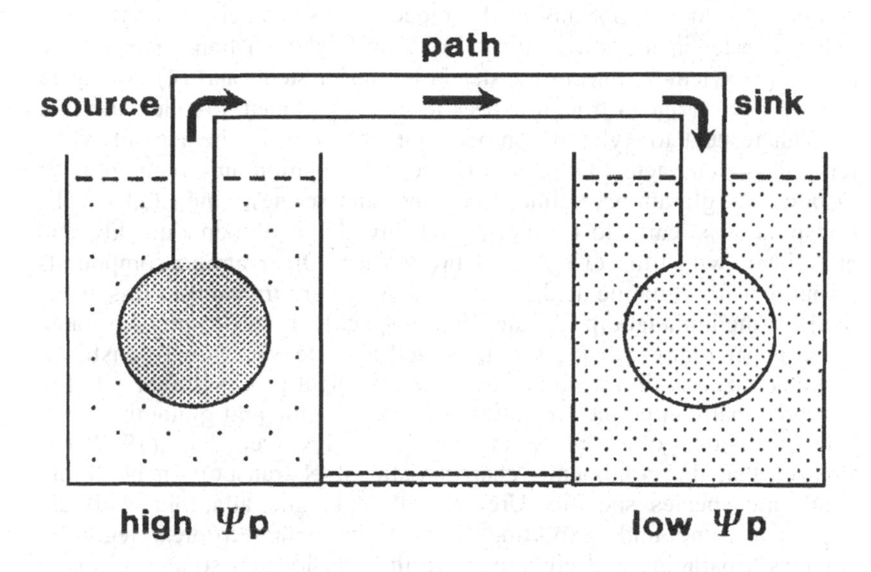

Source: Reproduced from Wolswinkel (1992), by kind permission from CAB International, Wallington, UK.

Note: This diagram (Münch, 1930) shows a high concentration of osmotically active solutes in the sink apoplast, leading to a further reduction of turgor pressure at the sink end of the phloem pathway. The concentration of solutes in the source apoplast is generally low. The xylem pathway between developing grains and source regions of the plant (representing a part of the apoplast) is much less developed than the phloem pathway.

Pate and co-workers have published a large amount of data on the exchange of solutes between xylem and phloem. This group has put a strong emphasis on the N-partitioning system of nodulated legumes (Pate, 1975, 1980, 1986, 1989, and references therein), but data on transfer of mineral solutes have also been collected (e.g., Hocking and Pate, 1977; Jeschke, Pate, and Atkins, 1987; Pate, 1989). In Pate's laboratory, a large-seeded annual grain legume, white lupin (*Lupinus albus* L.), has been the main plant species used for research. Other legume species have also been studied (e.g., *Pisum sativum* L. and *Vigna unguiculata* (L.) Walp.). With respect to short-distance exchanges of specific amounts of C and N between the adjacent transport channels of xylem and phloem, Pate (1989) has mentioned four components in the model for N, namely, (1) xylem-to-xylem transfer in the lower stem; (2) xylem-to-phloem transfer in mature leaves; (3) xylem-to-phloem transfer in the upper stem; and (4) cycling of shoot-derived C and N from root back to shoot by phloem-to-xylem transfer.

With respect to xylem-to-phloem transfer of N in the leaf of white lupin, it is a characteristic pattern for certain common amino compounds (asparagine, glutamine, valine, threonine, and serine), when fed via the transpiration stream, to be transferred directly to phloem, quickly and effectively, with little evidence of breakdown. Other amino compounds (aspartic acid, glutamic acid, arginine, glycine) are transferred less effectively in the unmetabolized state. With respect to xylem-to-phloem transfer in upper regions of the shoot, as well as in leaves, direct transfer of unmetabolized amino compounds is the dominant process involved. The two principal compounds transferred are asparagine and glutamine, with smaller amounts of valine, serine, and lysine involved. Pate (1980) has stressed that the compounds characteristic of N transport in N_2-fixing plants are species specific. Ureides (allantoin and allantoic acid) are important compounds exporting fixed N in certain tropical legumes, whereas asparagine and glutamine are the translocated solutes in many temperate legumes. For white lupin, a model for fruit growth was constructed based on extensive measurements by Pate and co-workers (Pate, 1986; Salisbury and Ross, 1992). According to this model, the phloem sap supplies about 98 percent of C, 89 percent of N, and 40 percent of water entering the fruit from the parent plant. The remaining N and water are supplied by the xylem sap.

Remobilization of Mineral Nutrients

With the exception of Ca, and presumably also Mn (e.g., Hill, Robson, and Loneragan, 1979; Nable and Loneragan, 1984), import of nutrients via the xylem and export (retranslocation) via the phloem are regular features throughout the life of an individual leaf (Marschner, 1995). Comprehensive studies on mineral nutrient cycling have been done in white lupin and castor bean by Jeschke and Pate (1991). Nutrient cycling is particularly important for N, and possibly P, nutrition of plants.

Remobilization of mineral nutrients is particularly important during reproductive growth, when seeds and fruits are formed (Marschner, 1995). At this growth stage, root activity and nutrient uptake generally decrease, mainly as a result of decreasing carbohydrate supply to the roots. Cereal grains, for example, are characterized by high concentration of N and P;

remobilization of these nutrients from leaves and the stem is an important nutrient source for grains during reproductive development.

In a detailed study on remobilization of macronutrients and micronutrients from several parts of the plant (leaf, pod wall, and seed coat) during reproductive development of grain legumes (pea and two species of lupin), Hocking and Pate (1977) demonstrated that remobilization of N, P, and K was very high, some 60 to 90 percent being remobilized from the leaf, pod, and seed coat during their respective periods of senescence, versus 20 to 60 percent of Mg, Zn, Mn, Fe, and Cu and less than 20 percent of Na and Ca. In an interesting paper on remobilization of Ca in bean during seed development, Mix and Marschner (1976) have suggested there may be a direct transfer of Ca from the pod tissue into the seed coat via the apoplast separating the seed coat and pod tissue (in contrast to the normal route for nutrient transport into developing seeds, in which the funiculus is the major route of transport). If this is an important mechanism for movement of Ca (and, possibly, also of some other mineral nutrients) into seed tissues, it may also be relevant for other types of fruits (e.g., the caryopsis of cereals).

In cereals such as wheat, up to about 90 percent of the total P accumulated in grains can be attributed to remobilization from vegetative parts. The remobilization of highly phloem-mobile mineral nutrients can lead to such a rapid decline in their content in the vegetative plant parts that rapid senescence is induced and plants accordingly behave as "self-destructing" systems (e.g., remobilization of P and senescence of the flag leaf in wheat; Batten and Wardlaw, 1987). Occurrence of deficiency symptoms (N, K, P, Mg) on older leaves reflects a high rate of remobilization via the phloem, whereas the occurrence of deficiency on only the young leaves (Ca, Fe, Mn, Cu, Zn, Mo, B) demonstrates insufficient rates of remobilization of these mineral elements via the phloem (Marschner, 1983). During the reproductive stage, the degree of remobilization of micronutrients and of Ca is often much higher than during vegetative growth (Marschner, 1995). The relatively high remobilization rates of micronutrients during reproductive growth, as compared with the vegetative growth stage, have been explained as the result of leaf senescence, in which changes in hormonal balance may play an important role (Marschner, 1995). The extent of remobilization of mineral nutrients may attract increasing attention in the future in connection with the selection and breeding of genotypes of high "nutrient efficiency." One factor involved may be high rates of remobilization from older leaves to actively growing parts of the plant.

PATHWAY OF NUTRIENT TRANSPORT INTO DEVELOPING GRAINS

Many data on the pathway of nutrient transport into developing grains have been compiled in several reviews on solute transport into developing seeds (Thorne, 1985; Wolswinkel, 1985, 1992; Bewley and Black, 1994; Patrick and Offler, 1995; Fisher and Oparka, 1996). The main emphasis in these reviews was on assimilate transport via the phloem. However, since transport of mineral nutrients into developing grains via the phloem is also very important, these data are relevant for a good understanding of the transport of mineral nutrients into developing grains.

In developing seeds of legumes and other taxonomic groups, the nurturing seed coat (maternal tissues; often called "testa") surrounds the cotyledons of the embryo. In the caryopses of cereals, the pericarp and rudimentary seed coat are fused. A protective function for the dry seed coat of mature seeds has long been recognized, but the seed coat is far more than a protective shell. In legumes, it is a multifunctional organ supplying nutrients to the embryo throughout seed development (Thorne, 1985; Murray, 1987). In grains of other species, for example, cereals, maternal tissues may have a comparable role.

No direct vascular connection exists between the vegetative plant and the embryo or endosperm. The vascular tissue ends in maternal tissues of the ovule, with no vascular penetration beyond its walls; filial tissues (embryo and endosperm) are isolated from the maternal vascular system. It is believed that as assimilates are transported from maternal tissues to filial tissues, they are delivered to the extracellular space (apoplast) separating the two generations, prior to uptake from the apoplast into the tissues of the embryo or endosperm. The symplastic discontinuity between the maternal and filial tissues in seeds necessitates membrane efflux from maternal tissues and subsequent membrane influx by filial tissues (Patrick and Offler, 1995).

Legumes

The transport of nutrients to the developing seeds occurs from two major pod vascular bundles along which seeds are alternately attached. Subsequently, nutrients are transported into the seed, entering the seed coat through a vascular strand in the funiculus that branches from the vascular tissue running through the pod wall. The seed coat of legumes is provided with bundles of xylem and phloem, with branches consisting of strands of phloem alone (Wolswinkel, 1992; Patrick and Offler, 1995). The phloem in the seed coat, through which the assimilates and other phloem-mobile solutes are distributed, may consist of only a few strands (e.g., in pea) or may exhibit an extensive, reticulate network (e.g., in soybean). Irrespective of the vascularization, photosynthate is delivered to the entire surface of the enclosed cotyledons; solute exchange occurs across the entire maternal/filial interface. Several studies provide support for the claim that symplastic transport via plasmodesmata occurs from phloem to the ground tissues (Patrick and Offler, 1995; Fisher and Oparka, 1996; Patrick and Offler, 1996). A specific tissue may act as the principal site of membrane efflux to the seed apoplast. In the cotyledons of developing legume seeds, the epidermal and subepidermal cells at the exchange interface form a dermal complex involved in active solute uptake from the apoplast separating maternal and filial tissues (Patrick and Offler, 1995).

Cereals

The filial storage cells are endospermic in origin; solute exchange between the maternal and filial tissues is restricted to a relatively short length of the maternal/filial interface (Patrick and Offler, 1995).

Corn and Several Other Tropical Cereals (Such As Sorghum and Millets)

Assimilates are unloaded from the phloem terminals located at the base of the grain called the pedicel (maternal tissues), where specialized transfer cells facilitate movement from this maternal tissue into the base of the developing endosperm (Felker and Shannon, 1980; Porter, Knievel, and Shannon, 1985). Assimilates entering the pedicel of corn are presumably transported symplastically from the phloem to the placenta-chalazal region, which is a maternal tissue (Felker and Shannon, 1980). Finally, assimilates may be passively transferred to the apoplast and diffuse apoplastically to the endosperm transfer cells, where they reenter the symplast and are translocated throughout the endosperm.

Temperate Cereals (Such As Wheat and Barley)

The importing vein extends along the length of the grain (Zee and O'Brien, 1970); assimilates are supplied via the vascular tissue running in the furrow (crease) and must first pass through the funiculus-chalazal region, then through the nucellar projection into the endosperm cavity. Patrick and Offler (1995) have described evidence for solute transfer from the sieve element-companion cell complex to the nucellar projection cells via the symplast (see also Wang and Fisher, 1994b). The use of apoplastic tracers has confirmed claims that lignification and suberization within the cell walls of the pigment strand prevent solute transport through the apoplast (Wang and Fisher, 1994b; Wang, Offler, and Patrick, 1995). Thus, solute flow is confined to the symplast of the pigment strand. This circumstance dictates that the nucellar projection cells must act as the exclusive cellular sites of sucrose efflux to the endosperm cavity. The filial tissues bordering the cavity differentiate in parallel with the nucellar projection cells to form a modified aleurone/subaleurone complex. The plasma membrane surface area amplification resulting from the development of wall ingrowths (transfer cell characteristics) has a role in supporting active membrane influx of sucrose at observed rates, and evidence exists for a symplastic route for subsequent transfer to the starchy endosperm (Patrick and Offler, 1995).

Rice

In the dorsal region of the caryopsis, a pericarp vascular bundle extends from one end of the grain to the other, showing a similarity with wheat. However, rice grains do not have a crease. Evidence suggests that, in rice, assimilates arrive in the phloem and move symplastically to the pigment strand and circumferentially via the nucellar epidermis, with high plasmodesmatal densities in the radial walls, before entering the apoplast at the nucellus-aleurone interface (Oparka and Gates, 1981a, b).

Xylem Discontinuity

A xylem discontinuity exists in the pedicel of wheat and barley grains (Zee and O'Brien, 1970). This point has been discussed in several reports (Frick and Pizzolato, 1987; Pizzolato, 1990; Wang and Fisher, 1994a, b; Pearson et al., 1995, 1996; Fisher and Oparka, 1996). More generally, in developing seeds, and in developing fleshy fruits as well, there is evidence for a restricted xylem connection between reproductive sinks and other parts of the plant (Wolswinkel, 1990, 1992, and references therein). Because the concentration of apoplastic solutes in seeds and fruits is often substantial, hydraulic isolation may be important in preventing the loss of these solutes via the xylem (Wolswinkel, 1990, 1992; Bradford, 1994; Fisher and Oparka, 1996). In developing grains of wheat and corn and seeds of soybean, water status is practically independent of whole-plant water status, particularly the leaf water potential (Wolswinkel, 1990, 1992). These data indicate that developing kernels are hydraulically uncoupled from the leaves and other parts of the plant. The xylem discontinuity in the pedicel of wheat and barley grains (Zee and O'Brien, 1970) may be a general feature of certain taxonomic groups of grasses (Pizzolato, 1990, and references therein; Fisher and Oparka, 1996).

Recent Research on Micronutrient Transport into Developing Wheat Grains

A need exists for better understanding of several factors involved in micronutrient transport into developing cereal grains: for example, (1) the route of transport from vegetative plant parts into developing grains; (2) the site(s) and the rate of transfer between xylem and phloem; (3) the timing and the rate of remobilization from vegetative plant parts; and (4) the function of different components of the wheat spikelet. Pearson and Rengel and co-workers (Pearson and Rengel, 1994; Pearson et al., 1995, 1996) have started a line of research in which labeled Zn and Mn are used to obtain more information on the mechanism(s) of Zn and Mn transport into developing wheat grains. Pearson and Rengel (1994) concluded that Zn was remobilized from wheat leaves during grain development and that Mn was not remobilized, confirming other data in the literature on a low mobility of Mn in the phloem. Xylem-to-phloem transfer of Zn may occur in the rachis and to a lesser extent in peduncle and lemma, while Mn may be predominantly translocated to the spikelets in the xylem (Pearson et al., 1995). By increasing the relative humidity surrounding the ear to approximately 100 percent, thus reducing transpiration and xylem transport, Zn and Mn transport into the grain was almost completely blocked (Pearson et al., 1996).

CONCLUSIONS

Xylem transport and phloem transport cooperate in long-distance nutrient transport into developing grains. The present knowledge on the relative importance of xylem and phloem transport is mainly based on detailed analyses of the phloem and xylem sap in different shoot parts of individual plants and the corresponding mineral element contents in the shoot parts at sequential harvests. The use of radioisotopes has been a valuable research tool. Our knowledge of loading of nutrients into xylem vessels and phloem sieve tubes is incomplete, particularly in the case of micronutrients. In the last two decades, research on assimilate transport into developing seeds has contributed to a better understanding of the pathway of nutrient transport into developing grains and the factors controlling phloem transport (e.g., the empty-ovule technique). Possibly, the techniques used and the data obtained may stimulate research on unsolved questions related to transport of mineral nutrients into developing grains.

In recent years, much progress has been made in the study of ion transport across cell membranes (e.g., the role of active and passive transport; carrier-mediated transport; the role of a plasma membrane-bound H^+-ATPase; the presence of proton pumps in both plasma membrane and tonoplast; measurements of membrane potentials and ion fluxes in isolated membrane vesicles; the existence of ion channels). In most of these studies, parenchymatous tissues or cells have been used. Much less is known about characteristics of ion transport across the plasma membrane of the sieve element-companion cell complex. These characteristics will influence loading of nutrient ions into, and unloading from, the sieve-tube system. With respect to the relative contribution of xylem and phloem to nutrient transport into developing grains, the effect of the transpiration rate on distribution within the shoot has to be stressed (e.g., a positive correlation can be observed between Ca distribution and the transpiration of shoot organs). Nutrient transport into developing grains will also be dependent on exchange between xylem and phloem and remobilization from senescent parts of the mother plant. Information on xylem transport and phloem transport of micronutrients is more incomplete than our knowledge of transport of macronutrients; metal complexation in transport fluids may be an important characteristic of micronutrient transport.

REFERENCES

Arnozis, P.A. and G.R. Findenegg (1986). Electrical charge balance in the xylem sap of beet and *Sorghum* plants grown with either NO_3 or NH_4 nitrogen. *Journal of Plant Physiology* 125: 441-449.

Baker, D.A. and J.A. Milburn, Eds. (1989). *Transport of Photoassimilates*. Harlow, Essex, UK: Longman Scientific and Technical.

Batten, G.D. and I.F. Wardlaw (1987). Senescence and grain development in wheat plants grown with contrasting phosphorus regimes. *Australian Journal of Plant Physiology* 14: 253-265.

Becker, R., A. Pich, G. Scholz, and K. Seifert (1989). Influence of nicotianamine and iron supply on formation and elongation of adventitious roots in hypocotyl cuttings of the tomato mutant "chloronerva" *(Lycopersicon esculentum)*. *Physiologia Plantarum* 76: 47-52.

Bewley, J.D. and M. Black (1994). *Seeds. Physiology of Development and Germination.* New York: Plenum Press.

Bradford, K.J. (1994). Water stress and the water relations of seed development: A critical review. *Crop Science* 34: 1-11.

Delrot, S. (1989). Loading of photoassimilates. In *Transport of Photoassimilates*, Eds. D.A. Baker and J.A. Milburn. Harlow, Essex, UK: Longman Scientific and Technical, pp. 167-205.

Engels, C. and H. Marschner (1993). Influence of the form of nitrogen supply on root uptake and translocation of cations in the xylem exudate of maize (*Zea mays* L.). *Journal of Experimental Botany* 44: 1695-1701.

Felker, F.C. and J.C. Shannon (1980). Movement of ^{14}C-labeled assimilates into kernels of *Zea mays* L. III. An anatomical examination and microautoradiographic study of assimilate transfer. *Plant Physiology* 77: 524-531.

Fisher, D.B. and K.J. Oparka (1996). Post-phloem transport: Principles and problems. *Journal of Experimental Botany* 47: 1141-1154.

Frick, H. and T.D. Pizzolato (1987). Adaptive value of the xylem discontinuity in partitioning of photoassimilate to the grain. *Bulletin of the Torrey Botanical Club* 114: 252-259.

Gamalei, Y. (1989). Structure and function of leaf minor veins in trees and herbs: A taxonomic review. *Trees* 3: 96-110.

Giaquinta, R.T. (1983). Phloem loading of sucrose. *Annual Review of Plant Physiology* 34: 347-387.

Grusak, M.A. and P.E.H. Minchin (1988). Seed coat unloading in *Pisum sativum*: Osmotic effects in attached versus excised empty ovules. *Journal of Experimental Botany* 39: 543-559.

Hamilton, D.A. and P.J. Davies (1988). Mechanism of export of organic material from the developing fruits of pea. *Plant Physiology* 86: 956-959.

Hayashi, H. and M. Chino (1990). Chemical composition of phloem sap from the uppermost internode of the rice plant. *Plant and Cell Physiology* 31: 247-251.

Hill, J., A.D. Robson, and J.F. Loneragan (1979). The effect of copper supply on the senescence and the retranslocation of nutrients of the oldest leaf of wheat. *Annals of Botany* 44: 279-287.

Hocking, P.J. (1980). The composition of phloem exudate and xylem sap from tree tobacco (*Nicotiana glauca* Groh.). *Annals of Botany* 45: 633-643.

Hocking, P.J. and J.S. Pate (1977). Mobilization of minerals to developing seeds of legumes. *Annals of Botany* 41: 1259-1278.

Jeschke, W.D. and J.S. Pate (1991). Cation and chloride partitioning through xylem and phloem within the whole plant of *Ricinus communis* L. under conditions of salt stress. *Journal of Experimental Botany* 42: 1105-1116.

Jeschke, W.D., J.S. Pate, and C.A. Atkins (1987). Partitioning of K^+, Na^+, Mg^{++}, and Ca^{++} through xylem and phloem to component organs of nodulated white lupin under mild salinity. *Journal of Plant Physiology* 128: 77-93.

Kirkby, E.A. and D.J. Pilbeam (1984). Calcium as a plant nutrient. *Plant, Cell and Environment* 7: 397-405.

Lang, A. and H. Düring (1991). Partitioning control by water potential gradient: Evidence for compartmentation breakdown in grape berries. *Journal of Experimental Botany* 42: 1117-1122.

Lang, A. and M.R. Thorpe (1989). Xylem, phloem and transpiration flows in a grape: Application of a technique for measuring the volume of attached fruits to high resolution using Archimedes' principle. *Journal of Experimental Botany* 40: 1069-1078.

Lee, D.R. and P.T. Atkey (1984). Water loss from the developing caryopsis of wheat (*Triticum aestivum*). *Canadian Journal of Botany* 62: 1319-1326.

Maas, F.M., D.A.M. Van de Wetering, M.L. Van Beusichem, and H.F. Bienfait (1988). Characterization of phloem iron and its possible role in the regulation of Fe-efficiency reactions. *Plant Physiology* 87: 167-171.

Marschner, H. (1983). General introduction to the mineral nutrition of plants. In *Inorganic Plant Nutrition. Encyclopedia of Plant Physiology.* New Series. Volume 15A, Eds. A. Laüchli and R.L. Bieleski. Heidelberg, Germany: Springer-Verlag, pp. 5-60.

Marschner, H. (1995). *Mineral Nutrition of Higher Plants*, Second Edition. London, UK: Academic Press.

Martin, P. (1982). Stem xylem as a possible pathway for mineral retranslocation from senescing leaves to the ear in wheat. *Australian Journal of Plant Physiology* 9: 197-207.

McGrath, J.F. and A.D. Robson (1984). The movement of zinc through excised stems of seedlings of *Pinus radiata* D. *Annals of Botany* 54: 231-242.

Milburn, J.A. (1975). Pressure flow. In *Phloem Transport. Encyclopedia of Plant Physiology.* New Series. Volume I, Eds. M.H. Zimmermann and J.A. Milburn. Heidelberg, Germany: Springer-Verlag, pp. 328-353.

Milburn, J.A. and D.A. Baker (1989). Physico-chemical aspects of phloem sap. In *Transport of Photoassimilates*, Eds. D.A. Baker and J.A. Milburn. Harlow, Essex, UK: Longman Scientific and Technical, pp. 345-359.

Milburn, J.A. and J. Kallarackal (1989). Physiological aspects of phloem translocation. In *Transport of Photoassimilates*, Eds. D.A. Baker and J.A. Milburn. Harlow, Essex, UK: Longman Scientific and Technical, pp. 264-305.

Mix, G.P. and H. Marschner (1976). Redistribution of calcium in bean fruits during seed development. *Zeitschrift für Pflanzenphysiologie* 80: 354-366.

Münch, E. (1930). *Die Stoffbewegungen in der Pflanze.* Jena, Germany: Fischer.

Murray, D.R. (1987). Nutritive role of seed coats in developing legume seeds. *American Journal of Botany* 74: 1122-1137.

Nable, R.O. and J.F. Loneragan (1984). Translocation of manganese in subterranean clover (*Trifolium subterraneum* L. cv. Seaton Park). II. Effects of leaf senescence and of restricting supply of manganese to part of a split root system. *Australian Journal of Plant Physiology* 11: 113-118.

Oparka, K.J. and P. Gates (1981a). Transport of assimilates in the developing caryopsis of rice (*Oryza sativa* L.). Ultrastructure of the pericarp vascular bundle and its connection with the aleurone layer. *Planta* 151: 561-573.

Oparka, K.J. and P. Gates (1981b). Transport of assimilates in the developing caryopsis of rice (*Oryza sativa* L.). The pathway of water and assimilated carbon. *Planta* 152: 388-396.

Pate, J.S. (1975). Exchange of solutes between phloem and xylem and circulation in the whole plant. In *Phloem Transport. Encyclopedia of Plant Physiology.* New

Series. Volume I, Eds. M.H. Zimmermann and J.A. Milburn. Heidelberg, Germany: Springer-Verlag, pp. 451-473.

Pate, J.S. (1980). Transport and partitioning of nitrogenous solutes. *Annual Review of Plant Physiology* 31: 313-340.

Pate, J.S. (1986). Xylem-to-phloem transfer—Vital component of the nitrogen-partitioning system of a nodulated legume. In *Phloem Transport*, Eds. J. Cronshaw, W.J. Lucas, and R.T. Giaquinta. New York: Alan R. Liss, pp. 445-462.

Pate, J.S. (1989). Origin, destination and fate of phloem solutes in relation to organ and whole plant functioning. In *Transport of Photoassimilates*, Eds. D.A. Baker and J.A. Milburn. Harlow, Essex, UK: Longman Scientific and Technical, pp. 138-166.

Pate, J.S. and B.E.S. Gunning (1972). Transfer cells. *Annual Review of Plant Physiology* 23: 173-196.

Pate, J.S. and W.D. Jeschke (1995). Role of stems in transport, storage, and circulation of ions and metabolites by the whole plant. In *Plant Stems: Physiology and Functional Morphology*, Ed. B.L. Gartner. San Diego, CA: Academic Press, pp. 177-204.

Pate, J.S., M.B. Peoples, A.J.E. Van Bel, J. Kuo, and C.A. Atkins (1985). Diurnal water balance of the cowpea fruit. *Plant Physiology* 77: 148-156.

Patrick, J.W. (1983). Photosynthate unloading from seed coats of *Phaseolus vulgaris* L. General characteristics and facilitated transfer. *Zeitschrift für Pflanzenphysiologie* 111: 9-18.

Patrick, J.W. (1990). Sieve element unloading: Cellular pathway, mechanism and control. *Physiologia Plantarum* 78: 298-308.

Patrick, J.W. and C.E. Offler (1995). Post-sieve element transport of sucrose in developing seeds. *Australian Journal of Plant Physiology* 22: 681-702.

Patrick, J.W. and C.E. Offler (1996). Post-sieve element transport of photoassimilates in sink regions. *Journal of Experimental Botany* 47: 1165-1177.

Pearson, J.N. and Z. Rengel (1994). Distribution and remobilization of Zn and Mn during grain development in wheat. *Journal of Experimental Botany* 45: 1829-1835.

Pearson, J.N., Z. Rengel, C.F. Jenner, and R.D. Graham (1995). Transport of zinc and manganese to developing wheat grains. *Physiologia Plantarum* 95: 449-455.

Pearson, J.N., Z. Rengel, C.F. Jenner, and R.D. Graham (1996). Manipulation of xylem transport affects Zn and Mn transport into developing wheat grains of cultured ears. *Physiologia Plantarum* 98: 229-234.

Pizzolato, T.D. (1990). Vascular system of the fertile floret of *Cynodon dactylon* (Gramineae, Eragrostoideae). *Botanical Gazette* 151: 477-489.

Porter, G.A., D.P. Knievel, and J.C. Shannon (1985). Sugar efflux from maize (*Zea mays* L.) pedicel tissue. *Plant Physiology* 77: 524-531.

Porter, G.A., D.P. Knievel, and J.C. Shannon (1987). Assimilate unloading from maize (*Zea mays* L.) pedicel tissues. I. Evidence for regulation of unloading by cell turgor. *Plant Physiology* 83: 131-136.

Salisbury, F.B. and C.W. Ross (1992). *Plant Physiology*. Belmont, CA: Wadsworth Publishing Company.

Scholz, G., R. Becker, U.W. Stephan, A. Rudolph, and A. Pich (1988). The regulation of iron uptake and possible functions of nicotianamine in higher plants. *Biochemie und Physiologie der Pflanzen* 183: 257-269.

Senden, M.H.M.N. and H.T. Wollerbeek (1990). Effect of citric acid on the transport of cadmium through xylem vessels of excised tomato stem-leaf systems. *Acta Botanica Neerlandica* 39: 297-303.

Stephan, U.W. and G. Scholz (1993). Nicotianamine: Mediator of transport of iron and heavy metals in the phloem? *Physiologia Plantarum* 88: 522-529.

Streeter, J.G. (1979). Allantoin and allantoic acid in tissues and stem exudate from field-grown soybean plants. *Plant Physiology* 63: 478-480.

Thorne, J.H. (1985). Phloem unloading of C and N assimilates in developing seeds. *Annual Review of Plant Physiology* 36: 317-343.

Thorne, J.H. and R.M. Rainbird (1983). An *in vivo* technique for the study of phloem unloading in seed coats of developing soybean seeds. *Plant Physiology* 72: 268-271.

Triplett, E.W., N.M. Barnett, and D.G. Blevins (1980). Organic acids and ionic balance in xylem exudate of wheat during nitrate or sulfate absorption. *Plant Physiology* 65: 610-613.

Van Beusichem, M.L., E.A. Kirkby, and R. Baas (1988). Influence of nitrate and ammonium nutrition and the uptake, assimilation, and distribution of nutrients in *Ricinus communis*. *Plant Physiology* 86: 914-921.

Wang, N. and D.B. Fisher (1994a). Monitoring phloem unloading and post-phloem transport by microperfusion of attached wheat grains. *Plant Physiology* 104: 7-16.

Wang, N. and D.B. Fisher (1994b). The use of fluorescent tracers to characterize the post-phloem transport pathway in maternal tissues of developing wheat grain. *Plant Physiology* 104: 17-27.

Wang, H.L., C.E. Offler, and J.W. Patrick (1995). The cellular pathway of photosynthate transfer in the developing wheat grain. II. A structural analysis and histochemical studies of the pathway from the crease phloem to the endosperm cavity. *Plant, Cell and Environment* 18: 373-388.

Welbaum, G.E. and F.C. Meinzer (1990). Compartmentation of solutes and water in developing sugarcane stalk tissue. *Plant Physiology* 93: 1147-1153.

Welch, R.M. (1995). Micronutrient nutrition of plants. *CRC Critical Reviews in Plant Sciences* 14: 49-82.

White, M.C., F.D. Baker, R.L. Chaney, and A.M. Decker (1981). Metal complexation in xylem fluid. II. Theoretical equilibrium model and computational computer program. *Plant Physiology* 67: 301-310.

White, M.C., R.L. Chaney, and A.M. Decker (1981). Metal complexation in xylem fluid. III. Electrophoretic evidence. *Plant Physiology* 67: 311-315.

White, M.C., A.M. Decker, and R.L. Chaney (1981). Metal complexation in xylem fluid. I. Chemical composition of tomato and soybean stem exudate. *Plant Physiology* 67: 292-300.

Wolswinkel, P. (1985). Phloem unloading and turgor-sensitive transport: Factors involved in sink control of assimilate partitioning. *Physiologia Plantarum* 65: 331-339.

Wolswinkel, P. (1990). Recent progress in research on the role of turgor-sensitive transport in seed development. *Plant Physiology and Biochemistry* 28: 399-410.

Wolswinkel, P. (1992). Transport of nutrients into developing seeds: A review of physiological mechanisms. *Seed Science Research* 2: 59-73.

Wolswinkel, P. and A. Ammerlaan (1983). Phloem unloading in developing seeds of *Vicia faba* L. The effect of several inhibitors on the release of sucrose and amino acids by the seed coat. *Planta* 158: 205-215.

Wolswinkel, P. and A. Ammerlaan (1984). Turgor-sensitive sucrose and amino acid transport into developing seeds of *Pisum sativum*. Effect of a high sucrose or mannitol concentration in experiments with empty ovules. *Physiologia Plantarum* 61: 172-182.

Wolswinkel, P., A. Ammerlaan, and J.W. Koerselman-Kooij (1992). Effect of the osmotic environment on K^+ and Mg^{2+} release from the seed coat and cotyledons of developing seeds of *Vicia faba* and *Pisum sativum*. Evidence for a stimulation of efflux from the vacuole at high cell turgor. *Journal of Experimental Botany* 43: 681-693.

Wolswinkel, P. and J.W. Koerselman-Kooij (1992). Effect of a pretreatment of seed coats with a low-osmolality solution on subsequent ^{14}C-sucrose transport into attached empty ovules of pea. *Plant, Cell and Environment* 15: 617-621.

Wolterbeek, H.T., J. Van Luipen, and M. De Bruin (1984). Non-steady state xylem transport of fifteen elements into the tomato leaf as measured by gamma-ray spectroscopy: A model. *Physiologia Plantarum* 61: 599-606.

Zee, S.Y. and T.P. O'Brien (1970). A special type of tracheary element associated with "xylem discontinuity" in the floral axis of wheat. *Australian Journal of Biological Sciences* 23: 783-791.

Ziegler, H. (1975). Nature of transported substances. In *Phloem Transport. Encyclopedia of Plant Physiology.* New Series. Volume I, Eds. M.H. Zimmermann and J.A. Milburn. Heidelberg, Germany: Springer-Verlag, pp. 59-100.

Zimmermann, M.H. and J.A. Milburn, Eds. (1975). *Phloem Transport. Encyclopedia of Plant Physiology.* New Series. Volume I. Heidelberg, Germany: Springer-Verlag.

Zimmermann, M.H. and H. Ziegler (1975). List of sugars and sugar alcohols in sieve-tube exudates. In *Phloem Transport. Encyclopedia of Plant Physiology.* New Series. Volume I, Eds. M.H. Zimmermann and J.A. Milburn. Heidelberg, Germany: Springer-Verlag, pp. 480-503.

Chapter 5

The Significance of Root Size for Plant Nutrition in Intensive Horticulture

Asher Bar-Tal

Key Words: nutrient uptake, root restriction, root size, transpiration rate.

INTRODUCTION

Deep-irrigated crops in intensive agriculture, and horticultural plants grown in soilless culture, generally develop small root systems. High yields with a high shoot growth rate and a high shoot/root ratio are common. However, the physiological capacity of the small root system to take up water and nutrients to meet the demands of the shoot may become the limiting factor. Under well-controlled experimental conditions, root restriction in a small container has been shown to limit plant growth in various species, as manifested in reductions of such growth parameters as leaf area, leaf number, plant height, and biomass production (Cooper, 1972; Richards and Rowe, 1977; Carmi and Heuer, 1981; Carmi and Van Staden, 1983; Carmi et al., 1983; Ruff et al., 1987; Robbins and Pharr, 1988; Peterson and Krizek, 1992).

The container volume in which the root can expand may affect plant growth either through plant nutrition and transpiration (Brouwer and De Wit, 1968; Hameed, Reid, and Rowe, 1987) or via root and shoot physiology (Aung, 1974; Jackson, 1993). In some cases, root confinement to a small container volume does not cause nutrient deficiency (Carmi and Heuer, 1981; Robbins and Pharr, 1988) or drought stress (Krizek et al., 1985; Ruff et al., 1987); instead, it has been suggested that the root restriction stress may decrease the supply of root-produced growth substances to shoots, causing an imbalance in growth regulators in roots and shoots (Carmi and Heuer, 1981; Peterson, Reinsel, and Krizek, 1991a, b). Root respiration was reduced by root confinement, with a decline in root respiration capacity reflecting a decline in root metabolism (Peterson, Reinsel, and Krizek, 1991a, b). Recently, Jackson (1993) suggested that the nutrient control hypothesis is not valid if the nutrient uptake rate per root unit volume, F_N, is not constant. The main subject of the following discussion is the effect of root size on nutrient uptake; the transfer of growth substances from roots to shoots, hormonal balance, and root metabolism are beyond the scope of this chapter.

THE ROLE OF ROOT SIZE AND CONTAINER VOLUME IN NUTRIENT UPTAKE

Plant nutrition may be affected by the container volume in which roots can grow, either indirectly via the quantities of water and nutrients available in the growth medium (Bar-Tal, Bar-Yosef, and Kafkafi, 1990, 1993) or directly via reducing the length and surface area of root that is absorbing nutrients (Humphries, 1958; Brouwer and De Wit, 1968). According to the nutrient control model (Brouwer and De Wit, 1968), a possible cause for the reduction in shoot growth by root restriction is shortage of nutrient supply (Mutsaers, 1983; Hanson, Dixon, and Dickson, 1987; Bar-Tal, Bar-Yosef, and Kafkafi, 1990) or drought stress (Hameed, Reid, and Rowe, 1987). Available information regarding the influence of the shoot/root ratio on total nutrient uptake and root activity is, to a great extent, equivocal; Jungk and Barber (1975) reported that total uptake of P (phosphorus) by corn plants depended on root length only and was not affected by the size of the shoot, whereas Edwards and Barber (1976) found that total P uptake by intact and trimmed (by 50 percent) soybean roots was similar, as P uptake per root unit weight was greater in plants with trimmed roots. Further data indicated an ability of the root to adjust its nutrient uptake activity upon trimming, for example, increased uptake of P by tomato (Jungk, 1974) and of N (nitrogen) by corn (Edwards and Barber, 1976). Schupp and Ferree (1990) found that a single pruning of roots in young apple trees had little effect on leaf mineral nutrient levels, whereas the second root pruning increased foliar levels of macro- and micronutrients.

The direct effect of root size on plant growth may be through water uptake. Carmi (1986) reported that the leaf stomatal resistance of beans (*Phaseolus vulgaris*), as measured one and six days after root pruning, increased considerably; such a change may have a direct influence on growth and photosynthesis. Schupp and Ferree (1990) reported that root pruning of young apple trees caused a temporary reduction in the transpiration rate for a few days, whereas a second pruning had no effect on that rate. According to Hameed, Reid, and Rowe (1987), root-restricted tomato plants exhibited drought stress, even though they were grown in solution culture; root hydraulic resistance increased for roots in a small container. The effect of root size on the transpiration rate depends greatly on environmental conditions. Downey and Mitchell (1971) reported that trimming the root system did not reduce the transpiration rate and leaf turgor, provided that soil and air moisture levels were maintained. Geisler and Ferree (1984) found that pruning 28 or 59 percent of the roots of young apple trees reduced the transpiration rate by 29 and 45 percent, respectively, one day after pruning, with recovery beginning ten days later; leaf water potential was less sensitive, and recovery was apparent after one day.

As the previous examples show, reduction of the volume of the growth medium occupied by roots may have two effects on plant nutrition: (1) root growth is physically restricted, resulting in decreased root volume, length, and surface area, and (2) a small volume of the growth medium leads to an increased need for frequent replenishment and to restriction of nutrient and water supplies. Separation of the aforementioned effects is quite difficult in most growing systems, and this may be the reason for the differing results and conclusions obtained from different experiments. The following examples examine the direct and indirect effects of root volume.

PLANT RESPONSE TO CONTAINER VOLUME AND NUTRIENT CONCENTRATIONS

Bar-Tal, Bar-Yosef, and Kafkafi (1990) studied the effects of container volume and N and P supply on N and P uptake by pepper seedlings. Elevating N and P concentrations in the nutrient solution in a given container volume (15 ml) resulted in significant increases in top and root weights and in top N and P contents. Similar results were obtained by increasing the growth medium volume from 5 to 15, 35, 65, and 700 ml (at constant N and P concentrations of 5.0 and 0.5 mM, respectively).

To determine whether nutrient supply in the various container volume treatments could be a limiting factor for N and P uptake, a comparison was made of the relationships between N and P consumption and supply rates using seedlings in containers of various volumes during the last day in the nursery. The highest N and P consumption rates were obtained in the 700-ml containers: 3 and 0.5 mg day^{-1}, respectively. Only 45 percent of this quantity was supplied daily via the irrigation solution to seedlings growing in 5-ml cells, while 120 percent and 300 percent quantities were supplied to plants in the 15- and 35-ml cells, respectively. These calculations show that supply could limit growth in the 5-ml volume, but not in the 15-, 35-, or 65-ml volumes. However, the relationships obtained between the container volume and the uptake rate per unit root volume (F_N and F_P for N and P, respectively) were curvilinear (see Figure 5.1), with a steep reduction in F as pot volume declined from approximately 100 ml to 5 ml per plant. These results conflict with the previous calculation that nutrient supply was not a limiting factor in container volumes above 5 ml per plant. It was suggested that in the experiment of Bar-Tal, Bar-Yosef, and Kafkafi (1990), the time-averaged nutrient concentration at the root surface between successive irrigations decreased as the container volume decreased. Since the relative effect of the container volume on F was greater than its effect on seedling weight, it appears that the main mechanism by which volume affected uptake of nutrients by the whole plant acted through the influence on the nutrient concentration at the root surface. This assumption was tested by Bar-Tal, Bar-Yosef, and Kafkafi (1993) using a mechanistic model for pepper seedlings. This model simulates seedling growth and uptake of water, N, and P as functions of N and P concentrations in the irrigation solution, rooting volume per plant, plant density, and substrate characteristics. The model predicted that changing the irrigation interval of seedlings grown in 5-ml volume to two hours should result in a shoot weight and N uptake equal to those of seedlings growing in 35-ml cells with a twelve-hour irrigation interval.

Bar-Yosef, Imas, and Levkovitz (1995) studied the effect of pot volume—10 and 20 l—on tomato plant growth and nutrient uptake. Plant growth was enhanced by the increase in pot volume from 10 to 20 l. The transpiration was also higher with an increase in the container volume, but the ratio of transpiration to dry matter production was not affected. The flux of water per unit mass of root decreased from 53 to 11 ml g^{-1} day^{-1} as plant age increased from twenty-three to forty-two days after planting, independently of container volume. The concentrations of N, P, K (potassium), Ca (calcium), and Mg (magnesium) in plant tissues were not affected by the container volume, but the uptake rates per plant were higher in the 20-l containers than in the 10-l ones. The increased N and P uptake rates stemmed from higher N and P uptake rates per unit root area and from the bigger root system in 20-l containers compared to 10-l ones. The maximal N uptake rates per unit root area were 60 and 83 mg m^{-2} day^{-1} in 10- and 20-l containers, respectively, while the maximal P uptake rates per unit root area were 10 and 13 mg m^{-2} day^{-1} in 10- and 20-l containers, respectively.

FIGURE 5.1. The Uptake Rate of P and N per Unit of Fresh Root Weight (F_P and F_N) As a Function of the Growth Medium Volume

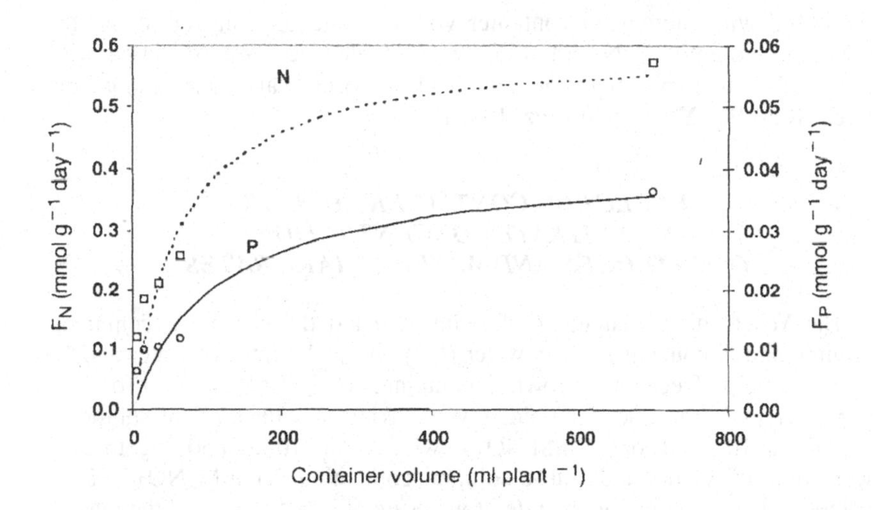

Source: Adapted from Bar-Tal, Bar-Yosef, and Kafkafi (1990).

Note: The curves are fitted using the equation: $F_i = F_i' V/(K_{Vi}+V)$. The subscript "i" stands for N or P. The best-fitted values of F_P' and F_N' are 0.6 mmol g^{-1} day^{-1} (8.4 mg g^{-1} day^{-1}, significant at P = 0.05) and 0.041 mmol g^{-1} day^{-1} (1.3 mg g^{-1} day^{-1}, P = 0.10), and of K_{VP} and K_{VN} are 60.6 and 112 ml per plant (P = 0.10), respectively.

In a more detailed study of the effect of container volume on tomato plant nutrition, Bar-Yosef and colleagues (1997) used a wider range of container volumes: 5, 10, 20, and 40 l. Three solution concentrations were studied in 10- and 20-l containers. Dry matter production and uptake of nutrients, especially of P and K, increased with an increasing container volume in the range studied. Increasing the nutrient concentration by 50 percent could compensate for a reduction in volume from 40 to 20 l, but not from 20 to 10 l. The reduction in P and K uptake as container volume decreased stemmed mainly from lower P and K uptake rate per unit root area. The maximal uptake rates of N, P, and K were 167, 30, and 83 mg plant^{-1} day^{-1}, respectively. The uptake rates of nutrients, especially P, were much lower in 5-l containers than in the other volumes. As the treatments had no significant effect on root weight, the reduction in P and

K uptake as container volume decreased resulted mainly from lower P and K uptake rates per unit root area.

In the previous experiments, the nutrient uptake rate per unit root area increased with increasing container volume. The opposite effect of the container volume on N and water uptake rates per unit root area was obtained in the experiments with apple (Bar-Yosef et al., 1988) and peach trees (Ran, Bar-Yosef, and Erez, 1992).

EFFECT OF CONTAINER VOLUME AND NITRATE CONCENTRATION ON NITROGEN AND WATER UPTAKE RATES

Bar-Yosef and colleagues (1988) investigated the effects of container volume and N concentration in water (C_N) on uptake rates of water and N by apple trees. Trees were grown in containers of 200, 50 and 10 l for five years. Trees in the 200-l containers were irrigated with nutrient solutions containing 10.7, 7.1, or 2.5 mM NO_3^-, whereas the 10- and 50-l treatments were uniformly supplied with a solution containing 7.1 mM NO_3^-. Elevated C_N had no effect on the rate of water uptake, but increased the rate of N uptake from 2.4 to 4.8 g N day^{-1} $tree^{-1}$ during July due to enhanced N uptake per unit of root. The C_N had negligible effects on root weight and permeability to water. Reducing the container volume decreased N and water uptake rates, but increased root permeability to NO_3^- and water (see Table 5.1). These data show that the plants with restricted roots were more efficient in taking up N and water, probably as a consequence of enhanced demand, elicited by the larger canopy-to-root ratios found in these treatments. Daily N uptake by trees grown in 50-l containers and irrigated with 7.5 mM NO_3^- solution was close to that of trees grown in 200-l containers and irrigated with 2.5 mM NO_3^- solution (data not shown). Thus, increased C_N and the enhanced root permeability in a 50-l container could compensate for the effect of the restricted root system on N nutrition.

Ran, Bar-Yosef, and Erez (1992) investigated the growth of peach trees for four years in containers holding 80, 40, 20, 10, or 5 l of coarse sand. All trees received an ample supply of NO_3^--based nutrient solution, so that the container volume was the predominant growth-limiting factor. As in the experiment with apple trees described previously, the container volume in which peach trees were grown significantly affected total dry matter and N uptake, but did not affect the N content in leaves. The container volume influenced the rate of water uptake per tree because of the effect on root weight per plant, but had no effect on water uptake per unit root weight. Daily N consumption increased with an increase in the container volume

TABLE 5.1. All-Season Average Flux of $NO3^-$ (F_N) and Water Uptake (F_W) per Unit Root Fresh Weight by Five-Year Golden Delicious Trees As a Function of Container Volume and N Concentration in the Irrigation Water (C_N)

Container volume	C_N	F_N	F_W
l	mM	mg N g^{-1} day^{-1}	g H_2O g^{-1} day^{-1}
10	7.1	0.49	5.0
50	7.1	0.29	3.8
200	2.5	0.11	2.8
200	7.1	0.21	2.8
200	10.7	0.27	2.8

Source: Adapted from Bar-Yosef et al. (1988).

because of the larger root weight, even though the average N uptake declined with an increase in the container volume. The enhancement of N uptake as the container volume decreased coincided with a greater fraction of roots with diameter less than 1 mm (millimeter) in the smaller containers than in the larger ones. On the assumption that the largest portion of uptake occurred through thin roots, the N uptake rate was correlated with the weight of roots with a diameter of less than 3 mm (see Figure 5.2). The 2.5-year-old trees showed a linear correlation, indicating that the average N uptake (0.49 g N kg^{-1} fresh root weight day^{-1}) was independent of changes in root weight. The relationships for 3.0- and 3.5-year-old trees were curvilinear, meaning that the average uptake decreased with increasing root weight. These results indicate that N uptake efficiency of the root is affected by the N demand of the whole plant or by the fraction of thin roots taking up N.

The last two examples confirm the ability of the whole plant root system to regulate N uptake. However, there was no clear-cut separation between the direct effect of container volume on the roots and the indirect effect of nutrient medium buffering capacity for water and nutrients. Also, only N was investigated in these studies.

To study the direct effect of container volume separately from the indirect effects, De Willigen and Van Noordwijk (1987) used containers of various sizes in a system of continuously circulating nutrient solution to eliminate differences in the amounts of nutrients and water available to the plant. They conducted an experiment with tomato, using container volumes of 0.5, 1.5, and 6 l, with sand and rockwool growth media. The corresponding pore volumes were 0.2, 0.6, and 2.4 l in the sand and 0.5,

FIGURE 5.2. Nitrate-N Uptake Rate (N_{uptake}) by Peach Trees As a Function of (1) the Fresh Weight (W) of Roots Measuring <3 mm in Diameter and (2) Tree Age

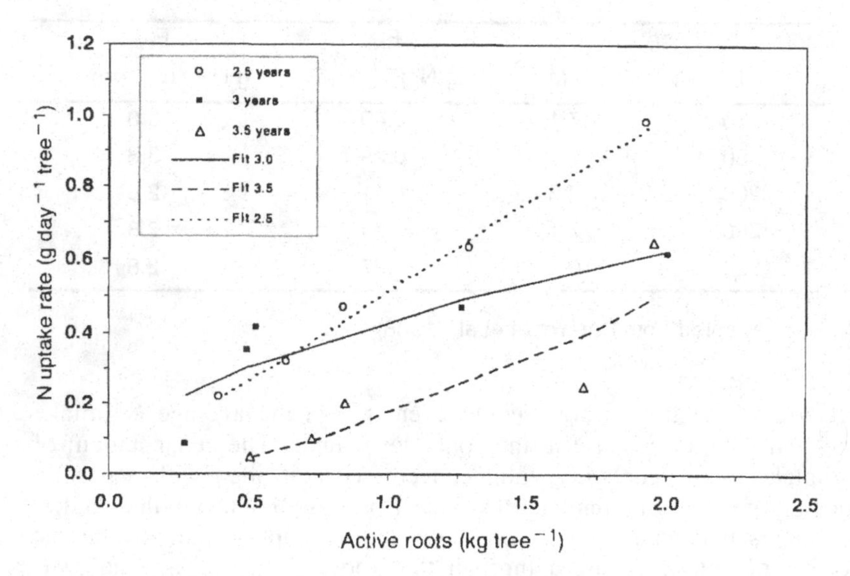

Source: Adapted from Ran, Bar-Yosef, and Erez (1992).

Note: Best-fit regression lines were:
$N_{uptake} = 0.159\ W^{1.71}$ ($r^2 = 0.82$) for 3.5-year-old trees
$N_{uptake} = 0.435\ W^{0.52}$ ($r^2 = 0.96$) for 3.0-year-old trees
$N_{uptake} = 0.021 + 0.490\ W$ ($r^2 = 0.99$) for 2.5-year-old trees

1.5, and 4.8 l in the rockwool. N and K contents in the shoots and fruits were not affected by container volume. P, Ca, and Mg contents were reduced in the smallest containers, but tissue concentrations of these nutrients were still much higher than the critical concentrations.

The following equation was employed by De Willigen and Van Noordwijk (1987) to determine the minimal root surface area ($A_{r,n}$, m^2) to meet plant demands for N, P, K, and Ca:

$$A_{r,n} = Q_m/(N_p \times F_{max}) \qquad \text{(Eq. 1)}$$

where Q_m = nutrient quantity required for optimal growth and yield (mg per plant); N_p = plant density (ha^{-1}); F_{max} = maximum nutrient uptake rate per unit root surface area (mg m^{-2} day^{-1})

The initial estimates of $A_{r,n}$ for N, P, K, and Ca are given in Table 5. 2. In a similar way, a required root surface area for water, $A_{r,w}$ (m^2 per plant), was proposed:

$$A_{r,w} = E_p/\{L_p \times (\Delta Hp - 2\pi_o\sigma_r^2/(1-\sigma_r))\} \quad \text{(Eq. 2)}$$

where E_p = transpiration rate (ml s^{-1} $plant^{-1}$); L_p = root conductance (ml m^{-2} s^{-1} MPa^{-1}); σ_r = reflection coefficient; π_o = osmotic potential of the solution; ΔH_p = difference in water potential between the root and the soil.

The estimate of $A_{r,w}$ for high E_p (2 l day^{-1} $plant^{-1}$) was 1.8 m^2 per plant (equivalent to 0.09 l root volume per plant) (De Willigen and Van Noordwijk, 1987), following direct measurements of the actual L_p. The critical root surface area for normal growth was 1.2 m^2 per plant. The required root surface areas for N, P, K, and Ca should be 0.7, 0.6, 1.6, and 2 m^2 per plant, respectively (see Table 5.2). De Willigen and Van Noordwijk (1987) concluded that no nutrient was limiting plant dry matter production and fruit yield in the smallest pots. The only possible exception was Ca uptake in the experiment with tomato plants, as evident from the occurrence of blossom end rot in the smallest pot. The observed minimum root surface area closely agrees with the root surface area required for water uptake under the experimental conditions, as evident from independent measurement of the potential for water uptake of the root system. It appeared that plants in smaller containers showed lower leaf water potentials in periods of high potential transpiration than those in larger containers, with no measurable differences in growth. Later, when differences in growth were apparent, no consistent variation in leaf water was observed.

TABLE 5.2. Estimates of Maximum Uptake Rate (F_{max}), Physiologically Required Root Surface Area ($A_{r,n}$) and Root Volume ($V_{r,n}$) for Tomato

Required uptake per plant in mg day^{-1}	N	P	K	Ca
	225	45	450	100
F_{max} (mg m^{-2} day^{-1})	84	15.5	136.5	24
$A_{r,n}$ (m^2)	2.7	2.9	3.2	4.0
$V_{r,n}$ (l)	0.14	0.15	0.16	0.22

Source: Adapted from De Willigen and Van Noordwijk (1987).

Note: A root diameter of 0.2 mm is assumed. Nutrient uptake was based on published data and growth rates of 3 g vegetative and 6 g generative dry matter production per plant per day.

TOMATO RESPONSE TO ROOT TREATMENTS IN HYDROPONICS

Although solution culture is a preferred system for studying the direct physiological effects of root restriction, only a few such studies have been carried out (Carmi and Heuer, 1981; Hameed, Reid, and Rowe, 1987; Robbins and Pharr, 1988; Peterson and Krizek, 1992), most of which were short-term experiments. Some of these studies analyzed nutrient contents in leaves, but none of them focused on the effects of root restriction on the nutrient uptake rate and distribution of nutrients among the organs of the entire plant. Bar-Tal and colleagues (1994a, 1994b, 1995) and Bar-Tal and Pressman (1996) used an airo-hydroponics system for investigating the effect of root size on uptake of nutrients and water by tomato plants to elucidate the effects of root pruning and restriction in relation to nutrient supply on transpiration and nutrient uptake.

In the first study, three root treatments were applied: intact (undisturbed root system), mild root pruning (pruning took place ten, seventeen, and forty-four days after transplanting, leaving two-thirds of the root system intact), and severe root pruning (pruning took place ten, seventeen, forty-four, fifty-seven, and eighty-six days after transplanting, leaving one-third of the root system intact). Equal proportions of primary and secondary roots were pruned. The root treatments were factorially combined with low and high concentrations of NO_3^- solution (1.5 and 9.0 mM N, respectively) (Bar-Tal et al., 1994a, b). In the second study, three root treatments were applied: unrestricted root system and severely (0.4 l) and mildly (1.0 l) restricted root systems. Low- and high-N treatment (C_N) were included: 1.0 and 9.0 mM (Bar-Tal et al., 1995). The third experiment was an investigation of the effects of root restriction on K and Ca uptake by tomato roots (Bar-Tal and Pressman, 1996). Treatments included unrestricted and restricted roots and three concentrations of K (2.5, 5, and 10 mM) and two concentrations of Ca (3 and 6 mM).

Root Growth

The actual progress of root growth, as affected by various treatments, is shown in Figure 5.3. Restriction of the root system into smaller containers resulted in a decrease in the root volume compared with unrestricted roots. Severe and mild root restriction almost stopped an increase in the root volume 70 and 130 days after transplanting, respectively. The growth curves of the unrestricted and restricted root systems in both nutrient solutions had a smooth "S" shape, which is described well by the Gompertz growth model:

FIGURE 5.3. Root Growth As Affected by Root Restriction

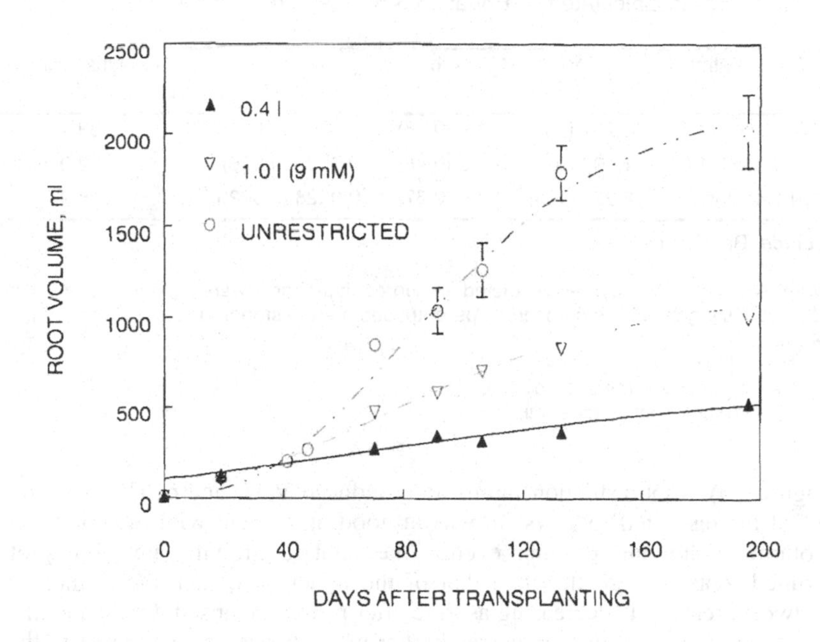

Source: Adapted from Bar-Tal et al. (1995)

Note: Each point represents a mean of NO_3^- solution concentrations (1.0 and 9.0 mM), except for the mild restriction that was studied only in 9 mM solution. The vertical bars represent $\pm LSD_{0.05}$.

$$V_t = V_f \times e^{(-b \times e^{-k \times t})} \qquad \text{(Eq. 3)}$$

where V_t is the root volume at time t (ml); V_f is the maximal root volume (ml); k is the rate coefficient (day^{-1}); b is a dimensionless parameter (for values, see Table 5.3)

The estimated parameters k and V_f were significantly affected by the solution NO_3^- concentration (data not shown). The rate coefficient values (k) of the intact root were in good agreement with published values (Ruff et al., 1987; Bar-Tal et al., 1994a). According to the model curves, the daily change in root volume of the low- and high-C_N solution treatments increased to maximum values of 18 and 20 ml day^{-1} on the forty-fifth and fifty-fifth day after transplanting, respectively, and decreased thereafter (see

TABLE 5.3. Statistical Analysis of Parameters Used in Figure 5.3 for the Curves of Root Growth Calculated by Equation 3: $V_t = V_f \times e^{(-b \times e^{-k \times t})}$

Root treatment	V_1 ml	b	k day^{-1}	SE of the estimate ml
Restricted (0.4 l)	701 (135)**	1.9 (0.16)**	0.0095 (0.0028)**	63.4
Restricted (1.0 l)	1,197 (93)**	3.5 (0.48)**	0.018 (0.0026)**	92.9
Unrestricted	2,225 (104)**	5.9 (0.87)**	0.0228 (0.0023)**	218.9

Source: Bar-Tal et al. (1995).

Note: The curves for severely restricted and unrestricted roots were fitted for 1 and 9 mM N solutions together. The numbers in parentheses are the standard errors of the parameters.

** P = 0.01

V_t is the root volume at time *t* (ml)

V_t is the final root volume (ml)

Figure 5.3). Root restriction significantly reduced V_f, *k*, and *b*. The resulting V_f of the restricted root system was in good agreement with the container volume. Following pruning events, the relative growth rate (RGR) of pruned roots was greater than that of the intact root, with the difference between treatments decreasing as time after pruning elapsed. Consequently, the proportion of young roots was higher in the pruned as compared to the intact root system. Initiation of young secondary roots was observed after pruning events.

Nutrient Content in Plant Tissue

Root pruning and restriction did not reduce N concentration in any of the plant organs. On the contrary, severe root pruning significantly increased N concentration of the younger leaves. Root pruning significantly decreased P concentration in the old leaves in the high-C_N treatments. Increasing C_N from 1.5 mM to 9 mM significantly increased N concentration in stems and leaves (see Table 5.4). Root restriction reduced K concentration in the measured plant organs, except for the roots and the leaves below the third truss (see Table 5.5).

Root restriction reduced Ca concentration in the roots, but in contrast to its effect on K concentration, it had no significant effect on Ca concentration in the stem, leaves, and fruits, except for the fruits of truss eight. Root restriction significantly reduced the K/Ca ratio in the leaves (data not shown). The effect of the K/Ca ratio in the fertilization management on blossom-end rot in tomato fruit is well known (Van der Boon, 1973). In the

TABLE 5.4. The Effect of Root Pruning and C_N on Nutrient Concentration in Tomato Leaves and Stem 143 Days After Transplanting

Root pruning	C_N mM	Stem			Old leaves[a]			Young leaves[b]		
		N	P	K	N	P	K	N	P	K
		mg g^{-1} DM								
severe	1.5	13.3	4.7	41.1	20.7	5.4	27.8	29.4	5.2	41.7
mild	1.5	13.7	4.9	42.9	21.8	5.9	31.7	27.7	5.6	38.3
intact	1.5	13.8	4.9	44.8	20.1	5.5	35.8	28.2	5.2	37.8
severe	9.0	15.3	5.1	40.5	25.1	5.8	34.0	32.1	5.4	47.3
mild	9.0	15.0	5.4	43.0	23.8	6.4	34.9	30.5	5.4	46.6
intact	9.0	15.6	5.2	42.2	24.5	6.7	37.0	29.9	5.1	45.3
$LSD_{0.05}$		1.2	0.54		1.6	0.58	2.7	1.9		4.3
F values:										
Root		NS	NS	NS	NS	4.6^{*}	17^{**}	4.2^{*}	NS	NS
C_N		24^{**}	6.8^{*}	NS	64^{**}	16^{**}	21^{**}	19^{**}	NS	34^{**}
Root × C_N		NS	NS	NS	3.4^{*}	NS	3.6^{*}	NS	NS	NS

Source: Adapted from Bar-Tal et al. (1994b).

[a] Old leaves—leaves of the first five strata that were detached when all the fruits of each stratum were picked.

[b] Young leaves—leaves of the sixth to tenth stratum analyzed at termination of the experiment, 143 days after transplanting.

* P = 0.05

** P = 0.01

NS, not significant at P ≤ 0.05

studies described here, root restriction reduced the incidence of blossom-end rot and a good correlation was established between the incidence of blossom-end rot and the K/Ca ratio in the leaves. This effect was more pronounced in solution containing 5.0 mM K and 1.5 mM (3 mmol(+) l^{-1}) Ca, where the incidence of blossom-end rot was reduced from 11.3 to 4.7 affected fruits per plant.

Nutrient Uptake

Total N accumulation by the severely pruned plants was 83 percent of the control in the high-C_N treatment throughout the experiment (Bar-Tal et al., 1995). Severe (0.4 l) and mild (1.0 l) root restriction also reduced total N accumulation significantly, as compared with intact-root plants. Increases

TABLE 5.5. The Effect of Root Restriction and C_N on Nutrient Concentration in Tomato Leaves and Stem 198 Days After Transplanting

C_N mM	Root treatment	Root	Stem	Leaves below trusses 3	5	7	Fruits
		mg g^{-1} DM					
		N concentration					
1.0	Restricted (0.4 l)	31	18	27	27	25	20
1.0	Unrestricted	30	16	28	26	24	19
9.0	Restricted (0.4 l)	36	22	30	30	27	20
9.0	Restricted (1.0 l)	36	22	32	29	28	22
9.0	Unrestricted	31	22	34	29	28	22
$LSD_{0.05}$		3.2	1.8	2.3	2.5	1.8	
F values:	Root	NS	NS	NS	NS	NS	NS
	C_N	5^{*}	32^{***}	18^{***}	6.4^{*}	13^{**}	NS
	Root × C_N	NS	NS	NS	NS	NS	NS
		K concentration					
1.0	Restricted (0.4 l)	37	33	33	28	24	38
1.0	Unrestricted	39	36	35	34	32	41
9.0	Restricted (0.4 l)	26	31	39	33	29	36
9.0	Restricted (1.0 l)	33	34	41	38	36	35
9.0	Unrestricted	39	35	36	40	39	40
$LSD_{0.05}$			1.9	3.9	4.1	3.5	3.7
F values:	Root	NS	11^{***}	NS	7^{**}	18^{***}	5^{*}
	C_N	NS	NS	4.3^{*}	13^{**}	21^{***}	NS
	Root × C_N	NS	NS	NS	NS	NS	NS

Source: Adapted from Bar-Tal et al. (1995).

* P = 0.05
** P = 0.01
*** P = 0.001
NS = not significant at $P \leq 0.05$

in the rate of N uptake (F_N) occurred after root pruning (at thirty-two, fifty-six, and eighty-six days after transplanting), as well as due to root restriction, indicating an adaptation of the smaller root system to the shoot demand (see Figure 5.4). Thus, the smaller root system was more efficient in uptake per unit root (see also Jungk, 1974; Jungk and Barber, 1975; Edwards and Barber, 1976; De Willigen and Van Noordwijk, 1987; Bar-Yosef et al., 1988).

Root restriction significantly reduced the K uptake rate per plant, but it had no effect on F_K. The values of the parameter F_{max} (0.97 and 0.35 mg

FIGURE 5.4. Nitrogen Uptake Rate per Root Unit Volume (F_N) As a Function of Time and As Affected by (a) Root Pruning and (b) Root Restriction

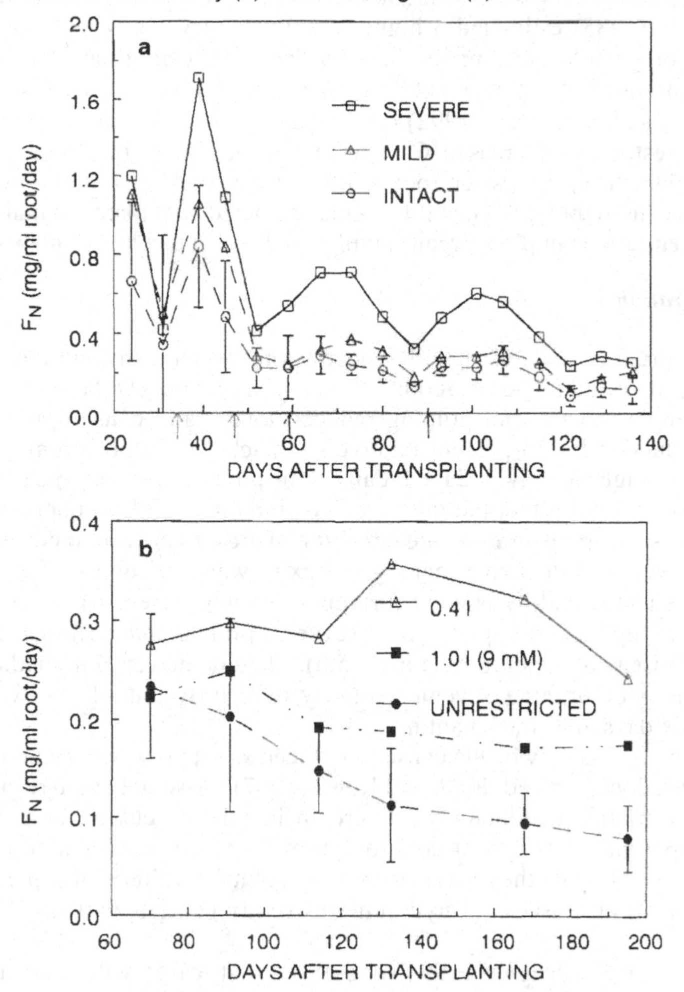

Source: Adapted from Bar-Tal et al. (1994b).

Part a: Each point represents a mean of two NO_3^- solution concentrations (1.5 and 9.0 mM). Arrows on the abscissa indicate pruning events.

Source: Adapted from Bar-Tal et al. (1995).

Part b: Each point represents a mean of two NO_3^- solution concentrations (1.0 and 9.0 mM), except for the mild restriction that was studied only in 9 mM solution. The vertical bars represent $\pm LSD_{0.05}$.

g^{-1} day^{-1} for the young and old plants, respectively) fell within the range of published data for four different species (Wild et al., 1974). Bar-Yosef and Sagiv (1985) estimated a higher maximal F_K value of 4.3 mg g^{-1} day^{-1} for greenhouse tomato (*Lycopersicon esculentum*, cv. Grandier) grown in soil, but a lower value, 0.34 mg g^{-1} day^{-1}, was found by Bar-Yosef and colleagues (1992) for another cultivar of tomato.

Root restriction decreased Ca uptake rate per plant, but there was a trend of increasing F_{Ca} when root restriction was combined with low C_K, which is consistent with the ability of the restricted-root plants to maintain Ca concentrations in plant organs similar to those in intact-root plants.

Transpiration

Root pruning and restriction reduced total water consumption by a similar proportion to the reduction in canopy dry matter (DM) production. For example, severe root pruning reduced total water consumption and DM production by 25 percent relative to intact plant. Root pruning and restriction significantly reduced daily transpiration (Figure 5.5). Root restriction did not affect the ratio of transpiration to DM production, but did increase the transpiration rate per unit leaf area, as measured during the last two weeks of the experiment. The flux of water per unit root volume (F_W) was unaffected by N concentration (data not presented), whereas it was much higher in the plants with severely pruned and restricted roots than in those with intact roots (Figure 5.6). The F_W increased immediately following root pruning seventeen, thirty-two, forty-four, fifty-six, and eighty-six days after transplanting.

Drought stress may be the cause of reduced shoot growth in response to root restriction. Hameed, Reid, and Rowe (1987) investigated the response to root restriction of tomato plants grown in solution culture in a short-term experiment. The restricted-root tomato plants exhibited drought stress, even though they were grown in solution culture, the possible reason being increased root hydraulic resistivity of roots in a small container.

Bar-Tal and colleagues (1994b) observed that leaves wilted for a day only after the first severe pruning, not after subsequent pruning events, in agreement with the results of Schupp and Ferree (1990). The reductions in total water consumption and daily transpiration caused by root pruning and root restriction could be attributed to a reduction in leaf area, meaning reduced transpiration would not be the cause, but a result of, retarded shoot growth. The increase in the transpiration rate per unit leaf area, as a result of root restriction, also indicated that root restriction did not limit water uptake. Krizek and colleagues (1985) showed that the effect of root

FIGURE 5.5. The Transpiration Rate per Plant As a Function of Time and As Affected by (a) Root Pruning and (b) Root Restriction

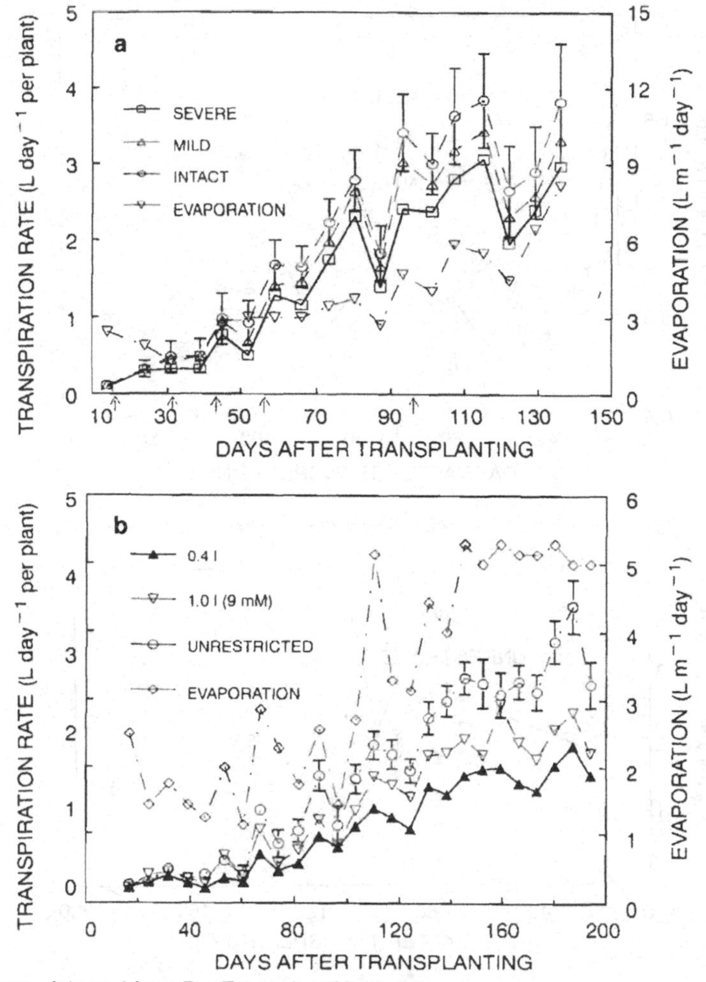

Source: Adapted from Bar-Tal et al. (1994b).

Part a: Each point represents a mean of two NO_3^- solution concentrations (1.5 and 9.0 mM) and outdoor evaporation data from a standard pan. Arrows on the abscissa indicate pruning events.

Source: Adapted from Bar-Tal et al. (1995).

Part b: Each point represents a mean of two NO_3^- solution concentrations (1.0 and 9.0 mM), except for the mild restriction that was studied only in 9 mM solution. The vertical bars represent $\pm LSD_{0.05}$.

FIGURE 5.6. The Transpiration Rate per Unit Root Volume (F_W) As a Function of Time and As Affected by (a) Root Pruning and (b) Root Restriction

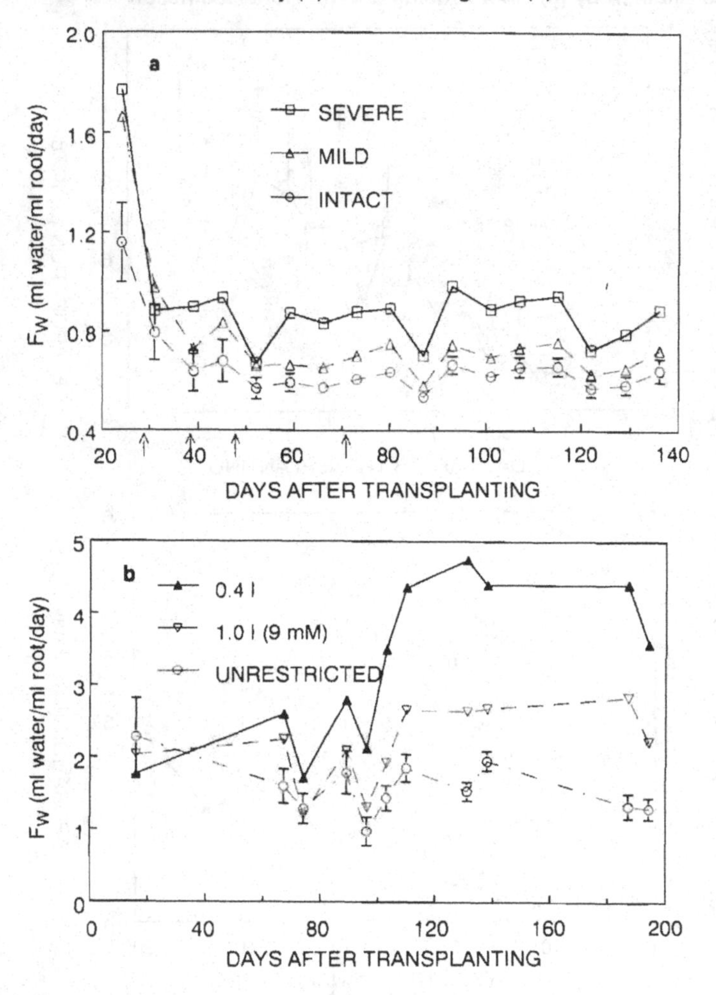

Source: Adapted from Bar-Tal et al. (1994b).

Part a: Each point represents a mean of two NO_3^- solution concentrations (1.5 and 9.0 mM). Arrows on the abscissa indicate pruning events.

Source: Adapted from Bar-Tal et al. (1995).

Part b: Each point represents a mean of two NO_3^- solution concentrations (1.0 and 9.0 mM), except for the mild restriction that was studied only in 9 mM solution. The vertical bars represent $\pm LSD_{0.05}$.

restriction on soybean growth and water status was different from the effect of soil moisture stress. Hameed, Reid, and Rowe (1987) reported an increase in the ratio of transpiration to dry matter production in root-restricted plants, although the net assimilation rate was decreased by root restriction. Ruff and colleagues (1987) failed to obtain such a decrease in the net assimilation rate in a long-term experiment. Bar-Tal and colleagues (1995) also obtained the enhanced ratio of transpiration to dry matter production in response to restriction of the tomato root system in a long-term experiment. Therefore, it seems that drought is not the main factor in reducing growth of the root-restricted plants.

INTERNAL REGULATION OF NUTRIENT UPTAKE

Most of the data on ion kinetics are based on the uptake of nutrients from solution by excised roots (Epstein, 1972). However, one of the factors that complicates the research on nutrient uptake by roots is the internal regulation of nutrient uptake. This phenomenon was recognized long ago. Williams (1948) stated that "the intake of phosphate was more controlled by internal factors than by external factors of supply." The existence of a regulation mechanism also follows from the ability of most plants to obtain a relatively constant nutrient composition in a wide range of nutrient solutions (Steiner, 1984). It is obvious that the roots do not function as a simple "pump" for nutrients, without regard for the rest of the plant, since plants would otherwise accumulate extremely large nutrient concentrations (Nye and Tinker, 1977); there must be a strong feedback control on root activity from the rest of the plant. Although the mechanism of this feedback is not known, it appears that the regulation of uptake is specific for each nutrient.

Clarkson (1985) suggested that there are short-term modulations of the activities of System I (high-affinity) carriers and long-term changes in carrier numbers. Short-term regulation of nutrient intake was obtained for K uptake by barley (Glass, 1976; Siddiqi and Glass, 1983) and Cl^- uptake by *Chara* (Sanders, 1980). Vale, Jackson, and Volk (1987) showed that the kinetics of K uptake by excised roots are affected by the K status in the roots. Long-term regulation of $H_2PO_4^-$ and SO_4^{2-} uptake by *Hordeum vulgare* during deficiency-stress-induced demand was achieved by increasing V_{max}, but with no change in K_m. In the same species, K deficiency stress induced both decrease in K_m and increase in V_{max} (Lee, 1982). It has been observed that uptake of NO_3^- and Cl^- is regulated to maintain a steady state "set" value for (NO_3^- + Cl^-) rather than for each separately. However, changes in the influx alone cannot explain the modulation in nutrient uptake by plants. It has been shown that regulation of the NO_3^- efflux is dominant in the control of net NO_3^- uptake by wheat (Minotti, Williams, and Jackson, 1969) and barley (Deane-Drummond and Glass, 1983).

Much of the data obtained from whole-plant experiments based on split-root systems, pruned roots, and application of fertilizers to a fraction of the root rhizosphere indicate that there is internal regulation of nutrient uptake by plants (Jungk, 1974; Jungk and Barber, 1975; Edwards and Barber, 1976; Richards and Rowe, 1977; Schupp and Ferree, 1988, 1990). However, the relevance of these experiments to the plethora of information based on excised roots is limited because the kinetics of nutrient uptake are affected by plant age (Clarkson, Sanderson, and Russel, 1968), the history of the plant, and the status of nutrients in the whole plant (Ben Zioni, Vaadia, and Lips, 1971; De Willigen and Van Noordwijk, 1987). The values of the parameters of the kinetics equation describing nutrient uptake at the root surface are different in whole plants as opposed to excised roots; the intact root system has lower K_m and V_{max} (Loneragan, 1978). The meaning of these differences is that intact roots can supply plant requirements from lower nutrient concentration, but their uptake from high-concentration solutions is lower than that of excised roots. Thus, the changes in the uptake rate are relatively small over a wide range of external concentrations. A model of shoot regulation of root uptake of NO_3^- and K, which involves the internal status of N and K in the shoot, and K circulation in the plant was proposed by Ben Zioni, Vaadia, and Lips (1971). This involvement of ion circulation is probably the reason that uptake of Ca, which is an immobile nutrient in plants, is less regulated. Clarkson (1985) concluded that "modulation of the kinetics is an early response to nutrient stress and occurs before changes in growth rate can be measured" (p. 100).

Another factor in the enhancement of nutrient and water uptake rates by restricted or pruned root systems may be changes in root morphology and in the average root age. Clarkson (1985) stated that "The second response to limiting supplies of N and P, possibly for S and K also, is the inhibition of overall growth and the relative enlargement of the root system" (p. 100). Bar-Tal and colleagues (1994b) suggested that root pruning enhanced root production, resulting in an increased number of active young roots. As such roots are very fine, they hardly contributed to the volume of the roots, but caused the average uptake rate per unit root volume to increase. Ruff and colleagues (1987) and Richards and Rowe (1977) reported that root-restricted tomato plants and peach seedlings developed more densely branched root systems than their unrestricted equivalents. Ran, Bar-Yosef, and Erez (1992) found that the proportion of fine roots of peach increased as the container volume decreased. The resultant change in morphology may affect the uptake properties of the roots in ways that cannot be adequately described on a weight and volume basis.

CONCLUDING REMARKS

The agricultural experience with intensively irrigated and fertilized crops indicates that small root systems may be sufficient for maximum plant growth under conditions of optimum supply of water and nutrients. The question arising from the previously described investigations is whether the size of the root system and its ability to absorb nutrients become the limiting factors in plant nutrition. In some of the studies discussed, plant nutrition was affected by the growth medium volume and root size, whereas in others such an effect was not obtained. In the former studies, the nutrient uptake rate per unit of root weight or volume was reduced as the container volume decreased, whereas the opposite result was obtained in the latter studies. These differing responses stemmed from the indirect and direct effects of the volume of soil or growth medium available for root expansion. As long as this volume does not affect the buffering capacity of the medium to maintain nutrient concentration and to satisfy the plant demand for nutrients, root efficiency increases as root size decreases. Theoretically, root restriction below a critical value should affect plant nutrition, but in practice, other factors such as water uptake and plant hormones restrict plant growth before root size can control nutrient uptake. According to De Willigen and Van Noordwijk (1987), root restriction might reduce Ca uptake because that uptake is less regulated by the plant. However, Bar-Tal and Pressman (1996) found that efficiency of Ca uptake by tomato roots increased as the root volume decreased, probably as a result of change in root architecture; the Ca uptake rate is more closely correlated with the number of root tips than with root length. Changes in root morphology, higher branching, and a greater proportion of small-diameter roots have been observed to result from root restriction.

When use of small volumes reduced the ability of the medium to supply nutrients at the required rate, there was a reduction in the apparent efficiency of the roots in absorbing nutrients. In fact, the latter effect occurs at container volumes and root sizes much bigger than the critical root size for absorption of nutrients from an unlimited source. Technically, the supply of sufficient nutrients via small volumes that are bigger than the theoretical lower border for absorption becomes impossible because of aeration problems associated with too frequent irrigation and the decrease in the soil or growth medium hydraulic coefficient, as root density increases above a threshold point.

De Willigen and Van Noordwijk (1987) found that smaller container sizes led to lower leaf water potentials in periods of high potential transpiration, as compared to plants grown in larger container sizes, before any differences in growth were noticed. Bar-Tal and colleagues (1994b) and Schupp and Ferree (1990) observed leaves wilting for a day only after the first pruning, but later, when a difference in growth was apparent, no consistent difference in leaf water was observed. Bar-Yosef, Stammers, and Sagiv (1980) showed that a bigger root mass, caused by altered irrigation frequency, affected fresh fruit yield by improving water uptake, but did not affect fruit dry weight; these results indicate a disturbance in the water supply to the plants with smaller root volumes. De Willigen and Van Noordwijk (1987) suggested that water uptake was the first limiting factor when root growth was restricted. The studies discussed here indicate that water uptake might be more sensitive to root restriction than nutrient uptake.

REFERENCES

Aung, L.H. (1974). Root-shoot relationships. In *The Plant Root and its Environment*, Ed. E.A. Carson. Charlottesville, VA: University of Virginia Press, pp. 29-61.

Bar-Tal, A., B. Bar-Yosef, and U. Kafkafi (1990). Pepper transplant response to root volume and nutrition in the nursery. *Agronomy Journal* 82: 989-995.

Bar-Tal, A., B. Bar-Yosef, and U. Kafkafi (1993). Modeling pepper seedling growth and nutrient uptake as a function of cultural conditions. *Agronomy Journal* 85: 718-724.

Bar-Tal, A., A. Feigin, I. Rylski, and E. Pressman (1994a). Root pruning and $N-NO_3$ solution concentration effects on tomato plant growth and fruit yield. *Scientia Horticulturae* 58: 91-103.

Bar-Tal, A., A. Feigin, I. Rylski, and E. Pressman (1994b). Root pruning and $N-NO_3$ solution concentration effects on nutrient uptake and transpiration by tomato. *Scientia Horticulturae* 58: 77-90.

Bar-Tal, A., A. Feigin, S. Sheinfeld, R. Rosenberg, B. Sternbaum, I. Rylski, and E. Pressman (1995). Root restriction and $N-NO_3$ solution concentration effects on nutrient uptake, transpiration and dry matter production of tomato. *Scientia Horticulturae* 63: 195-208.

Bar-Tal, A. and E. Pressman (1996). Root restriction and potassium and calcium solution concentration affect dry-matter production, cation uptake, and blossom end rot in greenhouse tomato. *Journal of the American Society for Horticultural Science* 121: 649-655.

Bar-Yosef, B., P. Imas, and I. Levkovitz (1995). The effects of pot volume, geometry and growth medium type on plant development. An annual report submitted to the Chief Scientist of the Ministry of Agriculture, Israel.

Bar-Yosef, B., E. Matan, I. Levkovitz, A. Assaf, and A. Dayan (1992). Response of two cultivars of greenhouse tomato (F.144 and 175) to fertilization in the Besor area. Final Report 301-170-92, submitted to the Chief Scientist of the Ministry of Agriculture, Israel.

Bar-Yosef, B. and B. Sagiv (1985). Potassium supply to field crops grown under drip irrigation and fertilization. In *Proceedings of Potassium Symposium*, Pretoria, South Africa. Bern, Switzerland: International Potash Institute, pp. 185-189.

Bar-Yosef, B., S. Schwartz, T. Markovich, B. Lucas, and R. Assaf (1988). Effect of root volume and nitrate solution concentration on growth, fruit yield, and temporal N and water uptake rates by apple trees. *Plant and Soil* 107: 49-56.

Bar-Yosef, B., C. Stammers, and B. Sagiv (1980). Growth of trickle-irrigated tomato as related to rooting volume and uptake of N and water. *Agronomy Journal* 72: 815-822.

Bar-Yosef, B., A. Zilber, A. Markovitz, M. Keinan, I. Levkovitz, and S. Suriano (1997). The effects of pot volume, growth medium type and fertilization on the development of greenhouse tomato. An annual report submitted to the Chief Scientist of the Ministry of Agriculture, Israel.

Ben Zioni, A., Y. Vaadia, and S.H. Lips (1971). Nitrate uptake by roots as regulated by nitrate reduction products of the shoot. *Physiologia Plantarum* 24: 288-290.

Brouwer, R. and C.T. De Wit (1968). A simulation model of plant growth with special attention to root growth and its consequences. In *Root Growth. Proceedings of Fifteenth Easter School in Agricultural Science, University of Nottingham*, ed. W.J. Whittington. London, UK: Butterworths, pp. 224-242.

Carmi, A. (1986). Effect of cytokinins and root pruning on photosynthesis and growth. *Photosynthetica* 20: 1-8.

Carmi, A., J.D. Hesketh, W.T. Enos, and D.B. Peters (1983). Interrelationships between shoot growth and photosynthesis as affected by root growth restriction. *Photosynthetica* 17: 240-245.

Carmi, A. and B. Heuer (1981). The role of roots in control of bean shoot growth. *Annals of Botany* 48: 519-527.

Carmi, A. and J. Van Staden (1983). The role of roots in regulating the growth rate and cytokinin content in leaves. *Plant Physiology* 73: 76-78.

Clarkson, D.T. (1985). Factors affecting mineral nutrient acquisition by plants. *Annual Review in Plant Physiology* 36: 77-115.

Clarkson, D.T., J. Sanderson, and R.S. Russel (1968). Ion uptake and root age. *Nature (London)* 220: 805-806.

Cooper, A.J. (1972). The influence of container volume, solution concentration, pH and aeration on dry matter partition by tomato plants in water culture. *Journal of Horticultural Science* 47: 341-347.

Deane-Drummond, C.E. and A.D.M. Glass (1983). Short term studies of nitrate uptake into barley plants using ion-specific electrodes and $^{36}ClO_3^-$. II. Regulation of NO_3^- efflux by NH_4^+. *Plant Physiology* 73: 105-110.

De Willigen, P. and M. Van Noordwijk (1987). Roots, Plant Production and Nutrient Use Efficiency. Doctoral Thesis. Agricultural University Wageningen, Wageningen, Netherlands, 282 pp.

Downey, L.A. and T.C. Mitchell (1971). Root length and vapor pressure deficit: Effect on relative water content in *Zea mays* L. *Australian Journal of Biological Science* 24: 811-814.

Edwards, J.H. and S.A. Barber (1976). Nitrogen flux into corn roots as influenced by shoot requirement. *Agronomy Journal* 68: 973-975.

Epstein, E. (1972). *Mineral Nutrition of Plants: Principles and Perspectives*. New York: John Wiley.

Geisler, D. and D.C. Ferree (1984). The influence of root pruning on water relations, net photosynthesis, and growth of young "Golden Delicious" apple trees. *Journal of the American Society for Horticultural Science* 109: 827-831.

Glass, A.D.M. (1976). Regulation of potassium absorption in barley roots. An allosteric model. *Plant Physiology* 58: 33-37.

Hameed, M.A., J.B. Reid, and R.N. Rowe (1987). Root confinement and its effects on the water relations, growth and assimilate partitioning of tomato (*Lycopersicon esculentum* Mill). *Annals of Botany* 59: 685-692.

Hanson, P.J., R.K. Dixon, and R.E. Dickson (1987). Effect of container size and shape on the growth of northern red oak seedlings. *HortScience* 22: 1293-1295.

Humphries, E.C. (1958). Effect of removal of a part of the root system on the subsequent growth of the root and shoot. *Annals of Botany* 22: 151-157.

Jackson, M.B. (1993). Are plant hormones involved in root to shoot communication? *Advances in Botanical Research* 19: 104-187.

Jungk, A. (1974). Phosphate uptake characteristics of intact root systems in nutrient solution as affected by plant species, age and P supply. In *Plant Analysis and Fertilizer Problems. Proceedings of the Seventh International Plant Nutrition Colloqium*, Volume 1, Ed. J. Wehrmann. Hannover, Germany, pp. 185-196.

Jungk, A. and S.A. Barber (1975). Plant age and the phosphorus uptake characteristics of trimmed and untrimmed corn root systems. *Plant and Soil* 42: 227-239.

Krizek, D.L., A. Carmi, R.M. Mirecki, F.W. Snyder, and J.A. Bunce (1985). Comparative effects of soil moisture stress and restricted root zone volume on morphogenetic and physiological responses of soybean [*Glycine max* (L.) Merr.]. *Journal of Experimental Botany* 36: 25-38.

Lee, R.B. (1982). Selectivity and kinetics of ion uptake by barley plants following nutrient deficiency. *Annals of Botany* 50: 429-449.

Loneragan, J.F. (1978). The interface in relation to root function and growth. In *The Soil-Root Interface*, Eds. J.L. Harley and R.S. Russell. London, UK: Academic Press, pp. 351-367.

Minotti, P.L., D.C. Williams, and W.A. Jackson (1969). Nitrate uptake by wheat as influenced by ammonium and other cations. *Crop Science* 9: 9-14.

Mutsaers, H.J.W. (1983). Leaf growth in cotton (*Gossypium hirsutum* L.). 2. The influence of temperature, light, water stress and root restriction on the growth and initiation of leaves. *Annals of Botany* 51: 521-529.

Nye, P.H. and P.B. Tinker (1977). *Solute Movement in the Soil-Root System*. Oxford, UK: Blackwell Scientific Publications.

Peterson, T. A. and D.T. Krizek (1992). A flow-through hydroponic system for the study of root restriction. *Journal of Plant Nutrition* 15: 893-911.

Peterson, T.A., M.D. Reinsel, and D.T. Krizek (1991a). Tomato (*Lycopersicon esculentum* Mill., cv. "Better Bush") plant response to root restriction. 1. Alteration of plant morphology. *Journal of Experimental Botany* 42: 1233-1240.

Peterson, T.A., M.D. Reinsel, and D.T. Krizek (1991b). Tomato (*Lycopersicon esculentum* Mill., cv. "Better Bush") plant response to root restriction. 2. Root respiration and ethylene generation. *Journal of Experimental Botany* 42: 1241-1249.

Ran, Y., B. Bar-Yosef, and A. Erez (1992). Root volume influence on dry matter production and partitioning as related to nitrogen and water uptake rates by peach trees. *Journal of Plant Nutrition* 15: 713-726.

Richards, D. and R.N. Rowe (1977). Effects of root restriction, root pruning and 6-benzylaminopurine on the growth of peach seedlings. *Annals of Botany* 41: 729-740.

Robbins, N.S. and D.M. Pharr (1988). Effect of restricted root growth on carbohydrate metabolism and whole plant growth of *Cucumis sativus* L. *Plant Physiology* 87: 409-413.

Ruff, M.S., D.L. Krizek, R.M. Mirecki, and D.W. Inouye (1987). Restricted root zone volume: Influence on growth and development of tomato. *Journal of the American Society for Horticultural Science* 112: 763-769.

Sanders, D. (1980). Control of Cl^- influx in *Chara* by cytoplasmic Cl^- concentration. *Journal of Membrane Biology* 52: 51-60.

Schupp, J.R. and D.C. Ferree (1988). Effects of root pruning at four levels of severity on growth and yield of "Melrose"/M.26 apple trees. *Journal of the American Society for Horticultural Science* 113: 194-198.

Schupp, J.R. and D.C. Ferree (1990). Influence of time of root pruning on growth, mineral nutrition, net photosynthesis and transpiration of young apple trees. *Scientia Horticulturae* 42: 299-306.

Siddiqi, M.Y. and A.D.M. Glass (1983). Studies of the growth and mineral nutrition of barley varieties. II. Potassium uptake and its regulation. *Canadian Journal of Botany* 61: 1551-1558.

Steiner A.A. (1984). The universal nutrient solution. In *Proceedings of the Sixth Congress of the International Society of Soilless Culture*. Wageningen, Netherlands: International Society of Soilless Culture, pp. 633-649.

Vale, F.R., W.A. Jackson, and R.J. Volk (1987). Potassium influx into maize root systems: Influence of root potassium concentration and ambient ammonium. *Plant Physiology* 84: 1416-1420.

Van der Boon, J. (1973). Influence of K/Ca ratio and drought on physiological disorders in tomato. *Netherlands Journal of Agricultural Science* 21: 56-67.

Wild, A., V. Skarlou, C.R. Clement, and R.W. Snaydon (1974). Comparison of potassium uptake by four plant species grown in sand and in flowing solution culture. *Journal of Applied Ecology* 11: 801-812.

Williams, R.F. (1948). The effects of phosphorus supply on the rates of intake of phosphorus and nitrogen and upon certain aspects of phosphorus metabolism in gramineous plants. *Australian Journal of Scientific Research (B)* 1: 333-361.

Chapter 6

Role of Mineral Nutrients in Photosynthesis and Yield Formation

Ismail Cakmak
Christof Engels

Key Words: carbohydrate, deficiency, nutrients, partitioning, phloem, photosynthesis, sink, transport, yield.

INTRODUCTION

The capacity of plants to assimilate carbon in the source organs and translocate and utilize photosynthates in the sink organs determines growth rate and productivity of crop plants. This capacity of crop plants is directly affected by several factors, such as water availability, light conditions, and mineral nutrients (Geiger, Koch, and Shieh, 1996). Mineral nutrients possess several roles in formation, partitioning, and utilization of photosynthates. Therefore, mineral nutrient deficiencies substantially impair production of dry matter and its partitioning between the plant organs (Marschner, Kirkby, and Cakmak, 1996; McDonald, Ericsson, and Larsson, 1996). Although total dry matter production is similarly affected by mineral nutrient deficiencies, partitioning of dry matter between plant organs (e.g., shoots and roots) is differentially influenced by different mineral nutrient deficiencies. In a study with bean plants, total dry matter production was similar under P (phosphorus), K (potassium), and Mg (magnesium) deficiencies. Under P deficiency, however, a greater proportion of dry matter was partitioned to roots, while under K and Mg deficiencies root growth was mark-

This study was supported by the Deutsche Forschungsgemeinschaft (DFG) and the Scientific and Technical Research Council of Turkey (TÜBYTAK). The authors thank Huriye Avsar for her excellent assistance in preparation of the manuscript.

edly decreased (Cakmak, Hengeler, and Marschner, 1994a). These changes were paralleled with the corresponding changes in the distribution of carbohydrates between roots and shoots, indicating differential effects of mineral nutrients on partitioning of photosynthates within plants. Similarly, in birch seedlings, shoot and root growth was differently diminished by deficiency of several nutrients. Under deficiency of N (nitrogen), K, and S (sulfur), a large proportion of total dry matter was allocated to the roots, but under deficiencies of Mg, K, and Mn (manganese) dry matter partitioning toward the roots was depressed severely (Ericsson, 1995).

Besides partitioning of photosynthates, mineral nutrients have strong effects on synthesis of photosynthates and their utilization in sink organs. The metabolic activity of sink organs shows a high dependency on mineral nutrient supply. Deficiencies of mineral nutrients severely limit flower initiation and development (Steer and Hocking, 1983), development and viability of pollen grains (Graham, 1975; Sharma et al., 1991), and development of vegetative sink organs such as tubers (Krauss, 1985). In this chapter, several examples are presented on the roles of mineral nutrients in photosynthetic CO_2 reduction, synthesis and partitioning of photosynthates, and metabolic activity (strength) of source and sink organs. For further information, the readers are referred to recent reviews (Marschner, Kirkby, and Cakmak, 1996; McDonald, Ericsson, and Larsson, 1996).

PHOTOSYNTHESIS AND RELATED PROCESSES

Photosynthetic Electron Transport

In the photosynthesis process, light energy is absorbed by chlorophyll and other pigments and converted into chemical energy in the forms of ATP and NADPH. Conversion of the light energy into chemical energy involves an electron flow between photosystem I (PS I) and photosystem II (PS II). This electron transport system is associated with regeneration of NADPH and establishment of an electrochemical gradient across the thylakoid membranes to produce ATP. NADPH and ATP are used as electron and energy sources at various steps of CO_2 reduction and carbohydrate synthesis.

Mineral nutrients influence photosynthetic electron flow in various ways, either as constituents of the light-harvesting complex or as ions facilitating electron flow. In this regard, Mg, as a central part of chlorophyll molecules in PS I and PS II, plays an important role in absorption of light energy. Depending on the Mg nutritional status of plants, between 6 percent and 35 percent of the total Mg is bound to chlorophylls (Scott and Robson, 1990). On the other side, in the water-splitting reaction and photosynthetic O_2 evolution, a Mn-containing enzyme system exerts an essential role (Rutherford, 1989). This enzyme system is associated with PS II and needed for its function in maintaining photosynthetic electron flow. Therefore, a decrease in photosynthetic electron flow or O_2 evolution can be expected in plants suffering from Mn deficiency. As shown in subterranean clover, Mn deficiency caused a marked depression in photosynthetic O_2 evolution, and a resupply of Mn to deficient plants for 24 h (hours) restored O_2 evolution up to the level of Mn-sufficient plants (Nable, Bar-Akiva, and Loneragan, 1984).

The photosynthetic electron transport system shows a high sensitivity to Fe (iron) deficiency. A most typical reaction of plants to Fe deficiency is the decrease in chlorophyll concentration in leaves (see Table 6.1). Iron has an important role in the biosynthesis of chlorophyll, by affecting the synthesis of chlorophyll precursor S-aminolevulinic acid (Pushnik and Miller, 1989). As with chlorophyll, several components of the photosynthetic electron transport system, particularly PS I, are decreased as a result of Fe deficiency.

The PS I components, such as the electron carrier cytochromes, the reaction center P700, and proteins, were decreased in tobacco leaves in response to Fe deficiency (see Table 6.1). Accordingly, photosynthetic electron transport capacities of PS I and PS II were restricted as a result of Fe deficiency. Iron deficiency also decreased the content of leaf ferredoxin (Alcaraz et al., 1986) that transmits electrons from PS I not only to $NADP^+$ but also to molecular O_2, nitrite, and sulfite. Therefore, under Fe deficiency, decreases in ferredoxin in leaves were accompanied by a reduced activity of NO_3^- reductase (Alcaraz et al., 1986).

TABLE 6.1. Effect of Fe Nutritional Status of Tobacco Leaves on Contents of Chlorophyll and Photosystem I (PS I) Components, and Photosynthetic Electron Transport Capacity of PS II and PS I

Treatment			PS I components			e^- transport capacity	
	Fe	Chlorophyll	P700	Cytochromes	Protein	PS II	PS I
	$\mu g\ cm^{-2}$ leaf			pmol cm^{-2}	$\mu g\ cm^{-2}$	$\mu eq\ cm^{-2}$ leaf h^{-1}	
+Fe	1.44	89	545	599	108	56	840
−Fe	0.25	26	220	201	38	30	390

Source: Adapted from Pushnik and Miller (1989).

Similar to Fe, Cu (copper) deficiency also impairs photosynthetic electron transport, possibly due to reduced levels of plastocyanin, which is a Cu-containing protein that transmits electrons to P700 in PS I (Ayala and Sandmann, 1989). In P-deficient plants, the ability of PS II pigments to absorb and transfer light energy to reaction centers was decreased, a phenomenon that was accompanied by an enhanced dissipation of thermal energy (Jacob, 1995). This reaction is considered a protective response of P-deficient plants against overexcitation of PS II and destruction of photosynthetic apparatus (Demming-Adams and Adams, 1992).

The synthesis of ATP (photophosphorylation) during photosynthetic electron flow is affected by mineral nutrient deficiencies, particularly those of Mg and K. As counter ions for H^+ (hydrogen) influx into thylakoid lumens, both K (Tester and Blatt, 1989) and Mg (Sugiyama et al., 1969) play a fundamental role in the establishment of a proton gradient across the thylakoid membrane. Magnesium is also required for synthesis of ATP as a bridging component between ADP and the enzyme (Lin and Nobel, 1971). According to Jacob and Lawlor (1993), P deficiency reduced ATP synthesis, possibly because of limited amount of inorganic P in the chloroplast stroma for the phosphorylation of ADP.

As a consequence of these impairments in different steps of photosynthetic electron transport, it can be expected that photosynthetic CO_2 reduction is also impaired by deficiency of mineral nutrients. As discussed next, mineral nutrient deficiency causes further impairments in reduction and conversion of CO_2 into carbohydrates.

CO_2 Fixation

In most plant species (C_3 plants), assimilation of CO_2 begins with its incorporation into ribulose bisphosphate (RuBP), the process which is catalyzed by the enzyme RuBP carboxylase, with the production of a C_3 compound phosphoglycerate (PGA). The production of PGA is followed by the formation of phosphoglyceraldehyde (PGAL) at the expense of ATP and NADPH. PGAL, a triosephosphate, is the key compound used in the starch synthesis in chloroplasts or, after leaving the chloroplasts to cytosol, in the sucrose synthesis in cytosol. In contrast, in C_4 plant species, such as corn and sugarcane, CO_2 is incorporated first into phosphoenol pyruvate (PEP) catalyzed by PEP carboxylase with the production of a C_4 compound oxaloacetate.

Mineral nutrients affect several steps of CO_2 assimilation. In C_4 plants, Na (sodium) is required for the conversion of pyruvate into PEP in mesophyll chloroplasts and, therefore, the amount of PEP is low in Na-deficient C_4 plants (Johnston, Grof, and Brownell, 1988). Usually, compared to C_3

plants, C_4 plants have higher CO_2 fixation capacity and produce more leaf area at lower leaf N contents (Sinclair and Horie, 1989), possibly because of a much higher proportion of soluble protein in RuBP carboxylase in C_3 plants than in C_4 plants (Sage, Pearcy, and Seemann, 1987).

Magnesium has a major role in CO_2 fixation (Terry and Ulrich, 1974; Fischer and Bremner, 1993), particularly by affecting activity of RuBP carboxylase in chloroplasts (Lin and Nobel, 1971). Magnesium exerts dual effects on RuBP carboxylase: (1) by binding to the enzyme complex, Mg activates the enzyme for a more efficient incorporation of CO_2 into RuBP (Sugiyama, Nakyama, and Akazawa, 1968), and (2) the stimulatory effect of Mg on RuBP carboxylase is also related to its role in the light-induced increase in pH of the chloroplast stroma. During electron flow between PS II and PS I, there is a light-driven pumping of protons into thylakoid lumens from stroma. This transfer of protons is counterbalanced by transport of Mg^{2+} into stroma, resulting in an increase in alkalinization and Mg concentration of stroma (Portis, 1981; Heineke and Heldt, 1988). With these changes in stroma upon illumination, the conditions become optimal for a high activity of RuBP carboxylase in chloroplast stroma. Decreases in CO_2 assimilation by Mg deficiency were also ascribed to Mg-deficiency-induced accumulation of photosynthates in source leaves (see the section "Phloem Export of Photosynthates") and, thus, reduction in activity of sink organs (Fischer and Bremner, 1993).

Similar to Mg, K is a counterion for H^+ flux across the thylakoid membranes and is involved in maintenance of high pH of chloroplast stroma (Fang, Mi, and Berkowitz, 1995). Accordingly, activity of RuBP carboxylase and rate of CO_2 fixation are depressed in K-deficient leaves (see Table 6.2). Potassium is also required for stomatal conductance and, thus, for an efficient CO_2 diffusion into leaves (Humble and Raschke, 1971). In alfalfa leaves, stomatal resistance was increased and CO_2 fixation decreased with the severity of K deficiency (see Table 6.2).

Decreases in CO_2 fixation are also well documented for N-deficient plants (Evans and Terashima, 1988). In jack pine seedlings, decreases in concentration of needle N were closely associated with impairments in carboxylation capacity of needles (Tan and Hogan, 1995). According to these results, N deficiency impaired carboxylation capacity of needles to a greater extent than it did photosynthetic electron transport and photophosphorylation. Photosynthetic CO_2 fixation is also decreased in plants suffering from P deficiency (Brooks, Woo, and Wong, 1988; Rao and Terry, 1989; Plesnicar et al., 1994). Decreases in CO_2 fixation by P deficiency have been associated with lower RuBP carboxylase activity caused by the limited regeneration of RuBP (see the section "Carbohydrate Metabolism").

TABLE 6.2. Relationships Between K Content in Leaves, Stomatal Resistance, CO_2 Fixation, and RuBP Carboxylase Activity in Alfalfa

K supply	Leaf K	Stomatal resistance	Photosynthesis	RuBP carboxylase activity
mM	mg g^{-1} DW	s cm^{-1}	mg CO_2 dm^{-2} h^{-1}	μmol CO_2 mg^{-1} protein h^{-1}
0.0	12.8	9.3	11.9	1.8
0.6	19.8	6.8	21.7	4.5
4.8	38.4	5.9	34.0	6.1

Source: Adapted from Peoples and Koch (1979).

Carbonic anhydrase (CA) is a Zn (zinc) -containing enzyme involved in fixation of CO_2 in higher plants, particularly in C_4 plants (Hatch and Burnell, 1990). As reviewed in Badger and Price (1994), CA affects photosynthetic CO_2 fixation through three primary functions: (1) converting HCO_3^- to CO_2 for fixation by RuBP carboxylase, (2) converting CO_2 to HCO_3^- for fixation by PEP carboxylase, and (3) providing a rapid equilibration between CO_2 and HCO_3^- and thus facilitating CO_2 diffusion to the active site of RuBP carboxylase. Activity of CA is sensitive to Zn deficiency. Zn-deficient cotton plants showed a sharp decrease in the activity of CA, but this decrease was not related to photosynthetic CO_2 assimilation (Ohki, 1976). In contrast, Zn-deficient wheat showed both a decrease in CA activity (Rengel, 1995) and a decrease in net photosynthetic CO_2 assimilation (Fischer, Thimm, and Rengel, 1997). However, since in C_4 plants PEP carboxylase uses HCO_3^- in preference to CO_2, and CA catalyzes the conversion of CO_2 to HCO_3^-, it is suggested that Zn deficiency may affect photosynthesis to a greater extent in C_4 than C_3 plants (Marschner, 1995).

Carbohydrate Metabolism

Triose phosphates produced during photosynthetic CO_2 reduction are the major compounds affecting biosynthesis of starch and sucrose. Triose phosphates are also required in the Calvin cycle for the regeneration of the CO_2 acceptor RuBP. Partitioning of triose phosphates between starch and sucrose biosynthesis is directly influenced by inorganic phosphate (Pi). Triose phosphates can be exported from the chloroplast to the cytosol by a phosphate translocator and used for the synthesis of sucrose, or they can be retained in the chloroplast stroma and used for starch synthesis (Foyer, 1987; Stitt and Quick, 1989). The translocator exchanges Pi from the cytoplasm for a triose phosphate from the chloroplast stroma and loses its activity if Pi is not available in the cytoplasm.

Triose phosphates are both substrate and activator of starch synthesis in stroma. As a result of Pi-induced depletion of triose phosphates in chloroplast stroma, synthesis of starch is decreased in plants supplied well with P (see Table 6.3). In addition, Pi has an inhibitory effect on the activity of ADP-glucose pyrophosphorylase, which is the key enzyme of starch synthesis (Rao, Freeden, and Terry, 1990). Therefore, the ratio of Pi to triose phosphates greatly affects the rate of starch synthesis in chloroplasts (Heldt et al., 1977). Because of these dual effects of Pi on starch synthesis, an accumulation of starch in P-deficient plants is typical. In soybean plants, Freeden, Rao, and Terry (1989) showed that leaf concentration of starch was enhanced by a factor of 30 in response to P deficiency (see Table 6.3). As a result of P deficiency, a smaller amount of triose phosphates is transferred from the chloroplasts; consequently, newly assimilated carbon, rather than sucrose, is used for starch synthesis. Therefore, in P-deficient leaves, a higher starch/sucrose ratio is usual and was reported in soybean (Fredeen, Rao, and Terry, 1989; Qui and Israel, 1994) and bean (*Phaseolus vulgaris*) plants (Cakmak, Hengeler, and Marschner, 1994a).

Because of enhanced utilization of fixed carbon in starch synthesis, P deficiency limits regeneration of RuBP and, thus, activity of RuBP carboxylase (Brooks, 1986; Freeden et al., 1990). In sugar beet plants, P deficiency caused a substantial increase in allocation of photosynthates to starch and sucrose, while sugar phosphates were markedly decreased (see Table 6.4). On P resupply to deficient plants, the allocation of photosynthates to starch and sucrose decreased and concentration of sugar phosphates increased rapidly (see Table 6.4). It is suggested that P-deficiency-induced decreases in photosynthesis were due to reduced levels of sugar phosphates and, thus, limited regeneration of RuBP (Rao and Terry, 1995). Impairments in RuBP regeneration by P deficiency were not a result of reduced supply of ATP and/or NADPH (Freeden et al., 1990; Rao and Terry, 1995).

Synthesis of sucrose by sucrose-phosphate synthase (SPS) is differentially affected by mineral nutrients. In general, the activity of SPS is correlated inversely with the accumulation of starch in leaves and directly with the synthesis of sucrose (Huber and Israel, 1982). The SPS activity and sucrose concentration were diminished in response to P deficiency in soybean leaves (Freeden, Rao, and Terry, 1989; Qui and Israel, 1992). By contrast, in sugar beet leaves, P deficiency enhanced concentration of sucrose and activity of SPS (Rao, Freeden, and Terry, 1990), indicating that P deficiency affects sucrose synthesis differently in different plant species. Since under P deficiency a decrease in leaf concentration of sucrose was accompanied by a marked increase in root concentration of sucrose, it appears that phloem export of sucrose is a key factor affecting

TABLE 6.3. Phosphorus, Starch, and Sucrose Contents of Leaves and Roots of Soybean Plants Grown with High or Low P Supply

Treatment	Total P content		Leaf		Root	
	Leaf	Root	Starch	Sucrose	Starch	Sucrose
	mg g^{-1} DW		g CH_2O m^{-2} leaf		mg CH_2O g^{-1} FW	
High P supply	6.9	10.7	0.4	0.70	23	16
Low P supply	0.9	1.3	12.9	0.20	160	177

Source: Adapted from Freeden, Rao, and Terry (1989).

TABLE 6.4. Changes in Concentrations of Leaf Sugar Phosphates (RuBP: Ribulose-1.5-Bisphosphate; PGA: 3-Phosphoglycerate) at 0, 3, and 6 Hours After Pi Resupply to P-deficient plants

Time after Pi resupply	P treatment	Leaf sugar phosphates		
		RuBP	PGA	Triose-P
h		μmol m^{-2}		
	Control	66	125	21
0	Deficient + Pi resupply	32	38	10
	Control	77	166	13
3	Deficient + Pi resupply	68	89	17
	Control	54	201	13
6	Deficient + Pi resupply	65	125	13

Source: Adapted from Rao and Terry (1995).

Note: The control treatment refers to plants growing with sufficient P supply.

sucrose status of P-deficient leaves (Rao, Freeden, and Terry, 1990; Cakmak, Hengeler, and Marschner, 1994b). An inhibition of phloem export of sucrose may result in accumulation of sucrose and, thus, inhibition of SPS activity (Stitt and Quick, 1989). Decreases in SPS activity have been reported for K-deficient soybean leaves (Huber, 1984). This decrease was paralleled with increases in sucrose concentration in leaves. Apparently, K deficiency caused an impairment in phloem export of sucrose (see the following section) and, therefore, a high accumulation of sucrose in leaves, leading to a feedback inhibition of SPS activity. Magnesium deficiency also resulted in a massive accumulation of sugars (sucrose and reducing sugars) and starch in source leaves, but decreased their amounts in the roots (Cakmak, Hengeler, and Marschner, 1994a). This result indicates that (1) Mg deficiency, as discussed next, impairs translocation of sucrose to sink organs, and (2) accumulation of sucrose in Mg-deficient leaves stimulates formation of starch by exerting an inhibitory effect on enzymes forming sucrose.

Phloem Export of Photosynthates

Mineral nutrition has pronounced effects on carbohydrate partitioning by affecting either phloem export of photosynthates (sucrose) or growth rate of sink and/or source organs. In this regard, K has an important effect on sucrose transport from source leaves to sink organs, as shown in tomato (Mengel and Viro, 1974) and bean plants (see Table 6.5). In source leaves of K-deficient bean plants, there was a distinct inverse relationship between leaf concentration and phloem export of sucrose: higher accumulation of sucrose in K-deficient leaves was accompanied with correspondingly lower phloem export of sucrose (see Table 6.5). The stimulatory effect of K on phloem export of photosynthates in tomato plants was not related to the CO_2 assimilation rate, indicating that K supply influences phloem export of photosynthates more than CO_2 assimilation rate (Mengel and Viro, 1974).

The enhancement effect of K on phloem export of sucrose is attributed to its roles in (1) maintenance of a high pH in the sieve-tube cells and (2) in establishment of a high osmotic potential in the phloem sap and, thus, enhancement of the rate of photosynthate flow into sink organs (Marschner, 1995). Stimulation of sucrose loading into sieve tubes by K was shown in willow (Peel and Rogers, 1982). The increasing effect of K on phloem transport of sucrose was also ascribed to a K-induced increase in the growth rate and, thus, in metabolic activity of sink organs to utilize or store photosynthates (Mengel and Viro, 1974; Conti and Geiger, 1982).

The role of Mg in phloem export of sucrose appeared to be more important than that of K (Cakmak, Hengeler, and Marschner, 1994b). In the studies with bean plants, Mg deficiency caused very rapid and severe inhibition of phloem export of sucrose from source leaves (see Table 6.5). Consistent with these results, concentration of photosynthates in the sink organs of Mg-deficient plants was reduced, for example, in cereal grains (Beringer and Forster, 1981) and in the pods (Fischer and Bussler, 1988) and roots (Cakmak, Hengeler, and Marschner, 1994a) of *Phaseolus vulgaris*. A decrease in sucrose export by Mg deficiency was evident before Mg deficiency had any depressive influence on root or shoot growth (Cakmak, Hengeler, and Marschner, 1994b). This result suggests that the impairment in sucrose export

TABLE 6.5. Effect of Sufficient (Control) and Deficient Supply of P, K, or Mg on Sucrose Concentration in Primary Leaves and Phloem Export of Sucrose from Primary Leaves of Bean Plants Grown for Twelve Days in Nutrient Solution

Treatment	6 days	9 days	12 days
		Sucrose concentration in leaves mg glucose equiv. g^{-1} DW	
Control	12	14	12
P Deficiency	11	15	19
K Deficiency	26	69	76
Mg Deficiency	41	119	108
		Phloem export of sucrose mg glucose equiv. g^{-1} FW $(8\ h)^{-1}$	
Control	3.3	3.8	3.4
P Deficiency	4.1	4.0	2.8
K Deficiency	1.7	2.9	1.6
Mg Deficiency	1.0	0.8	0.7

Source: Adapted from Cakmak, Hengeler, and Marschner (1994b).

from Mg-deficient source leaves is the reason for, and not the consequence of, the reduced sink activity.

A resupply of Mg to Mg-deficient plants for 12 h distinctly enhanced sucrose export from source leaves, and after a resupply for 24 h, the rate of sucrose export was comparable with the rate of Mg-sufficient plants (Cakmak, Hengeler, and Marschner, 1994b). These results indicate a particular and specific role of Mg in the sucrose export from source leaves. The mechanisms responsible for Mg-deficiency-induced inhibition of phloem export of sucrose are not well understood and need to be clarified.

In contrast to Mg and K deficiencies, P deficiency did not influence sucrose translocation from the source leaves into roots (see Table 6.5). Rao, Freeden, and Terry (1990) showed that in P-deficient sugar beet plants, the rates of carbon export were increased in darkness, but decreased in light. However, considering the higher concentrations of sucrose in P-deficient roots, it has been concluded that the export of sucrose to roots was stimulated by P deficiency (Rao, Freeden, and Terry, 1990). A high carbon transport from source leaves to roots caused by P deficiency was also observed in soybean plants (Freeden, Rao, and Terry, 1989). In that study, P deficiency enhanced root sucrose concentration eleven-fold and starch concentration seven-fold in comparison to P sufficiency.

As with P-deficient plants, an increased translocation of carbohydrates from source leaves to roots was also found in N-deficient plants, possibly because of decreased sucrose utilization in sink leaves (Rufty, Huber, and Volk, 1988) or enhanced sucrose utilization in roots (Aloni et al., 1991). Severe decreases in development of leaf canopy due to N deficiency (Radin and Boyer, 1982; Palmer et al., 1996) may decrease the sink strength in shoots, leading to a preferential allocation of photosynthates to roots. According to the results of Aloni and colleagues (1991), in pepper plants, transport of sucrose across the membranes, as well as sucrose loading into phloem veins, was not affected by N deficiency.

Photooxidation and Leaf Chlorosis

Impairments in CO_2 assimilation can cause decreases in utilization of NADPH and, thus, a limited pool of $NADP^+$ for acceptance of electrons from PS I. Under such conditions, particularly under high light intensity, reduced ferredoxin can use molecular O_2 as an electron acceptor, leading to photogeneration of toxic O_2 species such as superoxide radical ($O_2^{\cdot -}$), hydrogen peroxide (H_2O_2), and hydroxyl radical ($OH^{\cdot}$) (Polle, 1996). Additionally, impairments in conversion of absorbed light energy into chemical energy may result in transfer of excess energy to O_2, forming singlet oxygen ($^1O_2$) (Demming-Adams and Adams, 1992). These O_2 species are highly toxic to photosynthetic apparatus because they cause photooxidative damage to thylakoid membranes and chlorophyll (Cakmak, 1994; Foyer, Lelandais, and Kunert, 1994; Cakmak et al., 1995).

Based on inhibition of CO_2 assimilation in Mg- and K-deficient leaves, it can be suggested that the demand for NADPH in CO_2 reduction is decreased, and light is absorbed in excess of the capacity of chloroplasts to use it in CO_2 reduction. These conditions in Mg- and K-deficient leaves potentiate the conditions for photogeneration of toxic O_2 species and photooxidative damage to chloroplasts. Accordingly, it has been shown that Mg- or K-deficient leaves had high susceptibility to increased light intensity: the leaves became rapidly chlorotic and necrotic upon exposure to a high light intensity (Marschner and Cakmak, 1989). Susceptibility of Mg- or K-deficient leaves to high light intensity was attributed to enhanced light-driven generation of destructive O_2 species (Cakmak and Marschner, 1992; Cakmak, 1994). Also, in forest ecosystems, Mg deficiency chlorosis was found to be pronounced on sides of trees exposed to light (Mies and Zöttl, 1985; Polle et al., 1992).

In Mg- and K-deficient plants, increases in severity of leaf chlorosis and necrosis were closely associated with increases in the activity of enzymes scavenging $O_2^{\cdot -}$ and H_2O_2 (Cakmak and Marschner, 1992; See Table 6.6).

This result supports the idea that production of toxic O_2 species is intensified in Mg- or K-deficient leaves at the expense of CO_2 reduction, particularly under high light intensity (Marschner and Cakmak, 1989). In contrast to Mg- or K-deficient plants, activities of the antioxidative enzymes were not influenced in P-deficient plants, where sucrose export is not inhibited, carbohydrates are not accumulated, and neither chlorosis nor necrosis appeared in leaves (see Table 6.6). It seems likely that photooxidative damage to chloroplasts is a major contributing factor to development of leaf chlorosis and necrosis in Mg- and K-deficient leaves. Accordingly, a partial shading of deficient leaves either prevented or markedly delayed development of chlorosis and necrosis, while concentrations of Mg or K were not affected by shading (Marschner and Cakmak, 1989). In P-deficient leaves, photooxidative damage does not seem to be important, and therefore, leaf chlorosis or necrosis is usually absent in P-deficient plants (Bergmann, 1992; Cakmak, 1994).

Zinc deficiency also leads to enhanced photooxidative damage to thylakoid constituents. In citrus trees, Zn-deficiency-induced leaf chlorosis and necrosis are generally more severe on the branches exposed directly to sunlight (Cakmak et al., 1995). As with Mg-deficient plants, concentration of carbohydrates, particularly sucrose, increased in source leaves of Zn-deficient plants, while root concentrations of carbohydrates were decreased (Marschner and Cakmak, 1989). Besides the disturbance in sucrose translocation, activity of Zn-containing superoxide dismutase enzyme was depressed in Zn-deficient leaves (Cakmak and Marschner, 1993). Since superoxide dismutase is responsible for detoxification of $O_2^{\cdot -}$, Zn-deficient plants are highly light sensitive and become rapidly chlorotic with increases

TABLE 6.6. Concentration of Chlorophyll and Activities of Ascorbate-Dependent H_2O_2-Scavenging Enzymes in Primary Leaves of Twelve-Day-Old Bean Plants Grown in Nutrient Solution with Sufficient (Control) or Deficient Supply of P, K, or Mg

Treatment	Chlorophyll mg g^{-1} DW	Ascorbate peroxidase	Monodehydro-ascorbate reductase	Glutathione reductase
		μmol ascorbate g^{-1} FW min^{-1}		μmol NADPH g^{-1} FW min^{-1}
Control	11.2	4.2	1.2	0.27
P deficiency	12.0	2.6	0.4	0.28
K deficiency	3.7	9.5	3.5	0.34
Mg deficiency	3.6	12.9	4.8	0.82

Source: Adapted from Cakmak (1994).

in light intensity. A partial shading of Zn-deficient leaves did not affect tissue Zn concentration but was effective in preventing leaf chlorosis and necrosis in bean plants (Marschner and Cakmak, 1989) or in field-grown citrus trees (Cakmak et al., 1995).

Shoot and Root Growth (Partitioning of Carbohydrates)

Mineral nutritional status of plants has a considerable impact on partitioning of carbohydrates and dry matter between shoots and roots (Marschner, 1995; Marschner, Kirkby, and Cakmak, 1996). Under N and P deficiencies, a considerably larger proportion of dry matter (photosynthates) is partitioned to roots than shoots, leading to reduced shoot/root dry weight ratios (Cakmak, Hengeler, and Marschner, 1994a; Engels and Marschner, 1995). Lower shoot/root dry weight ratios under N deficiency were associated with an increased translocation of carbohydrates from source leaves to roots (Rufty, Huber, and Volk, 1988). Stimulation of carbohydrate transport into roots and corresponding enhancement in root growth were ascribed to the limited sink activities in shoots for utilization of carbohydrates. According to the results of Aloni and colleagues (1991), in N-deficient pepper plants, roots were more competitive sinks for photoassimilates than young leaves and shoot meristems. Recently, a severe decrease in leaf expansion was found in response to N deficiency in sunflower plants (Palmer et al., 1996), possibly due to adverse effects on cell wall properties and activity of enzymes involved in cell wall expansion, such as xyloglucan endotransglycosylase (XET) (Palmer and Davies, 1996). According to McDonald, Ericsson, and Larsson (1996), N-deficiency-induced inhibition of shoot expansion may be a result of insufficient supply of N substrates to loosening and synthesis processes in the cell walls of leaf cells. These processes are considered major components of the sink strength. Consequently, decreases in average cell length caused by N deficiency were less in the root cortex than in leaf epidermal cells of sunflower plants, causing partitioning of a greater proportion of photoassimilates into roots (McDonald, Ericsson, and Larsson, 1996). The decrease in shoot/root dry weight ratio can also be the result of an induced N export from shoots into roots (Peuke, Hartung, and Jeschke, 1994). An enhancement in export of N from shoots to roots under N-deficient conditions can increase the sink strength for carbohydrates in roots.

As with N-deficient plants, shoot growth in P-deficient plants is more severely affected than root growth. In bean (*Phaseolus vulgaris*) and soybean plants, the shoot/root dry weight ratios were 4.9 and 4.1 for P-sufficient, and 1.8 and 1.0 for P-deficient, plants (Freeden, Rao, and Terry, 1989; Cakmak, Hengeler, and Marschner, 1994a). The decreases in shoot/

root dry weight ratios were closely related to enhanced carbohydrate translocation from leaves to roots (Freeden, Rao, and Terry, 1989; see Table 6.7). A severe decrease in the leaf expansion rate of P-deficient plants, possibly because of decreases in root hydraulic conductivity (Radin and Eidenbock, 1984), can result in a higher sink strength for carbohydrates in roots than in shoots. Phosphorus-deficiency-induced decreases in the leaf expansion rate might also be, in part, a consequence of enhanced P translocation from shoots to roots (Smith, Jackson, and Van den Berg, 1990).

Partitioning of carbohydrates between shoots and roots and growth rate of shoots and roots are substantially affected in K-deficient and, particularly, in Mg-deficient plants (see Table 6.7). However, the effects of Mg or K deficiency on shoot and root growth are quite different from those caused by N and P deficiencies. Impairments of root growth and enhancement in shoot/root dry weight ratios are typical for Mg-deficient plants (see Table 6.7; Cakmak, 1994). Under Mg deficiency, severe decreases in translocation of carbohydrates from source leaves into roots (see Table 6.5) were seen as a major reason for impairment of root growth (Cakmak, Hengeler, and Marschner, 1994b). In bean plants, shoot/root dry weight ratios were about 10 for Mg-deficient plants, which was much higher than the ratios for nutrient sufficient (control) and P-deficient plants (see Table 6.7). Accordingly, as compared with control and P-deficient plants, almost 30 times lower proportion of carbohydrates was partitioned to the Mg-deficient roots (see Table 6.7). Potassium-deficient plants behaved similarly to Mg-deficient plants with respect to carbohydrate partitioning and shoot/root growth. Only 3 percent of total carbohydrates per plant were found in K-deficient roots, with reduced carbohydrate translocation from shoot to roots, resulting in limited root growth and enhanced shoot/root dry weight ratio (see Table 6.5).

TABLE 6.7. Shoot/Root Dry Weight Ratios and Relative Distribution of Carbohydrates (Sum of Reducing Sugars, Sucrose, and Starch) Between Shoot and Roots of Twelve-Day-Old Bean Plants Grown in Nutrient Solution with Sufficient (Control) or Deficient Supply of P, K, or Mg

Treatment	Shoot/root dry weight ratio	Relative distribution of carbohydrates, %	
		Roots	Shoots
Control	4.9	16	84
P deficiency	1.8	23	77
K deficiency	6.9	3	97
Mg deficiency	10.2	1	99

Source: Adapted from Cakmak, Hengeler, and Marschner (1994a).

ROLE OF MINERAL NUTRIENTS IN YIELD FORMATION

It is well known from numerous fertilizer experiments that the yield of agricultural crops is strongly dependent on the supply of mineral nutrients such as N (Greenwood et al., 1980) and P (Barry and Miller, 1989). The relationship between yield and mineral nutrient supply may be described by an optimum curve, which is dependent on genotype, nutrient supply, and the availability of other growth factors, for example, water (Frederick and Camberato, 1995; Marschner, 1995). Yield may be defined as the product of the number and weight of sink organs, such as fruits, grains, or tubers. Often, the number of sink organs is the yield component that is affected most by mineral nutrients, for example, the number of kernels in wheat (Fischer, 1993), corn (Uhart and Andrade, 1995), soybean (Grabau, Blevins, and Minor, 1986), or sunflower (Steer et al., 1984). The number of generative sink organs is controlled by a sequence of processes, including spikelet or flower initiation and development, fertilization, and kernel abortion (Steer et al., 1984; Jacobs and Pearson, 1991); each of these processes may be affected by mineral nutrition (see Table 6.8). Mineral nutrient supply strongly affects leaf area and the rate of photosynthesis (as discussed previously) and, thus, the ability of the plant to deliver photosynthates to the sink sites. As shown by the defoliation treatments in Table 6.8, the number of sink organs is regulated also by the photosynthetic capacity of the plants. Thus, the positive effect of mineral nutrient supply on the number of sink organs may result not only from an increase in mineral nutrient supply but also from an increase in photosynthate supply to the sink sites or from hormonal effects (Michael and Beringer, 1980).

Flower Initiation and Development

Mineral nutrient supply affects flower initiation and development, for example, K in *Solanum sisymbrifolium* (Wakhloo, 1975), P in tomato (Menary and Van Staden, 1976), and N in sunflower (Steer and Hocking, 1983) and various fruit trees (Sanchez et al., 1995). At least for sunflower, the number of seeds per plant is strongly influenced by the N supply before floret initiation and during the phase of floret development, indicating the importance of optimal N nutrition before anthesis (see Table 6.9).

In apple trees, flowering is influenced not only by the amount but also by the time and form of N fertilization (Grasmanis and Edwards, 1974). In apple trees supplied well with NO_3^-, floral initiation was substantially enhanced by NH_4^+ supply for only 24 h in the preceding summer (Rohozinski, Edwards, and Hoskyns, 1986). The positive effect of short-term

TABLE 6.8. Effect of N Supply and Defoliation on the Spikelet and Kernel Number, Single Kernel Weight, and the Grain Yield of Corn

Parameter	Treatment		
	Control	N deficiency	Defoliation
Grain yield, g plant^{-1}	224	77	268
Number of spikelets, plant^{-1}	799	602	729
Number of kernels, ear^{-1}	669	345	515
Spikelet abortion, %	16	57	29
Kernel weight, mg	305	273	293
Ears, plant^{-1}	1.06	1.0	1.0

Source: Adapted from Jacobs and Pearson (1991).

Note: Control = undefoliated plants that received 15 kg N ha^{-1} week^{-1} to give a total of 270 kg N ha^{-1}; N deficiency = undefoliated plants that received 15 kg N ha^{-1} only at sowing; Defoliation = removal of half of the lamina of all fully expanded leaves from floral initiation to give the same leaf area as in N deficiency treatment, while N fertilization was as in control treatment.

TABLE 6.9. Number of Seeds per Plant in Sunflower As Affected by N Supply Before and After Floret Initiation (Experiment 1) and Before and After Anthesis (Experiment 2)

	Experiment 1 N supply changed at the end of floret initiation			Experiment 2 N supply changed at the anthesis in the first floret		
N supply after[a]	N supply before[a]			N supply before[a]		
	1.2	8.7	21.0	1.2	8.7	21.0
1.2	59	150	233	61	145	225
8.8	95	196	193	48	175	264
21.0	84	183	247	53	228	314

Source: Adapted from Steer et al. (1984).

[a]mg NO_3^--N plant^{-1}d^{-1}

exposure to NH_4^+ was associated with an increase in the stem content of arginine, a precursor for polyamine synthesis. Direct application of polyamines enhanced percentage of flowering to a similar extent as did NH_4^+ supply, indicating that polyamines are involved in the NH_4^+-induced enhancement of flowering (Rohozinski, Edwards, and Hoskyns, 1986). In another study with apple trees, increased flowering of NH_4^+-fed plants was associated with higher cytokinin concentrations in the xylem sap (Gao, Motosugi, and Sugiura, 1992). The positive effect of cytokinins on flower development is documented well for other plant species as well (Herzog, 1981; Mullins, 1986). Cytokinins are presumably also involved in the regulation of P effects on flowering in tomato (Menary and Van Staden, 1976). Thus, there is good evidence that mineral nutrients exert their influence on flower development not only by increasing the rate of mineral or photosynthate supply to the sites of flower initiation but also by altering the production of growth-regulating substances.

Fertilization and Grain Set

Mineral nutrients are involved in the formation of growing sink organs by their effects on fertilization and abortion after fertilization. Development and viability of pollen grains is adversely affected by deficiencies of micronutrients such as Mo (molybdenum) (Agarwala et al., 1979), Cu (Dell, 1981), B (boron) (Agarwala et al., 1981), Zn (Sharma et al., 1990), and Mn (Sharma et al., 1991). Viability of pollen is often more depressed by nutrient deficiency than viability of the ovules, as shown for Cu (Graham, 1975), Mn (Sharma et al., 1991), and Zn (see Table 6.10) by cross-pollination experiments in which flowers from nutrient-sufficient plants were pollinated with pollen from deficient plants and vice versa. In B-deficient corn plants, however, fertilization of silks with pollen from high-B plants also severely decreased grain set, indicating that the silks of B-deficient plants were nonreceptive (Vaughan, 1977).

As compared with N and K, micronutrients have relatively low phloem mobility (Marschner, 1995). For high pollen viability and grain set, it is therefore necessary to ensure sufficient supply of micronutrients in the "critical" stages of pollen development. As shown by the low grain number of the cross-pollinations $DN_2 \times N$ and $ND_1 \times N$ in Table 6.10, adequate Zn supply was particularly important during microsporogenesis, whereas yield reductions were less severe when Zn deficiency occurred before the pollen mother cell stage ($DN_1 \times N$) or after completion of microsporogenesis ($ND_2 \times N$). These results indicate the great potential for yield increases by foliar application of micronutrients in critical growth phases.

TABLE 6.10. Effect of Zn Supply to the Pollen (Male) and Pollen Receiver (Female) Plants on Yield Formation in Corn

Cross male × female	Cob weight g plant^{-1}	Grain weight g cob^{-1}	Grain number grains cob^{-1}	100-grain weight g
N × N	154	117	385	30.4
D × N	30	0	0	0.0
DN_1 × N	89	54	171	31.7
ND_2 × N	71	43	138	30.8
DN_2 × N	36	2	4	43.3
ND_1 × N	28	0	0	0.0
N × ND_1	84	54	252	21.6
N × ND_2	95	65	220	29.3

Source: Adapted from Sharma et al. (1990).

Note: N = continuously adequate Zn supply; D = continuously inadequate Zn supply; ND_1 = Zn supply withheld from day 36 on (appearance of tassel initials); ND_2 = Zn supply withheld from day 45 on (one day after release of microspores from pollen tetrads); DN_1 = adequate Zn supply from day 36 on; DN_2 = adequate Zn supply from day 45 on.

The loss of pollen viability induced by micronutrient deficiencies is associated with various modifications of pollen characteristics, for example, lack of starch accumulation under Cu deficiency (Graham, 1975) or a decrease in the activities of starch phosphorylase and invertase, in case of deficiencies of Mo (Agarwala et al., 1979) and B (Agarwala et al., 1981). Deficiency of Cu leads to reduced thickening or lignification in the endothecial layer of the anthers in wheat (Dell, 1981) and to abnormal development of the tapetum (Jewell, Murray, and Alloway, 1988). The specific effects of micronutrient deficiencies on pollen and ovule viability may result in severe yield reductions, even when the vegetative growth is not yet strongly depressed (Sharma et al., 1991; Mozafar, 1993), indicating a strong sink limitation of yield.

In white clover, B supply increases flower fertilization indirectly by increasing the amount of nectar and changing its sugar composition, thus making flowers more attractive for pollinating insects (Smith and Johnson, 1969; Eriksson, 1979). In contrast, low N supply in corn delays the development of ears relative to the tassel and may thus affect the coincidence of pollen shed and silking of individual spikelets (Jacobs and Pearson, 1991). Furthermore, in semiprolific genotypes, low N supply increases the time interval between silking of the first and second ear (Anderson et al., 1984). Asynchronous flowering decreases the percentage of plants that develop two ears (Anderson et al., 1984) and the kernel number per cob (Jacobs and Pearson, 1991).

Nitrogen is the nutrient that has the greatest practical importance for yield formation in most agricultural areas. In field experiments with irrigated spring wheat, in which the amount and timing of N fertilization were varied, differences in grain yield of 170 to 750 g m^{-2} were mainly due to differences in the kernel number (Fischer, 1993). The number of kernels was closely related to the crop dry weight accumulation in the spike during the phase of spike growth (from one week before flag leaf emergence to anthesis), indicating that the effects of N on the grain number were mainly the consequence of modification of the net assimilation rate during this "critical" growing phase (Fischer, 1985, 1993). However, at least at low levels of N supply, the reduction in grain number may not always be explained only by a decrease in assimilate supply to the spike, as has been shown for corn (Abbate, Andrade, and Culot, 1995). Floral survival may be directly affected by the supply of reduced N to the spike (Below et al., 1981; Abbate, Andrade, and Culot, 1995).

As a rule, highest growth rates and plant yields are obtained by combined supply of both NH_4^+ and NO_3^- (Marschner, 1995). In field-grown plants, the positive effect of mixed N nutrition can be achieved by application of NH_4^+ fertilizer, together with nitrification inhibitors (Smiciklas and Below, 1992a, b). In comparison to NO_3^- nutrition, mixed-N nutrition increases yield, for example, in wheat and corn, mainly by increasing the grain number, whereas grain weight either remains the same or decreases. Furthermore, yield increases are obtained only if mixed N is supplied before anthesis (Below and Gentry, 1992). This indicates that mixed-N-induced yield increases are associated with physiological events that occur during ovule initiation and pollination rather than with processes occurring during the grain-filling period.

In a field study with corn, depending on genotype, the yield increase induced with mixed-N nutrition was the result of either increased partitioning of dry matter to the grain, increased total dry matter production, decreased percentage of aborted kernels, or an increased number of ovules at anthesis (Smiciklas and Below, 1992a). During vegetative growth, mixed-N-fed plants had higher concentrations of endogenous cytokinins (zeatin, zeatin riboside) in root tips than NO_3^--fed plants (Smiciklas and Below, 1992b). Furthermore, cytokinin application to NO_3^--fed plants enhanced dry matter partitioning to the grain and decreased kernel abortion (Smiciklas and Below, 1992b), indicating that the positive effects of mixed-N-nutrition may be partly associated with an increase in endogenous cytokinin supply.

Tuberization

In plant species that form vegetative sink organs, such as sugar beet and potato, the effect of mineral nutrients on yield formation is not entirely confined to the effects on leaf area duration and photosynthetic rate, that is, source capacity. Particularly for potato, there is good evidence that tuberization and the sink strength of the growing tubers are affected by the rate of N supply (Krauss, 1985). It is known from field experiments that high rates of N supply delay tuberization in potato (Ivins and Bremner, 1965), thus reducing the phase of tuber bulking when the growing season is shortened, for example, by early frost or drought (Clutterbuck and Simpson, 1978). In nutrient solution, where the rate of N supply can be regulated well, tuberization is completely prevented by continuous high N supply, even under environmental conditions (short day length, low temperatures) that would favor tuber initiation (Krauss and Marschner, 1971). Excessive N supply after tuberization may decrease tuber growth rate and lead to cessation of tuber growth with subsequent "regrowth," that is, outgrowth of stolons from the apical eyes of the tubers (Krauss and Marschner, 1982). The effects of N supply on tuber growth rate and "regrowth" are caused by N-induced changes in the phytohormone balance in the plants, such as an increase in the cytokinin export from roots to the shoot (Sattelmacher and Marschner, 1978), high gibberellic acid and low abscisic acid contents in the shoot (Krauss and Marschner, 1976) and in the tubers (Krauss, 1978). In potato, above-ground shoot organs and tubers compete for assimilates. Presumably, the N-induced changes in the phytohormone balance in plants favor assimilate partitioning to the above-ground shoot organs by virtue of their high sink strength (Booth and Lovell, 1972), as opposed to the decreased sink strength of the tubers (Mares, Marschner, and Krauss, 1981).

CONCLUSIONS

Most mineral nutrients are essential for photosynthetic CO_2 assimilation and are involved directly in partitioning of photosynthates between source and sink organs. Mineral nutrients also affect metabolic activity (strength) of source and sink organs, for example, cell expansion and flowering and, consequently, synthesis and utilization of photosynthates. When deficient, mineral nutrients behave more or less similarly in their final effect on total dry matter production of plants, but distinctly differently in their final effect on dry matter partitioning between source and sink organs of shoot and root systems. Phosphorus and magnesium cause stimulation and depression of dry matter allocation to roots, respectively. In the case of some mineral nutrient deficiencies, vegetative growth is little affected, but pollen viability and grain set are substantially depressed, causing a strong limitation in sink strength and, thus, plant productivity. It is concluded that mineral nutritional status of plants is a major determinant of dry matter production and productivity of crop plants.

REFERENCES

Abbate, P.E., F.H. Andrade, and J.P. Culot (1995). The effects of radiation and nitrogen on number of grains in wheat. *Journal of Agricultural Science (Cambridge)* 124: 351-360.

Agarwala, S.C., C. Chatterjee, P.N. Sharma, C.P. Sharma, and N. Nautiyal (1979). Pollen development in maize plants subjected to molybdenum deficiency. *Canadian Journal of Botany* 57: 1946-1950.

Agarwala, S.C., P.N. Sharma, C. Chatterjee, and C.P. Sharma (1981). Development and enzymatic changes during pollen development in boron deficient maize plants. *Journal of Plant Nutrition* 3: 329-336.

Alcaraz, C.F., F. Martinez-Sánchez, F. Sevilla, and E. Hellin (1986). Influence of ferredoxin levels on nitrate reductase activity in iron deficient lemon leaves. *Journal of Plant Nutrition* 9: 1405-1413.

Aloni, B., T. Pashkar, L. Karni, and J. Daie (1991). Nitrogen supply influences carbohydrate partitioning in pepper seedlings and transplant development. *Journal of the American Society for Horticultural Science* 116: 995-999.

Anderson, E.L., E.J. Kamprath, R.H. Moll, and W.A. Jackson (1984). Effect of N fertilization on silk synchrony, ear number, and growth of semiprolific maize genotypes. *Crop Science* 24: 663-666.

Ayala, M.B. and G. Sandmann (1989). Activities of Cu-containing proteins in Cu-depleted pea leaves. *Physiologia Plantarum* 72: 801-806.

Badger, M. and D.G. Price (1994). The role of carbonic anhydrase in photosynthesis. *Annual Review of Plant Physiology and Plant Molecular Biology* 45: 369-392.

Barry, D.A.J. and M.H. Miller (1989). Phosphorus nutritional requirement of maize seedlings for maximum yield. *Agronomy Journal* 81: 95-99.

Below, F.E., L.E. Christensen, A.J. Reed, and R.H. Hageman (1981). Availability of reduced N and carbohydrates for ear development of maize. *Plant Physiology* 68: 1186-1190.

Below, F.E. and L.E. Gentry (1992). Maize productivity as influenced by mixed nitrogen supplied before and after anthesis. *Crop Science* 32: 163-168.

Bergmann, W. (1992). *Nutritional Disorder of Plants—Development, Visual and Analytical Diagnosis.* Jena, Germany: Fischer Verlag.

Beringer, H. and H. Forster (1981). Einfluss variierter Mg-Ernährung auf Tausendkorngewicht und P-Fraktionen des Gerstenkorns. *Zeitschrift für Pflanzenernährung und Bodenkunde* 144: 8-15.

Booth, A. and P.H. Lovell (1972). The effect of pre-treatment with gibberellic acid on the distribution of photosynthate in intact and disbudded plants of *Solanum tuberosum* L. *New Phytologist* 71: 795-804.

Brooks, A. (1986). Effects of phosphorus nutrition on ribulose-1, 5-bisphosphate carboxylase activation, photosynthetic quantum yield and amounts of some Calvin-cycle metabolites in spinach leaves. *Australian Journal of Plant Physiology* 13: 221-237.

Brooks, A., K.C. Woo, and S.C. Wong (1988). Effects of phosphorus nutrition on the response of photosynthesis to CO_2 and O_2, activation of ribulose bisphosphate carboxylase and amounts of ribulose bisphosphate and 3-phosphoglycerate in spinach leaves. *Photosynthesis Research* 15: 133-141.

Cakmak, I. (1994). Activity of ascorbate dependent H_2O_2-scavenging enzymes and leaf chlorosis are enhanced in magnesium- and potassium-deficient leaves, but not in phosphorus-deficient leaves. *Journal of Experimental Botany* 45: 1259-1266.

Cakmak, I., M. Atly, R. Kaya, H. Evliya, and H. Marschner (1995). Association of high light and zinc deficiency in cold induced leaf chlorosis in grapefruit and mandarin trees. *Journal of Plant Physiology* 146: 355-360.

Cakmak, I., C. Hengeler, and H. Marschner (1994a). Partitioning of shoot and root dry matter and carbohydrates in bean plants suffering from phosphorus, potassium and magnesium deficiency. *Journal of Experimental Botany* 45: 1245-1250.

Cakmak, I., C. Hengeler, and H. Marschner (1994b). Changes in phloem export of sucrose in leaves in response to phosphorus, potassium and magnesium deficiency in bean plants. *Journal of Experimental Botany* 45: 1251-1257.

Cakmak, I. and H. Marschner (1992). Magnesium deficiency and high light intensity enhance activities of superoxide dismutase, ascorbate peroxidase and glutathione reductase in bean leaves. *Plant Physiology* 98: 1222-1227.

Cakmak, I. and H. Marschner (1993). Effect of zinc nutritional status on activities of superoxide radical and hydrogen peroxide scavenging enzymes in bean leaves. *Plant and Soil* 155/156: 127-130.

Clutterbuck, B.J. and H. Simpson (1978). The interactions of water and fertilizer nitrogen in effects on growth pattern and yield of potatoes. *Journal of Agricultural Science (Cambridge)* 91: 161-172.

Conti, T.R. and D.R. Geiger (1982). Potassium nutrition and translocation in sugar beet. *Plant Physiology* 70: 168-172.

Dell, B. (1981). Male sterility and anther wall structure in copper-deficient plants. *Annals of Botany* 48: 599-608.

Demming-Adams, B. and W.W. Adams (1992). Photoprotection and other responses of plants to high light stress. *Annual Review of Plant Physiology and Plant Molecular Biology* 43: 599-626.

Engels, C. and H. Marschner (1995). Plant uptake and utilization of nitrogen. In *Nitrogen Fertilization in the Environment*. Ed. P.E. Bacan. New York: Marcel Dekker, Inc., pp. 41-81.

Ericsson, T. (1995). Growth and shoot: Root ratio of seedlings in relation to nutrient availability. *Plant and Soil* 168/169: 205-214.

Eriksson, M. (1979). The effects of boron on nectar production and seed setting of red clover (*Trifolium pratense* L.). *Swedish Journal of Agricultural Research* 9: 37-41.

Evans, J.R. and I. Terashima (1988). Photosynthetic characteristics of spinach leaves grown with different nitrogen treatments. *Plant and Cell Physiology* 29: 157-165.

Fang, Z., F. Mi, and G.A. Berkowitz (1995). Molecular and physiological analysis of a thylakoid K^+ channel protein. *Plant Physiology* 108: 1725-1734.

Fischer, E.S. and E. Bremner (1993). Influence of magnesium deficiency on rates of leaf expansion, starch and sucrose accumulation, and net assimilation in *Phaseolus vulgaris*. *Physiologia Plantarum* 89: 271-276.

Fischer, E.S. and W. Bussler (1988). Effects of magnesium deficiency on carbohydrates in *Phaseolus vulgaris*. *Zeitschrift für Pflanzenernährung und Bodenkunde* 151: 292-298.

Fischer, E.S., O. Thimm, and Z. Rengel (1997). Zinc nutrition influences gas exchange in wheat. *Photosynthetica* 33: 505-508.

Fischer, R.A. (1985). Number of kernels in wheat crops and the influence of solar radiation and temperature. *Journal of Agricultural Science (Cambridge)* 105: 447-461.

Fischer, R.A. (1993). Irrigated spring wheat and timing and amount of nitrogen fertilizer. II. Physiology of grain yield response. *Field Crops Research* 33: 57-80.

Foyer, C.H. (1987). The basis for source-sink interaction in leaves. *Plant Physiology and Biochemistry* 25: 649-657.

Foyer, C.H., M. Lelandais, and K.J. Kunert (1994). Photooxidative stress in plants. *Physiologia Plantarum* 92: 696-717.

Frederick, J.R. and J.J. Camberato (1995). Water and nitrogen effects on winter wheat in the Southeastern coastal plain. I. Grain yield and kernel traits. *Agronomy Journal* 87: 521-526.

Freeden, A.L., T.K. Raab, I.M. Rao, and N. Terry (1990). Effects of phosphorus nutrition on photosynthesis of *Glycine max* (L.) Merr. *Planta* 181: 399-405.

Freeden, A.L., I.M. Rao, and N. Terry (1989). Influence of phosphorus nutrition on growth and carbon partitioning in *Glycine max*. *Plant Physiology* 89: 225-230.

Gao, Y-P., H. Motosugi, and A. Sugiura (1992). Rootstock effects on growth and flowering in young apple trees grown with ammonium and nitrate nitrogen. *Journal of the American Society for Horticultural Science* 117: 446-452.

Geiger, D.R., K.E. Koch, and W.-J. Shieh (1996). Effect of environmental factors on whole plant assimilate partitioning and associated gene expression. *Journal of Experimental Botany* 47: 1229-1238.

Grabau, L.J., D.G. Blevins, and H.C. Minor (1986). Phosphorus nutrition during seed development. Leaf senescence, pod retention, and seed weight of soybean. *Plant Physiology* 82: 1008-1012.

Graham, R.D. (1975). Male sterility in wheat plants deficient in copper. *Nature* 254: 514-515.

Grasmanis, V.O. and G.E. Edwards (1974). Promotion of flower initiation in apple trees by short exposure to the ammonium ion. *Australian Journal of Plant Physiology* 1: 99-105.

Greenwood, D.J., T.J. Cleaver, M.K. Turner, J. Hunt, K.B. Niendorf, and S.M.G. Loguens (1980). Comparison of the effects of nitrogen fertilizer on the yield, nitrogen content and quality of 22 different vegetables and agricultural crops. *Journal of Agricultural Science (Cambridge)* 95: 441-456.

Hatch, M.D. and J. N. Burnell (1990). Carbonic anhydrase activity in leaves and its role in the first step of C_4 photosynthesis. *Plant Physiology* 93: 825-828.

Heineke, D. and H.W. Heldt (1988). Measurement of light-dependent changes of the stromal pH in wheat leaf protoplasts. *Botanica Acta* 101: 45-47.

Heldt, H.W., C.J. Chan, D. Maronde, A. Herold, Z.S. Stankovic, D.A. Walker, A. Kraminer, M.R. Kirk, and U. Heber (1977). Role of orthophosphate and other factors in the regulation of starch formation in leaves and isolated chloroplasts. *Plant Physiology* 59: 1146-1155.

Herzog, H. (1981). Wirkungen von zeitlich begrenzten Stickstoff- und Cytokiningaben auf die Fahnenblatt- und Kornentwicklung von Weizen. *Zeitschrift für Pflanzenernährung und Bodenkunde* 144: 241-253.

Huber, S.C. (1984). Biochemical basis for effects of K deficiency on assimilate export rate and accumulation of soluble sugars in soybean leaves. *Plant Physiology* 76: 424-430.

Huber, S.C. and D.W. Israel (1982). Biochemical basis for partitioning of photosynthetically fixed carbon between starch and sucrose in soybean (*Glycine max*. (L.) Merr.) leaves. *Plant Physiology* 69: 691-696.

Humble, G.D. and K. Raschke (1971). Stomatal opening quantitatively related to potassium transport. *Plant Physiology* 48: 447-453.

Ivins, J.D. and P.M. Bremner (1965). Growth, development and yield in the potato. *Outlook in Agriculture* 4: 211-217.

Jacob, J. (1995). Phosphate deficiency increases the rate constant of thermal dissipation of excitation energy by photosystem II in intact leaves of sunflower and maize. *Australian Journal of Plant Physiology* 22: 417-424.

Jacob, J. and D.W. Lawlor (1993). *In vivo* photosynthetic electron transport does not limit photosynthetic capacity in phosphate deficient sunflower and maize leaves. *Plant, Cell and Environment* 16: 785-795.

Jacobs, B.C. and C.J. Pearson (1991). Potential yield of maize as determined by rates of growth and development of ears. *Field Crops Research* 27: 281-298.

Jewell, A.W., B.G. Murray, and B.J. Alloway (1988). Light and electron microscope studies on pollen development in barley (*Hordeum vulgare* L.) grown under copper-sufficient and deficient conditions. *Plant, Cell and Environment* 11: 273-281.

Johnston, M., C.P.L. Grof, and P.F. Brownell (1988). The effect of sodium nutrition on the pool sizes of intermediates of the C_4 photosynthetic pathway. *Australian Journal of Plant Physiology* 15: 449-457.

Krauss, A. (1978). Tuberization and abscisic acid content in *Solanum tuberosum* as affected by nitrogen nutrition. *Potato Research* 21: 183-193.

Krauss, A. (1985). Interaction of nitrogen nutrition, phytohormones, and tuberization. In *Potato Physiology*, Ed. P.H. Li. Orland, CA: Academic Press, pp. 209-230.

Krauss, A. and H. Marschner (1971). Einfluss der Stickstoffernährung der Kartoffeln auf Induktion und Wachstumsrate der Knolle. *Zeitschrift für Pflanzenernährung und Bodenkunde* 128: 153-168.

Krauss, A. and H. Marschner (1976). Einfluss von Stickstoffernährung und Wuchsstoffapplikation auf die Knolleninduktion bei Kartoffelpflanzen. *Zeitschrift für Pflanzenernährung und Bodenkunde* 139: 143-155.

Krauss, A. and H. Marschner (1982). Influence of nitrogen nutrition, daylength and temperature on contents of gibberellic and abscisic acid and on tuberization of potato plants. *Potato Research* 25: 13-21.

Lin, D.C. and P.S. Nobel (1971). Control of photosynthesis by Mg^{2+}. *Archives of Biochemistry and Biophysics* 145: 622-632.

Mares, D.J., H. Marschner, and A. Krauss (1981). Effect of gibberellic acid on the growth and carbohydrate metabolism of developing tubers of potato (*Solanum tuberosum* L.) *Physiologia Plantarum* 52: 267-274.

Marschner, H. (1995). *Mineral Nutrition of Higher Plants*, Second Edition. London, UK: Academic Press.

Marschner, H. and I. Cakmak (1989). High light intensity enhances chlorosis and necrosis in leaves of zinc, potassium, and magnesium deficient bean (*Phaseolus vulgaris* L.) plants. *Journal of Plant Physiology* 134: 308-315.

Marschner, H., E.A. Kirkby, and I. Cakmak (1996). Effect of mineral nutritional status on shoot-root partitioning of photoassimilates and cycling of mineral nutrients. *Journal of Experimental Botany* 47: 1255-1263.

McDonald, A.J.S., T. Ericsson, and C.M. Larsson (1996). Plant nutrition, dry matter gain and partitioning at the whole-plant level. *Journal of Experimental Botany* 47: 1245-1253.

Menary, R.C. and J. Van Staden (1976). Effect of phosphorus nutrition and cytokinins on flowering in the tomato, *Lycopersicon esculentum* Mill. *Australian Journal of Plant Physiology* 3: 201-205.

Mengel, K. and M. Viro (1974). Effect of potassium supply on the transport of photosynthates to the fruits of tomatoes (*Lycopersicon esculentum*). *Physiologia Plantarum* 30: 295-300.

Michael, G. and H. Beringer (1980). The role of hormones in yield formation. In *Proceedings of the 15th Colloquium of the International Potash Institute*, Bern, Switzerland: International Kali-Institut, pp. 85-116.

Mies, F. and H.W. Zöttl (1985). Zeitliche Veränderung der Chlorophyll- und Elementgehalte in den Nadeln eines gelb-chlorotischen fichtenbestandes. *Forstwirtschaft Centralblatt* 104: 1-8.

Mozafar, A. (1993). Role of boron in seed production. In *Boron and Its Role in Crop Production*, Ed. U.C. Gupta. Boca Raton, FL: CRC Press, pp. 185-206.

Mullins, M.G. (1986). Hormonal regulation of flowering and fruit set in the grapevine. *Acta Horticulturae* 179: 309-315.

Nable, R.O., A. Bar-Akiva, and J.F. Loneragan (1984). Functional manganese requirement and its use as a critical value for diagnosis of manganese deficiency in subterranean clover (*Trifolium subterraneum* L. cv. Seaton Park). *Annals of Botany* 54: 39-49.

Ohki, K. (1976). Effect of zinc nutrition on photosynthesis and carbonic anhydrase activity in cotton. *Physiologia Plantarum* 38: 300-304.

Palmer, S.J., D.M. Berridge, A.J.S. McDonald, and W.J. Davies (1996). Control of leaf expansion in sunflower (*Helianthus annuus* L.) by nitrogen nutrition. *Journal of Experimental Botany* 47: 359-368.

Palmer, S.J. and W.J. Davies (1996). An analysis of relative elemental growth rate, epidermal cell size and xyloglucan endotransglycosylase activity through the growing zone of ageing maize leaves. *Journal of Experimental Botany* 47: 339-347.

Peel, A.J. and S. Rogers (1982). Stimulation of sugar loading into sieve elements of willow by potassium and sodium salts. *Planta* 154: 94-96.

Peoples, T.R. and D.W. Koch (1979). Role of potassium in carbon dioxide assimilation in *Medicago sativa* L. *Plant Physiology* 63: 878-881.

Peuke, A.D., W. Hartung, and W.D. Jeschke (1994). The uptake and flow of C, N and ions between roots and shoots in *Ricinus communis* L. II. Grown with low or high nitrate supply. *Journal of Experimental Botany* 45: 733-740.

Plesnicar, M., R. Kastori, N. Petrovic, and D. Pankovic (1994). Photosynthesis and chlorophyll fluorescence in sunflower (*Helianthus annuus* L.) leaves affected by phosphorus nutrition. *Journal of Experimental Botany* 45: 919-924.

Polle, A. (1996). Mehler reaction: Friend or foe in photosynthesis? *Botanica Acta* 109: 84-89.

Polle, A., K. Chakrabarti, S. Chakrabarti, F. Seifert, P. Schramel, and H. Rennenberg (1992). Antioxidant and manganese deficiency in needles of Norway spruce (*Picea abies* L.) trees. *Plant Physiology* 99: 1084-1089.

Portis Jr., A.R. (1981). Evidence of a low stromal Mg^{2+} concentration in intact chloroplasts in the dark. I. Studies with the ionophore A 23187. *Plant Physiology* 67: 985-989.

Pushnik, J.C. and E.W. Miller (1989). Iron regulation of chloroplast photosynthetic function: Mediation of PS I development. *Journal of Plant Nutrition* 12: 407-421.

Qui, J. and D.W. Israel (1992). Diurnal starch accumulation and utilization in phosphorus-deficient soybean plants. *Plant Physiology* 98: 316-323.

Qui, J. and D.W. Israel (1994). Carbohydrate accumulation and utilization in soybean plants in response to altered phosphorus nutrition. *Physiologia Plantarum* 90: 722-728.

Radin, J.W. and J.S. Boyer (1982). Control of leaf expansion by nitrogen nutrition in sunflower plants: Role of hydraulic conductivity and turgor. *Plant Physiology* 69: 771-775.

Radin, J.W. and M.P. Eidenbock (1984). Carbon accumulation during photosynthesis in leaves of nitrogen- and phosphorus-stressed cotton. *Plant Physiology* 82: 869-871.

Rao, I.M., A.L. Freeden, and N. Terry (1990). Leaf phosphate status, phototosynthesis and carbon partitioning in sugar beet. III. Diurnal changes in carbon partitioning and carbon export. *Plant Physiology* 92: 29-36.

Rao, I.M. and N. Terry (1989). Leaf phosphate status, photosynthesis, and carbon partitioning in sugar beet. I. Changes in growth, gas exchange, and Calvin cycle enzymes. *Plant Physiology* 90: 814-819.

Rao, I.M. and N. Terry (1995). Leaf phosphate status, photosynthesis, and carbon partitioning in sugar beet. IV. Changes with time following increased supply of phosphate to low-phosphate plants. *Plant Physiology* 107: 1313-1321.

Rengel, Z. (1995). Carbonic anhydrase activity in leaves of wheat genotypes differing in Zn efficiency. *Journal of Plant Physiology* 147: 251-256.

Rohozinski, J., G.R. Edwards, and P. Hoskyns (1986). Effects of brief exposure to nitrogenous compounds on floral initiation in apple trees. *Physiologie Vègètale* 24: 673-677.

Rufty, T.W., S.C. Huber, and R.J. Volk (1988). Alterations in leaf carbohydrate metabolism in response to nitrogen stress. *Plant Physiology* 88: 725-730.

Rutherford, A.W. (1989). Photosystem II, the water-splitting enzyme. *Trends in Biochemical Science* 14: 227-232.

Sage, R.F., R.W. Pearcy, and J.R. Seemann (1987). The nitrogenase efficiency of C_3 and C_4 plants. III. Leaf nitrogen effects on the activity of carboxylating enzymes in *Chenopodium album* (L.) and *Amaranthus retroflexus* (L.). *Plant Physiology* 85: 355-359.

Sanchez, E.E., H. Khemira, D. Sugar, and T.L. Righetti (1995). Nitrogen management in orchards. In *Nitrogen Fertilization in the Environment*, Ed. P.E. Bacon. New York: Marcel Dekker, Inc., pp. 327-380.

Sattelmacher, B. and H. Marschner (1978). Nitrogen nutrition and cytokinin activity in *Solanum tuberosum*. *Physiologia Plantarum* 42: 185-189.

Scott, B.J. and A.D. Robson (1990). Changes in the content and form of magnesium in the first trifoliate leaf of subterranean clover under altered or constant root supply. *Australian Journal of Agricultural Research* 41: 511-519.

Sharma, P.N., C. Chatterjee, S.C. Agarwala, and C.P. Sharma (1990). Zinc deficiency and pollen fertility in maize (*Zea mays*). *Plant and Soil* 124: 221-225.

Sharma, C.P., P.N. Sharma, C. Chatterjee, and S.C. Agarwala (1991). Manganese deficiency in maize affects pollen viability. *Plant and Soil* 138: 139-142.

Sinclair, T.R. and T. Horie (1989). Leaf nitrogen, photosynthesis, and crop radiation use efficiency: A review. *Crop Science* 29: 90-98.

Smiciklas, K.D. and F.E. Below (1992a). Role of nitrogen form in determining yield of field-grown maize. *Crop Science* 32: 1220-1225.

Smiciklas, K.D. and F.E. Below (1992b). Role of cytokinin in enhanced productivity of maize supplied with NH_4^+ and NO_3^-. *Plant and Soil* 142: 307-313.

Smith, F.W., W.A. Jackson, and P.J. Van den Berg (1990). Internal phosphorus flows during development of phosphorus stress in *Stylosanthes hamata*. *Australian Journal of Agricultural Research* 17: 451-464.

Smith, R.H. and W.C. Johnson (1969). Effect of boron on white clover nectar production. *Crop Science* 9: 75-76.

Steer, B.T. and P.J. Hocking (1983). Leaf and floret production in sunflower (*Helianthus annuus* L.) as affected by nitrogen supply. *Annals of Botany* 52: 267-277.

Steer, B.T., P.J. Hocking, A.A. Kortt, and C.M. Roxburgh (1984). Nitrogen nutrition of sunflower (*Helianthus annuus* L.): Yield components, the timing of their establishment and seed characteristics in response to nitrogen supply. *Field Crops Research* 9: 219-236.

Stitt, M. and W.P. Quick (1989). Photosynthetic carbon partitioning: Its regulation and possibilities for manipulation. *Physiologia Plantarum* 77: 633-641.

Sugiyama, T., C. Matsumoto, T. Akazawa, and S. Miyachi (1969). Structure and function of chloroplast proteins. VII. Ribulose-1,5-diphosphate carboxylase of *Chlorella ellipsoida*. *Archives of Biochemistry and Biophysics* 129: 597-602.

Sugiyama, T., N. Nakyama, and T. Akazawa (1968). Structure and function of chloroplast proteins. V. Homotropic effect of bicarbonate in RuBP carboxylase relation and the mechanism of activation by magnesium ions. *Archives of Biochemistry and Biophysics* 126: 734-745.

Tan, W. and D.G. Hogan (1995). Limitations to net photosynthesis as affected by nitrogen status in jack pine (*Pinus banksiana* Lamb.) seedlings. *Journal of Experimental Botany* 46: 407-413.

Terry, N. and A. Ulrich (1974). Effects of Mg deficiency on the photosynthesis and respiration of leaves of sugar beet. *Plant Physiology* 54: 379-381.

Tester, M. and M.R. Blatt (1989). Direct measurement of K^+ channels in thylakoid membranes by incorporation of vesicles into planar lipid bilayers. *Plant Physiology* 91: 249-252.

Uhart, S.A. and F.H. Andrade (1995). Nitrogen deficiency in maize. II. Carbon-nitrogen interaction effects on kernel number and grain yield. *Crop Science* 35: 1384-1389.

Vaughan, A.K.F. (1977). The relation between the concentration of boron in the reproductive and vegetative organs of maize plants and their development. *Rhodesian Journal of Agricultural Research* 15: 163-171.

Wakhloo, J.L. (1975). Studies on the growth, flowering and production of female sterile flowers as affected by different levels of foliar potassium in *Solanum sisymbrifolium* Lam. I. Effect of potassium content of the plant on vegetative growth and flowering. *Journal of Experimental Botany* 26: 425-432.

Chapter 7

The Role of Nutrition in Crop Resistance and Tolerance to Diseases

Don M. Huber
Robin D. Graham

Key Words: bacteria, cultural disease control, defense, disease, fungi, nematodes, nutrition, resistance, tolerance, viruses.

INTRODUCTION

Nutrients are part of the "environment" for plant and microbial growth and, although frequently unrecognized, always have been an important factor in disease control. All the essential nutrients are reported to influence the incidence or severity of some disease (see Table 7.1); however, no single nutrient controls all diseases or favors disease control for any one group of plants (Graham, 1983; Engelhard, 1989; Huber, 1991). A particular element may decrease severity of some diseases, but increase others (see Figure 7.1), and have an opposite effect in a different environment. The effects of N (nitrogen), P (phosphorus), and K (potassium) on disease are reported more frequently than the effects of minor nutrients because of the limited availability of the macronutrients in many soils relative to the large quantity required for optimum plant growth. Nevertheless, it is through an understanding of the disease interactions with each specific nutrient that the roles of the plant, pathogen, and environment can be effectively modified for improved disease control.

The authors acknowledge Purdue Agricultural Research Programs, Journal Paper No. 15355.

TABLE 7.1. Reported Effects of Mineral Nutrient Fertilization on Disease

Mineral element	Disease is:		
	Decreased	Increased	Variable effect
Nitrogen ($N/NH_4/NO_3$)	168	233	17
Phosphorus (P)	82	42	2
Potassium (K)	144	52	12
Calcium (Ca)	66	17	4
Manganese (Mn)	68	13	2
Copper (Cu)	49	3	0
Zinc (Zn)	23	10	3
Boron (B)	25	4	0
Iron (Fe)	17	7	0
Sulfur (S)	11	3	0
Magnesium (Mg)	18	12	2
Silicon (not essential) (Si)	15	0	0
Chloride (Cl)	9	2	8
Other	27	4	0

Note: Table is based on 1,180 reports in the literature.

Nutrient manipulation through fertilization, or modification of the soil environment to influence nutrient availability, is an important cultural control for plant disease and an integral component of production agriculture. Cultural practices used for disease control (such as crop sequence, organic amendment, liming for pH adjustment, tillage and seedbed preparation, and irrigation) frequently influence disease through increasing or decreasing the availability of various nutrients, either directly or through altered microbial activity. The availability of inorganic fertilizers for crop production provides a more direct means of using nutrition to reduce the severity of many diseases than previously available through the more generalized practices of crop rotation or organic amendment. Several reviews of this topic are available (Huber and Watson, 1974; Huber, 1980, 1989, 1991; Graham, 1983; Huber and Arny, 1985; Huber and Wilhelm, 1988; Engelhard, 1989; Graham and Webb, 1991; Marschner, 1995; Savant, Snyder, and Datnoff, 1997). Integrating the effects of specific mineral nutrients with genetic resistance, sanitation, other cultural practices, and chemical treatments has provided effective control of many diseases.

FIGURE 7.1. Opposite Effects of the Form of N (175 kg ha^{-1}) on Two Potato Diseases

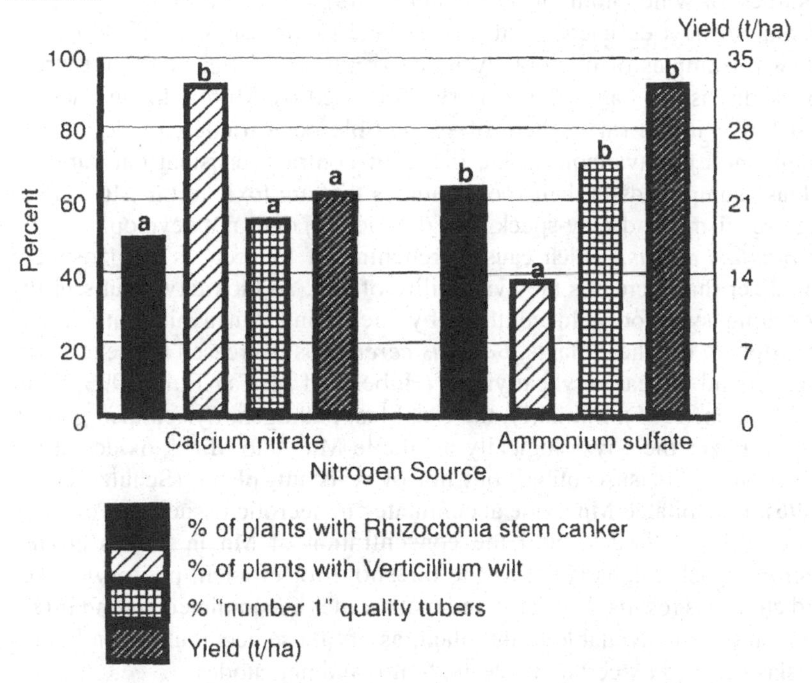

Source: Adapted from Huber (1989).

Note: Comparable means bars having the same letter above them are not significantly different.

EFFECT OF DISEASE ON NUTRIENT STATUS OF PLANTS

Nutrient deficiency often predisposes the plant to infection that, once established, can stress the plant further by impairing nutrient acquisition and/or utilization. This section discusses the latter effect, and the following sections, the former. Mineral nutrition is one of the basic plant processes impaired by disease. Pathogens alter the nutrition of the plant in diverse ways that are reflected in the symptoms of disease. Some pathogens may immobilize nutrients in the rhizosphere (the soil surrounding plant roots) or in infected tissues; others interfere with translocation or utilization efficiency, while still others cause hyperaccumulation and nutrient toxicity (Huber, 1978). Soilborne organisms, which utilize nutrients for their own metabolism, especially during decomposition of organic residues, or which immobilize the nutrient(s) in the rhizosphere, can produce a nutrient-deficient plant predisposed to disease. One of the better-known examples of microbially induced nutrient deficiency is the grey-speck disease of oats (Timonin, 1965) caused by Mn-oxidizing bacteria that thrive in the rhizosphere of susceptible oat varieties, rendering Mn (manganese) unavailable to the plant. In contrast, resistant oat varieties release compounds in their root exudates that are toxic to the Mn-oxidizing organisms, and grey-speck (Mn deficiency) does not develop.

Bacillus cereus, which causes frenching of tobacco, is a rhizosphere colonizer that increases the availability of Mn, and toxicity results in the frenching symptoms. In contrast, by increasing Mn availability in the rhizosphere of wheat plants, *Bacillus cereus* has decreased the severity of take-all and increased wheat yields (Huber and McCay-Buis, 1993; Scott and Wilken, 1995). Similarly, the rice blast pathogen, *Pyricularia grisea*, also oxidizes the physiologically available Mn^{2+} to Mn^{4+} oxides at the infection site of susceptible, but not of resistant, plants (Schulze et al., 1996). Precipitated Mn oxide accumulates in necrotic tissues as it diffuses from adjacent tissues, until the concentration of Mn in tissues around necrotic blast lesions is below the threshold for efficient photosynthesis, and chlorosis results. Nutrients and metabolites accumulated around infection sites are unavailable to the plant, as are those accumulated in hyperplasias induced by certain bacteria, fungi, and nematodes.

Root destruction is a common symptom induced by many soilborne fungi and nematodes. The effect of root rots is generally more severe in marginally sufficient soils or for relatively immobile nutrients for which an extensive functional root system is needed to provide the full nutritional needs of the plant. Restricted root growth from stem girdling or acropetal infection can indirectly influence nutrient absorption. A common effect of viral infection is restricted root growth and predisposition of plants to root-infecting fungi (Huber, 1978).

Although a malfunctioning vascular system caused by fungal or bacterial wilt diseases impairs the translocation and utilization of nutrients directly, changes in membrane permeability, or mobilization toward infection sites, can induce a nutrient deficiency or excess at the cellular level, even though the concentration in the plant may remain unchanged. In this way, *Fusarium oxysporum* f. *vasinfectum* decreases the concentration of N, K, Ca (calcium), and Mg (magnesium), but raises the concentration of P in leaves of infected cotton plants (Haag, Balmer, and De Carvalho, 1967). Leaching losses of nutrients in root exudates or from infected leaf tissue are also increased by disease and may deplete the level of nutrients in the plant below that required for efficient metabolism.

CHARACTERIZING NUTRIENT INTERACTIONS WITH DISEASE

The recognition that mineral nutrients affect disease has been based on (1) the observed effects of fertilization on disease severity; (2) comparisons of mineral concentrations in tissues of resistant and susceptible cultivars or diseased and nondiseased tissues; and/or (3) common patterns of conditions influencing mineral availability with disease incidence or severity (Huber and Watson, 1974; Huber, 1980, 1991; Graham, 1983; Huber and Arny, 1985; Huber and Wilhelm, 1988; Engelhard, 1989; Graham and Webb, 1991; Huber and Graham, 1992; Marschner, 1995). The evidence is discussed next.

The Observed Effects of Fertilization on Disease Severity

The effect of many nutrients on disease has been observed under deficiency conditions in which nutrient amendments were made to optimize plant growth. The effect of nutrients on disease was incidental to fertilizer programs, and the observed field responses were subsequently studied in the laboratory under more controlled conditions (Huber and Watson, 1970; Smiley and Cook, 1973; Smiley, 1975; Huber, 1994). Observations that barnyard manuring and other organic fertilizers decreased the severity of various soilborne diseases have frequently been attributed to increased microbial activity and stimulated "biological control," without recognition of the direct influence of the nutrients applied on the plant. The striking reduction of *Rhizoctonia cerealis*' "winterkilling" of wheat after fertilization with barnyard manure has been shown to be an effect of Zn (zinc) contained in the manure applied to the low-N soil (Huber et al., 1993) rather than some unknown biological phenomenon. The use of inorganic fertilizers allows direct observation of the effects of a specific nutrient that could only be generalized when organic manuring or crop rotation was the means of achieving the response.

Early observations that take-all of cereal crops was decreased by N fertilization have led to a standard recommendation to maintain N sufficiency when take-all is a potential problem; however, a deficiency of most essential mineral elements also will increase take-all (Reis, Cook, and McNeal, 1982, 1983; Huber and McCay-Buis, 1993). When P fertilization is optimized, both take-all and Pythium root rot decline. In contrast to take-all, foliar diseases of cereal crops such as rust and mildew, which are caused by obligate pathogens, are frequently increased as the rate of N is increased. Potassium has been observed to decrease the severity of various diseases, but the effect is frequently from the Cl (chlorine) ion rather than K when this salt is used.

Rhizoctonia and Fusarium cortical rots are each less severe under alkaline soil conditions in which a high availability of Ca prevails. On the other hand, Streptomyces scab of potato and Verticillium wilt are less severe in acid soils in which the availability of Mn and various other micronutrients is enhanced (Mortvedt, Berger, and Darling, 1963; Graham, 1983; Keinath and Loria, 1989).

Nutrient Concentration in Tissues of Resistant and Susceptible Plants or Diseased and Nondiseased Plants

Differences in the concentration of nutrients in tissues have been correlated with susceptibility or resistance to fungal pathogens (Wilhelm, Graham, and Rovira, 1988; Sparrow and Graham, 1988; Thongbai et al., 1993), as well as viruses and bacteria. Manganese concentration is usually lower in susceptible than resistant tissues (Bruck and Manion, 1980; Brain and Whittington, 1981), although exceptions have been noted with several virus diseases (Huber and Wilhelm, 1988) and Phytophthora root rot of avocado (Falcon, Fox, and Trujillo, 1984). Resistance of rice to blast, sheath blight, brown spot, and stem rot is correlated with high Si (silicon) content in plant tissues (Savant, Snyder, and Datnoff, 1997). There is an inverse correlation of Ca in tissues with severity of macerating-type diseases caused by *Sclerotium rolfsii*, *Pseudomonas solanacearum* (bacterial wilt), *Erwinia carotovora* and *E. chrysanthemi* (soft rots), *Pythium myriotylum* (peanut pod rot), *Rhizoctonia solani*, *Cylindrocladium crotalariae*, *Sclerotinia minor*, and *Fusarium solani* (Bateman and Basham, 1976; Kelman, McGuire, and Tzeng, 1989; Huber, 1994). A wilt-resistant flax took up more K than a susceptible variety (Dastur and Bhatt, 1964), and resistance of white pines to Peridermium rust has been associated with higher tissue concentrations of K than those found in susceptible varieties (Hutchinson, 1935). Rice cultivars susceptible to blast caused by *Pyricularia grisea* accumulate oxidized Mn in necrotic tissues as a result of pathogenic activity (Schulze et al., 1996). The concentration of Zn was higher in wheat tissues where *Rhizoctonia cerealis* (winterkill) was less severe than where it was more severe (see Table 7.2).

TABLE 7.2. Correlation of Rhizoctonia Winterkill with the Concentration of Zn in Wheat Leaves

Treatment	Tissue Zn		Plants killed	
	Without[a]	With	Without[a]	With
	mg kg^{-1}		%	
Barnyard manure applied	20[a]	41[b]	80[b]	18[a]
Sediment area of field	17[a]	27[b]	100[b]	45[a]
Tree leaf-drop area	19[a]	34[b]	65[b]	20[a]
Plant sufficiency level is 20 to 150 mg kg^{-1}				

Source: Adapted from Huber et al. (1993).

Note: Tissue zinc or plants killed means followed by a different letter in a row are significantly different from each other.

[a]Without the treatment or outside the area of influence.

Common Patterns in Nutrient-Disease Interactions

Diseases have been categorized as "high" or "low" pH types (Smiley, 1975), "ammonium" or "nitrate" types (Huber and Watson, 1974), "high" or "low" moisture diseases, etc., depending on the cultural conditions favoring severity (see Table 7.3). Inhibition of nitrification is also important in these effects (see Table 7.4). The effect of specific crop residues on nitrification was consistent with the effect of the form of N available in the soil on the severity of Fusarium and Rhizoctonia root rot of bean, *Phaseolus vulgaris* L. (Huber, Watson, and Steiner, 1965) (see Figure 7.2). Residues and crop sequences that stimulated nitrification to provide a predominantly NO_3^- form of N available in the soil decreased the severity of bean root rot, while those residues which inhibited nitrification increased the severity of bean root rot, independently of the population of *Fusarium solani* in soil (Huber, 1963) (see Figure 7.3). Residues inhibiting nitrification decreased the severity of Streptomyces scab of potato and take-all of wheat. Many cultural conditions influencing the form of N (Huber and Watson, 1974; Huber, 1991; Huber and McCay-Buis, 1993) also have an effect on rhizosphere pH (Smiley, 1975; Falcon, Fox, and Trujillo, 1984).

Although such cultural conditions influence the availability of nutrients, the recent focus has been on specific nutrients as a key to understanding the environmental or cultural effects on disease (Huber and Watson, 1970, 1974; Graham, 1983). Several diseases (Streptomyces scab of potatoes, soil pox of sweet potato, Verticillium wilt of vegetables, Phymatotrichum root rot of cotton, take-all of cereals, and Thielaviopsis root rot of tobacco) are associated with alkaline soils (Huber and Watson, 1970, 1974; Graham and

TABLE 7.3. Some Diseases Influenced by the Form of N and pH

Crop	Disease	Pathogen
Diseases decreased by NO_3^- fertilization and alkaline pH:		
Asparagus	Wilt	*Fusarium oxysporum*
Bean (*Phaseolus vulgaris*)	Chocolate spot	*Botrytis*
	Root and hypocotyl rot	*Fusarium solani, Rhizoctonia solani*
Beet	Damping-off	*Pythium spp.*
Cabbage	Club root	*Plasmodiophora brassica*
	Yellows	*Fusarium oxysporum*
Celery	Yellows	*Fusarium oxysporum*
Cucumber	Wilt	*Fusarium oxysporum*
Pea (*Pisum sativum*)	Damping-off	*Rhizoctonia solani*
Pepper	Wilt	*Fusarium oxysporum*
Potato	Stem canker	*Rhizoctonia solani*
Tomato	Gray mold	*Sclerotinia*
	Sclerotium blight	*Sclerotium rolfsii*
	Wilt	*Fusarium oxysporum*
Wheat	Eye spot	*Pseudocercosporella herpotrichoides*
Diseases decreased by NH_4^+ fertilization and acid pH:		
Bean (*P. vulgaris*)	Root rot	*Thielaviopsis basicola*
	Root knot	*Meloidogyne*
Carrot	Root rot	*Sclerotium rolfsii*
Corn	Stalk rot	*Gibberella zeae*
Eggplant	Wilt	*Fusarium oxysporum*
Onion	White rot	*Sclerotium rolfsii*
Pea	Root rot	*Pythium spp.*
Potato	Scab	*Streptomyces scabies*
	Wilt	*Verticillium dahliae*
	Virus	*Potato virus X*
Rice	Blast	*Pyricularia grisea*
Tomato	Southern wilt	*Pseudomonas solanacearum*
	Anthracnose	*Colletotrichum*
	Wilt	*Verticullium dahliae*
	Virus	*Potato virus X*
Wheat	Take-all	*Gaeumannomyces graminis*

Sources: Adapted from Huber and Watson (1974), Smiley (1975), Engelhard (1989), and Huber (1991; 1994).

Rovira, 1984; Huber and Arny, 1985; Huber and Wilhelm, 1988). Sulfur, which itself has little direct effect on *Streptomyces scabies*, has been used to reduce soil pH and scab for almost 100 years. In contrast, Ca, K, NO_3^-, or manure may increase scab. In addition to S (sulfur), NH_4^+ and other acidifying fertilizers also are commonly observed to reduce potato scab (Rogers, 1969; Huber and Watson, 1970; Smiley, 1975; Graham, 1983; Huber and Wilhelm, 1988; Keinath and Loria, 1989; Graham and Webb, 1991; Huber, 1991). Inhibiting nitrification of both inorganic and organic fertilizers enhances their ability to control scab (Huber, 1991), take-all of wheat (Huber and McCay-Buis, 1993), and stalk rot of corn (Sutton et al., 1995).

TABLE 7.4. Factors Affecting Nitrification, Mn Availability, and Take-All

Factor	Nitrification	Mn availability	Take-all
Liming ($CaCO_3$)	Increase	Decrease	Increase
Nitrate N	—	Decrease	Increase
Short (2-4 years) wheat monocropping	Increase	Decrease	Increase
Plant stress	Increase	Decrease	Increase
Manuring	Increase	Decrease	Increase
Loose seedbed	Increase	Decrease	Increase
Soybean precrop to wheat	Increase	Decrease	Increase
Alfalfa precrop to wheat	Increase	Decrease	Increase
Moderate soil moisture	Increase	Decrease	Increase
Alkaline pH soils	Increase	Decrease	Increase
High plant population	Increase	Decrease	Increase
Ammonium nitrogen	Increase	Increase	Decrease
Take-all-tolerant plant cultivars	—	Increase	Decrease
Nitrification inhibitors	Decrease	Increase	Decrease
Acid pH soils	Decrease	Increase	Decrease
Oat precrop to wheat	—	Increase	Decrease
Lupin precrop to wheat	Decrease	Increase	Decrease
Rice (paddy) precrop to wheat	Decrease	Increase	Decrease
Wheat monoculture (6+ years)	Decrease	Increase	Decrease
Late seeding	—	Increase	Decrease
Manganese fertilization	—	Increase	Decrease
Chloride fertilization	Decrease	Increase	Decrease

Sources: Adapted from Graham (1983); Huber and Wilhelm (1988), Huber (1989, 1991), Graham and Webb (1991), and Huber and McCay-Buis (1993).

The transformation from insoluble Mn^{3+} or Mn^{4+} oxides to soluble Mn^{2+} is highly dependent on environmental factors such as soil pH, moisture, nutrients, Cl^-, nitrification inhibitors, organic matter, and microbial activity (Graham, 1983). Thus, many of the factors predisposing plants to disease influence the availability of Mn in soil and may influence disease severity through their regulation of Mn uptake by the plant (Huber and Wilhelm, 1988). The greater resistance of paddy-grown rice to blast (*Pyricularia grisea*) compared to upland rice has been attributed to the increased uptake of Mn by rice under paddy conditions (Choong-Hoe, 1986; Pearson and Jacobs, 1986). Incorporation of rice straw decreases Mn and available N, and increases blast (Choong-Hoe, 1986). The correlation of factors influencing Mn availability with take-all of cereals (see Table 7.4), Strepto-

FIGURE 7.2. The Effect of Specific Crop Residues on Nitrification

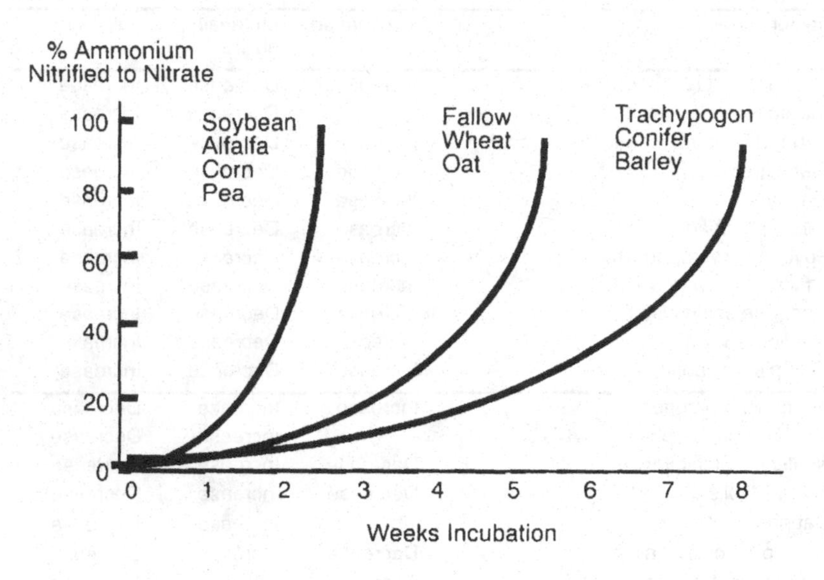

Source: Adapted from Huber, Watson, and Steiner (1965).

myces scab of potatoes, and blast of rice illustrates how cultural conditions affect both the availability of Mn and disease severity. Scab occurs on acid soils only when the Mn content is very low (McGregor and Wilson, 1964, 1966). The addition of "press wheels" to the planter to firm the seedbed during planting of winter wheat increased the concentration of Mn in wheat tissues by 5 to 12 mg g^{-1} and decreased the take-all index on a 0 to 10 scale from 6.0 to 4.5 (Huber and McCay-Buis, 1993).

STRATEGIES TO REDUCE DISEASE SEVERITY BY IMPROVING NUTRITION

Considerations in managing plant diseases by nutrition include (1) the level of genetic resistance (highly susceptible, tolerant, resistant, or immune) of the cultivar grown; (2) the nutrient status of the soil (deficient, sufficient, or excess); (3) the predominant form and biological stability of a nutrient that is available or applied; (4) the rate, time, and method of

FIGURE 7.3. Effect of Crop Sequence on the Population of *Fusarium Solani* f.sp. *Phaseoli* in Soil (Based on Frequency of Isolation in the Plate-Profile), Severity of Root Rot (% Roots Infected at Anthesis), and Seed Yield (% of Corn/Bean Rotation Yield)

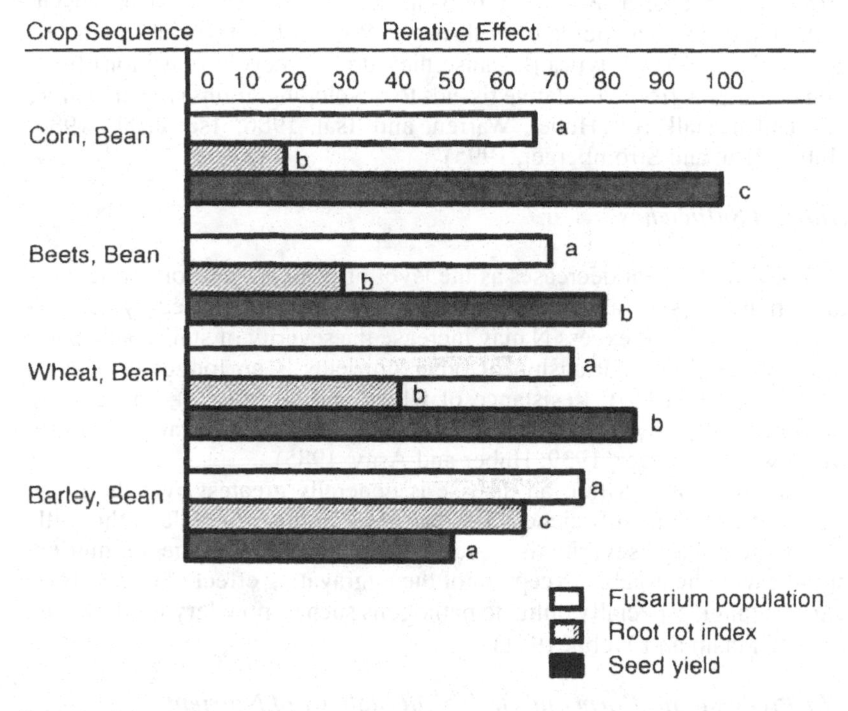

Source: Adapted from Huber (1963).

Note: Comparable bars followed by the same letter are not significantly different (P = 0.05).

nutrient application; (5) the source of an element and associated ions; and (6) integration of fertilizers and other cultural practices.

Level of Plant Resistance

The greatest disease suppression with nutrient amendments generally is with tolerant or resistant cultivars; cultivars immune to a particular disease may be highly efficient in nutrient uptake or function. Rye, considered resistant to take-all because of its rapid response to infection by "walling off" ("lignitubers") penetration hyphae of *Gaeumannomyces graminis*, is one of the most efficient cereal plants in nutrient absorption (Graham, 1984, 1988; Cooper, Graham, and Longnecker, 1988). Wheat, by contrast, is inefficient in micronutrient uptake and is susceptible to take-all.

Corn cultivars resistant to Diplodia and Gibberella stalk rots are either efficient in uptake of N or they prevent the cannibalization of photosynthetic enzymes as a source of N throughout grain-fill. Hybrids characterized as "stay green" types (because they do not recycle [partition] functional nutrients from vegetative tissues to developing grains) are also more resistant to stalk rot (Huber, Warren, and Tsai, 1986; Tsai et al., 1986; Huber, Tsai, and Stromberger, 1995).

Nutrient Sufficiency

Take-all of wheat decreases as the level of N increases from deficiency to sufficiency (see Table 7.5), as does stalk rot of corn caused by *Gibberella zeae*; however, excess N may increase the severity of stalk rot because the physiological relationship of other nutrients is no longer in balance (Warren et al., 1980). Resistance of wheat and flax to rust, and corn to Stewart's wilt, may be lost under K-deficient conditions (Sharvelle, 1936; McNew and Spencer, 1939; Huber and Arny, 1985).

The effect of nutrition on disease is generally greatest over the range from deficiency to sufficiency. The corollary to this principle is that little effect on disease severity is observed in the luxury range of nutrient supply, with the notable exception of the aggravating effect of excess N on various fungi, especially obligate pathogens such as powdery mildews and rusts (Graham and Webb, 1991).

The Predominant Form and Biological Stability of Nutrients

Oxidized and reduced forms of a nutrient often have opposing effects on disease because of differences in availability or the use of different metabolic pathways (Huber and Watson, 1974; Smiley, 1975; Huber and Wilhelm, 1988). This is especially true for N, Mn, S, and Fe (iron). The biological mineralization of organic N to inorganic NH_4^+ and its subsequent nitrification to NO_3^- are dynamic processes in most soils. Both the cation (NH_4^+) and the anion (NO_3^-) forms of N may be assimilated by plants, but they frequently have opposite effects on disease (Huber and Watson, 1974). Ammonium (NH_4^+) decreases the severity of take-all of cereals, Verticillium wilt of potatoes and other crops, Streptomyces scab of potatoes, and blast of rice, while similar rates of nitrate (NO_3^-) increase these diseases. Nitrate, in turn, decreases the severity of Fusarium and Rhizoctonia cortical rot of beans and potatoes, Fusarium wilt of many crops, and clubroot of crucifers.

TABLE 7.5. Effect of N Rate and Inhibiting Nitrification on Grain Yield and Take-All Severity of Wheat Grown in Natural Pathogen-Infested Soils

Nitrogen rate kg ha^{-1}	Nitrification inhibitor	Soil type					
		Sandy		Sandy loam		Silt loam	
		Yield kg ha^{-1}	White heads %	Yield kg ha^{-1}	White heads %	Yield kg ha^{-1}	White heads %
None	None	1270	30	938	50	2410	8
45	None	1540	26	2350	23	2950	3
45	Yes	1876	16	3080	11	3690	2
90	None	1680	18	3150	19	4020	1
90	Yes	2750	9	4220	7	4420	1
135	None	2550	12	3220	20	4420	1
135	Yes	3080	3	4760	6	4560	1
LSD		312	5	480	5	308	5

Source: Adapted from Huber and McCay-Buis (1993).

The beneficial effect of NH_4^+ on potato scab was not observed when high levels of residual NO_3^- were present in the soil. Growing an unfertilized cereal crop prior to the potatoes to remove excess residual N resulted in significantly lower scab when NH_4^+, rather than NO_3^-, was the source of N. The beneficial effect of NH_4^+ was increased by inhibiting nitrification to maintain applied N fertilizers in the NH_4^+ form for longer. Inhibiting nitrification of NH_4^+ fertilizers also improves the control of take-all of cereals through the effect of the form of N. The benefit of NH_4^+ in reducing some diseases results from the increased availability of Mn (Huber and Wilhelm, 1988).

Soil fumigation, which inhibits nitrification, has been a common practice for controlling Verticillium wilt of potato and other high-value crops, even though the population of the pathogen may increase in the soil following fumigation (Easton, 1964; Huber and Watson, 1970). The benefits of fumigation in reducing wilt were enhanced when NH_4^+ fertilizers were applied and decreased with NO_3^- sources of N (see Figure 7.4). Inhibiting nitrification and increasing rates of N also decreased the severity of tan-spot of wheat caused by *Pyrenophora tritici repentis* (see Table 7.6). Inhibiting nitrification of swine manure, applied as a fertilizer for corn, increased grain yield and eliminated the predisposition to Gibberella stalk rot generally observed with manure applications (Sutton et al., 1995) (see Figure 7.5).

FIGURE 7.4. Interaction of the Form of N with Dichloropene Fumigation on Verticillium Wilt of Potato As Measured by Tuber Yield

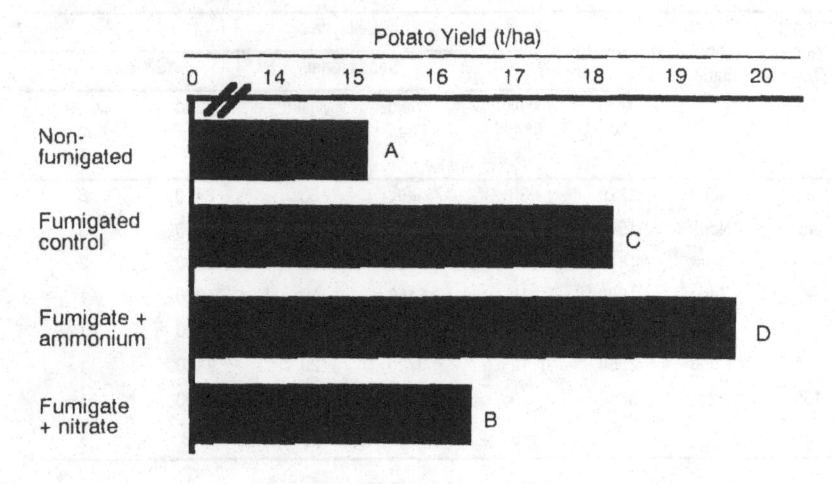

Source: Adapted from Huber (1996).

Note: Bars followed by various letters are significantly different (P = 0.05).

TABLE 7.6. Effect of N Rate and Inhibiting Nitrification on Tan-Spot (*Pyrenophora tritici repentis*) of Winter Wheat

Nitrogen rate	Nitrification inhibitor[a] kg ha^{-1}	Leaf necrosis[b] %
None	None	75[a]
50	None	30[b]
50	Yes	10[c]
100	None	20[bc]
100	Yes	3[c]

Source: Adapted from Huber (1996).

[a] Applied with NH_3 at the rate of 55 kg ha^{-1}.
[b] Means followed by the same letter are not significantly different.

FIGURE 7.5. Stalk Rot of Corn Fertilized with Liquid Swine Manure Without and with Nitrapyrin, a Nitrification Inhibitor

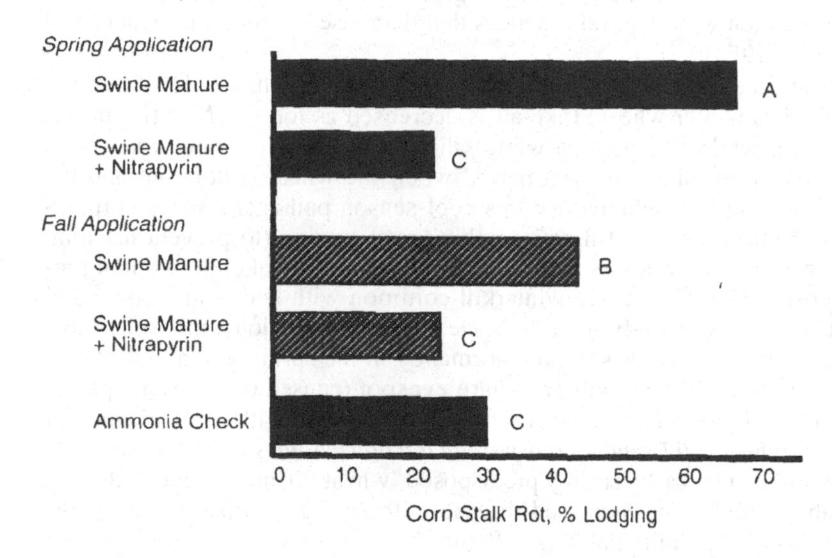

Sources: Adapted from Huber, Warren, and Tsai (1986) and Sutton et al. (1995).

Note: Bars followed by the same letter are not significantly different (P = 0.05).

Rate, Time, and Method of Nutrient Application

The potential nutritional needs for efficient crop production must be met economically and be environmentally sound. Nutrient availability to the plant will depend on the level of residual nutrient in the soil, the rate and time of fertilizer application, specific microbial activity, seasonal losses, efficiency of the plant, and general health of the plant. Nutrient losses can be decreased by applying only the amount required during various growth stages of the plant, avoiding applications during high loss periods, or by modifying the chemical or biological environment influencing nutrient availability (inhibiting nitrification, changing pH, or modifying soil moisture). In the earlier example of the opposing effects of NO_3^- and NH_4^+ on Rhizoctonia canker and Verticillium wilt of potato, the most damaging of the two diseases on yield was the Verticillium wilt. Increasing the size of the potato "seed piece," modifying tillage, or crop rotation can prevent damage from *Rhizoctonia*. Few alternatives are available to decrease the severity of Verticillium wilt, so the ability to maintain a low incidence of Verticillium wilt through the form of N chosen is useful in combination with cultural practices that decrease Rhizoctonia (Huber and Watson, 1974).

The time of fertilization is especially important when N fertilizers are applied to winter wheat. Take-all is decreased as long as N deficiency is avoided, while Rhizoctonia winterkill is a potential problem when N is applied during the cool, wet period when the wheat is dormant and the environment is conducive for this cool-season pathogen. Soil incorporation of all N with a nitrification inhibitor at sowing (to prevent leaching and denitrification losses during winter) will reduce take-all, without predisposing plants to severe winterkill common with late-winter-applied N (Huber and McCay-Buis, 1993). Delaying fertilization until soils have dried and the wheat has broken dormancy in the spring will avoid predisposition to both winterkill and sharp eyespot (caused by the same pathogen) (see Table 7.7). Similar reactions are observed with N fertilization of winter wheat and *Pseudocercosporella herpotrichoides* infection, where N applied in the early spring predisposed wheat to more severe disease (Huber, 1968). The increased disease with spring fertilization offset the benefits of the additional N (see Figure 7.6).

Source of Element and Associated Ions

Nutrient imbalance may be just as detrimental as nutrient deficiency or excess. Take-all is increased by K if N and P are deficient, but decreased by K if N and P are sufficient (Butler, 1961; Stetter, 1971). Potassium decreased the severity of late blight of potato (*Phytophthora infestans*), tobacco wildfire, and cereal rusts on plants receiving high rates of N. An imbalance of K relative to other elements increases the severity of Fusarium yellows of cabbage, Fusarium wilt of tomato, Stewart's wilt of corn, and downey mildew of tobacco (Huber and Arny, 1985).

TABLE 7.7. Effect of Time of Fertilization on Sharp Eyespot of Wheat

Applied kg N ha^{-1}	Disease severity[a]	Percent lodging	Yield kg ha^{-1}
October	2.1a	3a	3100a
March	3.2b	47b	2660b

Note: Means in a column followed by various letters are significantly different.

[a] Disease index based on a 0 to 5 scale with 0 = healthy

FIGURE 7.6. Effect of Time of Fertilization on Foot Rot (Pseudocercosporella Eyespot) and Yield of Winter Wheat

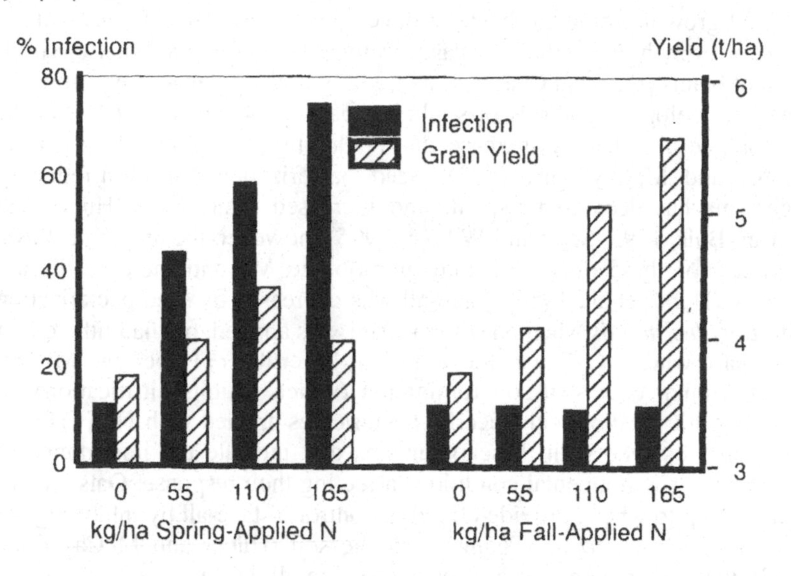

Sources: Adapted from Huber (1968) and Huber and Watson (1974).

The ratio of various elements may be more important than the absolute amount of each element for some diseases. Fusarium wilt of cotton has been correlated with the K/Mg ratio; Streptomyces scab of potato, brown rot gummosis (*Phytophthora parasitica*) of orange, and clubroot of crucifers with the Ca/K ratio; and late blight of potato (*P. infestans*) with the N/K ratio (Huber, 1980).

The associated ion applied with an organic or inorganic nutrient fertilizer may have an effect on disease that is independent from the primary ion. The Zn in barnyard manure, rather than the N, K, or P, was the key to its control of Rhizoctonia, as already mentioned. In some soils, the reduction in severity of Diplodia and Gibberella stalk rots of corn, take-all of cereals, northern leaf blight of corn, and stripe and leaf rust of wheat following the application of high rates of K has been shown to result from the accompanying Cl^- anion suppressesing NO_3^- uptake rather than from K directly (Huber and Arny, 1985). In these cases, effects on stalk rot were reproduced with NaCl and NH_4Cl, but not with K_2SO_4 or K-phosphate (Nelson, 1963; Huber and Arny, 1985).

Biological Control

Plant-growth-promoting bacteria have been applied to seed or plants to increase growth and reduce disease by modifying the microbial environment of plants (Burr and Ceasar, 1984). Many of the organisms promoted as potential biological controls of soilborne pathogens can reduce Mn in the rhizosphere and thereby increase the availability of Mn for plant uptake (Huber and McCay-Buis, 1993). Seed bacterization with Mn-reducing organisms has decreased take-all and increased grain yield (Huber and McCay-Buis, 1993; Scott and Wilken, 1995); however, the use of peat as a "carrier" for these organisms may immobilize Mn and increase disease severity (Huber et al., 1986). Take-all was decreased by seed bacterization with *Bacillus subtilis* when NH_4^+ fertilizer was applied, but had little effect on disease when NO_3^- was the source of N fertilizer (Huber and Mburu, 1984). Similar effects of soil fumigation (which inhibits nitrification), or inhibiting nitrification with specific bactericides applied with NH_4^+ fertilizers, are observed with these organisms and indicate the importance of recognizing environmental conditions affecting their response. Oats, grown as a precrop to wheat, provide effective control of take-all by inhibiting the activity of Mn-oxidizing organisms in the soil (Huber and McCay-Buis, 1993), while a precrop of rye, resistant to take-all, has no effect on take-all severity in a subsequent wheat crop (Rothrock and Cunfer, 1991).

Integration with Other Cultural Practices

Crop rotation and early fallowing practices made crop production possible in many areas of the world by increasing the supply of readily available nutrients and controlling weeds that compete for nutrients and moisture. Both factors markedly contributed to disease reduction and remain important practices for controlling many soilborne diseases. Integration of nutrient amendment with cultural practices such as tillage, crop rotation, seeding rate, and pH adjustment can accentuate the benefits of the nutrient amendment by modifying the environment for plant growth or microbial activity. Nutrition should be used in conjunction with disease resistance, crop rotation, weed control, and insect management to maximize plant productivity.

MECHANISMS OF DISEASE CONTROL WITH NUTRITION

Disease resistance is a property of the plant that describes the relative incompatibility of the plant-pathogen interaction, while tolerance describes the ability of the plant to produce in a compatible relationship. These terms allude to different mechanisms of disease control and, along with disease escape, effects on pathogen virulence, survival, or multiplication, and biological control, describe ways pathogen activity is regulated and the impact of disease decreased.

Nutrient Effects on Plant Resistance

Disease resistance is heritable and may be specific (to a particular pathogen) or general (controlling a range of pathogen groups), and it may be either single gene or multigenic. Some forms of general resistance may be identical with nutrient efficiency genes that confer on a plant genotype the ability to yield well in soils too deficient for a "standard" or reference genotype. Such genotypes are able to acquire adequate amounts of a limiting nutrient, while the standard genotype suffers from nutrient deficiency that compromises its resistance mechanisms in the same way as already described for deficiency. Host-pathogen systems studied in this respect include wheat and its pathogens, *G. graminis* var. *tritici* and *F. graminearum* (take-all fungus in Mn-deficient soil, and crown rot fungus in Zn-deficient soil) (Wilhelm, Graham, and Rovira, 1990; Pedler, 1994; Rengel, Pedler, and Graham, 1994; Grewal, Graham, and Rengel, 1996). In these systems, cultivars selected only for their efficiency in absorbing the limiting nutrient were more resistant to the fungus inoculated into the soil. In other words, heritable resistance to these two soilborne pathogens of wheat is expressed only in soils deficient in the nutrient in question; the genetic advantage is eliminated when the limiting nutrient is available in the soil in sufficient amounts.

Accumulation of Inhibitory Compounds Around the Infection Site

Many compounds produced by the secondary metabolic pathway of plants (shikimate) are formed prior to infection and provide a general type of resistance to disease. Thus, inhibitory phytoalexins, phenols, flavonoids, and auxins accumulate around infection sites of resistant plants, depending on the level of various nutrients. Specific steps at which plant defense mechanisms are regulated by micronutrients have been detailed by Graham (1983) and Graham and Webb (1991). Manganese, Cu (copper), Fe, B (boron), and Co (cobalt) are either integral parts of the enzyme molecules or cofactors required for enzyme activity. As an example, phenolic compounds in the vascular system of potato are inhibitory to the Verticillium wilt pathogen (see Figure 7.7). Nitrate inhibits phenol metab-

FIGURE 7.7. Comparison of the Concentration of Inhibitory Phenolic Compounds (Chlorogenic Acid) in the Vascular Systems of Verticillium Wilt-Resistant (P141256) and Wilt-Susceptible Potato Cultivars (Early Gem and Russet Burbank) During Growth

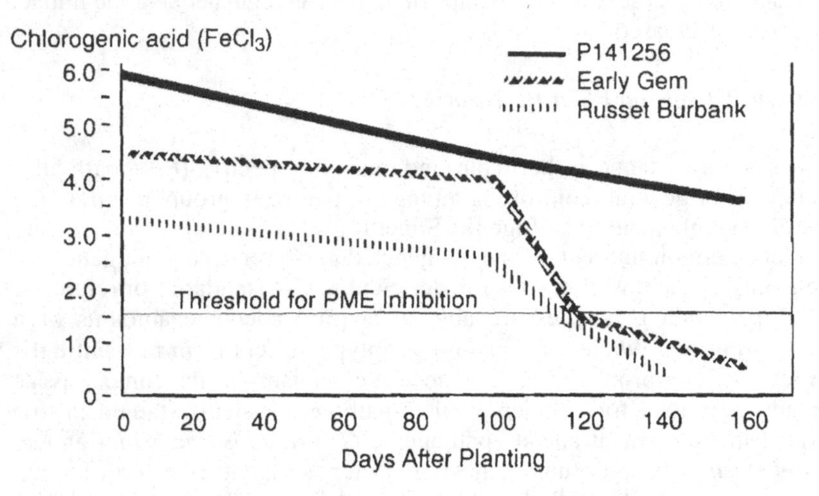

Sources: Adapted from Huber (1989) and Huber and Watson (1974).

Note: Disease onset is observed around days 90 to 100 in susceptible cultivars fertilized with NO_3^- and delayed until days 120 to 140 in susceptible cultivars fertilized with NH_4^+ that has been stabilized with a nitrification inhibitor. As the concentration of chlorogenic acid in xylem vessels drops below the amount required to inhibit pectin methyl esterase (PME) activity of the fungus (horizontal line), wilt symptoms are initiated.

olism (Chaudhry and Loneragan, 1970) and hastens wilt, while NH_4^+ increases uptake of Mn required for phenol production; an inhibitory concentration of phenolic materials is maintained in the vascular system to delay symptom expression (Huber, 1989). The concentration of inhibitory phenols remained above the threshold needed to inhibit *Verticillium* until harvest in fully resistant plants. Decreased concentrations of fungistatic phenols when K is deficient are also associated with susceptibility of rice to bacterial blight, grape to *Botrytis cinerea*, and wheat to rusts and mildew (Kiraly, 1976).

Resistance of potato to *Phytophthora* is associated with the K-induced accumulation of fungistatic levels of arginine in leaves (Alten and Orth, 1941). In contrast, high levels of glutamine and glutamic acid in leaves of tobacco susceptible to *Alternaria*, *Cercospora*, and *Sclerotinia* were decreased by K, and plants became more resistant to disease (Klein, 1957).

Glycoproteins (lectin) associated with resistance of sweet potato to *Ceratocystis fimbriota* (black rot) and potato to *Phytophthora infestans* (late blight) require Mn for activity (Garas and Kuc, 1981). Pathogen-induced peptidase activity, which provided specific amino acids required for the obligate rust and mildew pathogen's growth, was stimulated in wheat cultivars susceptible to rust and mildew, but was inhibited by Mn-containing fungicides and in resistant cultivars (Huber and Keeler, 1977). Hypersensitive cultivars were characterized by an initial stimulation of peptidase activity, as the pathogens penetrated the plant, followed by a rapid drop in plant aminopeptidase activity (see Figure 7.8).

Degradation of structural proteins (e.g., glycoproteins in cell walls) causes stalk tissues of corn to become susceptible to the macerating enzymes of the *Gibberella* and *Diplodia* fungi. Increasing rates of N, or inhibiting nitrification of applied NH_4^+ fertilizer, prevents these degradative processes and maintains resistance to stalk rot (Huber, Warren, and Tsai, 1986). Corn hybrids ("stay green") that genetically block degradation of RuBisCo and PEP-carboxylase are more resistant to stalk rot than normal hybrids. The accumulation of nutrients around infection sites also could have a direct toxic effect on pathogens similar to the "limited

FIGURE 7.8. Aminopeptidase Activity of Wheat Leaves After Inoculation with *Puccinia recondita*

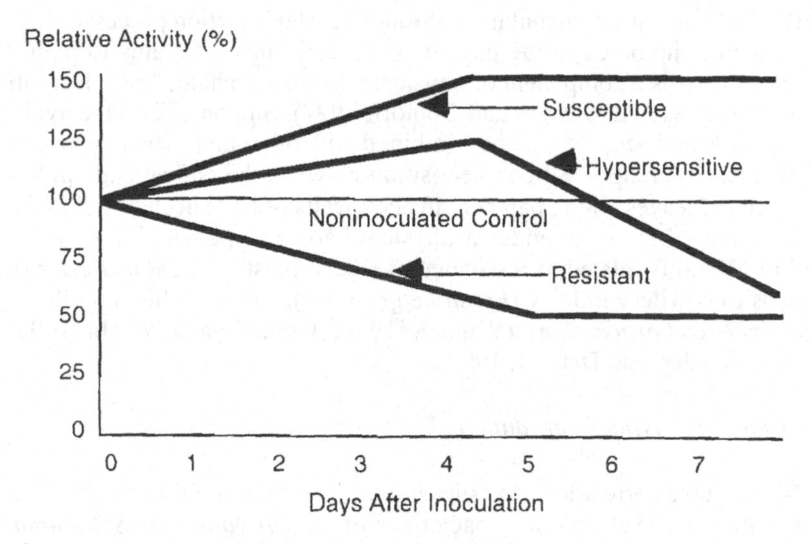

Source: Adapted from Huber and Keeler (1977).

growth" phenomenon of fungi (Kliejunas and Ko, 1975), for which the excess accumulation of nutrients becomes toxic to further growth.

Physical Barriers to Penetration and Infection

Physical barriers to penetration or development of pathogens include lignification, cicatrix formation, and callose development, and these also depend on various nutrient elements as cofactors, activators, or regulators of the biosynthetic pathways involved (Graham, 1983; Graham and Webb, 1991). Lignin is recognized as a partial barrier to penetration of many pathogens (Beckman, 1980) and may account for the importance of Cu in adult plant resistance to various pathogens (Graham, 1983). The increased thickness and speed of formation of wound periderm in potatoes is related to the sufficiency of Mn and provides a partial defense to Streptomyces scab (Graham, 1983). Lignitubers (Skou, 1981; Rengel, Pedler, and Graham, 1994) develop around penetrating hyphae of *Gaeumannomyces graminis* only when Mn and Cu are available, and autofluorescence of halos associated with callose tissue limiting penetration of *Erysiphe graminis* is enhanced with high, but not low, Mn availability (Kunoh, Kuno, and Ishizaki, 1985). Penetration of *G. graminis* hyphae to vascular elements was decreased by 80 percent when there was a sufficiency of Mn prior to fungal encounter of the roots, compared to Mn-deficient conditions (Wilhelm, 1991). Penetration of *G. graminis* was slowed even more when a sufficiency of Mn was maintained throughout the infection process.

Silicon is involved in the physiological resistance of plants to pathogens and also as a component of physical barriers (Graham, 1983; Datnoff et al., 1991; Savant, Snyder, and Datnoff, 1997). Silicon affects the availability of K and Mn and may be combined with other materials to give cell walls greater strength. Silicon deposition between the cuticle and epidermal cells of leaves, along parenchyma cell walls of leaf sheaths, and in the endodermis of stems provides a physical barrier to penetration of fungi and pests and is related to resistance of rice to blast (*Pyricularia grisea*), cereals to powdery mildew (*Erysiphe graminis*), and cucurbits to mildew (*Erysiphe cichoracearum*) (Wagner, 1940; Graham and Webb, 1991; Savant, Snyder, and Datnoff, 1997).

Resistance to Tissue Degradation

The inverse correlation of tissue calcium concentrations and severity of macerating diseases caused by bacterial soft rots (*Erwinia* spp.) *Sclerotium rolfsii, Pythium myriotylum, Rhizoctonia solani, Cylindrocladium crotala-*

riae, *Sclerotinia minor*, and *Fusarium solani* has been explained by the greater structural integrity of Ca-pectate in the middle lamella, cell wall components, and cell membranes induced by Ca (Bateman and Basham, 1976; Kelman, McGuire, and Tzeng, 1989). The activities of both polygalacturonase and pectin lyase are influenced by Ca, and oxalic acid produced by some fungal pathogens serves as a chelating agent to increase the activity of these enzymes (Kelman, McGuire, and Tzeng, 1989). Magnesium is also involved in structural components of the cell and may serve to reduce susceptibility to pathogen-produced macerating enzymes; however, both K and Mg reduce the Ca content of peanut pods and predispose them to pod breakdown by *Pythium* and *Rhizoctonia* (Csinos and Bell, 1989).

Disease Escape

Nutrient deficiency (or nutrient-stimulated plant growth) often shortens the time to maturity and, in the process, may decrease the length of the physiological stage for susceptibility, allowing the plant to escape disease (Rengel and Graham, 1995). Supplying a number of nutrients, both macro- and micronutrients, has this effect when the soil is deficient; however, in cool temperate climates, a high dose of N can delay maturity through the stimulation of tillering, thus increasing the risk from some foliar pathogens.

Tolerance

Tolerance is the ability of plants to "outgrow" the debilitating effects of disease or compensate for the loss of infected tissue through new growth. Nitrogen fertilization of barley increased the leaf area covered by powdery mildew from 10 to 20 percent, yet the N also increased the grain yield by 50 percent, showing that the vigorous, N-fertilized plants were highly tolerant of the increased disease burden (Last, 1962). Last's data showed that P also increased the tolerance to mildew and increased the yield as much as N, but that the addition of K, which was sufficient to begin with, did not affect tolerance or resistance. Phosphorus and N also stimulate root growth of cereal plants so that plants are able to compensate for tissue lost through root rots such as take-all. Browning root rot of cereals caused by *Pythium* spp. is of limited economic significance in many areas because of adequate P fertilization, which stimulates root growth to compensate for tissue lost by pathogenesis (Huber, 1980). Increasing rates of NO_3^-, while increasing the amount of root tissue damaged by *Gaeumannomyces*, also will increase grain yield because the increased availability of N compensates somewhat for the decreased root absorption efficiency caused by the disease. Tolerance may also be expressed by plants that have a decreased requirement for tissues damaged by disease. An example of this type of tolerance is expressed in some hard red winter wheat cultivars that actively take up N and store an adequate amount in vegetative tissues prior to anthesis, after which it is retranslocated to the developing grain.

Interactions with the Microbial Environment

Microorganisms in the environment may enhance (synergize) or inhibit disease (biological control) through their effects on nutrient availability. Microorganisms can immobilize nutrients in their own cells and change the availability of nutrients for plant uptake. Mineralization (the conversion of organic to inorganic forms), nitrification (oxidation of NH_4^+ to NO_3^-), and denitrification (reduction of nitrite and NO_3^- to gaseous N) are all microbial processes common in most agricultural environments. Take-all of wheat is generally more severe in nonsterile than in inoculated sterile soil because other Mn-oxidizing soil organisms complement the Mn-oxidizing activity of the pathogen. Soybeans growing in areas where take-all was severe on the previous wheat crop have lower tissue levels of Mn and may exhibit Mn-deficiency symptoms, compared with plants growing in adjacent areas where take-all was less severe (Huber and McCay-Buis, 1993).

Biological control may function either through direct effects on the pathogen to reduce its virulence or activity or by increased plant resistance, if the biocontrol agent increases the availability of specific nutrients. Mycorrhizae increase the uptake of P and Zn required for plant resistance. Most biological control agents reported to decrease take-all are Mn reducers that can increase the availability of Mn needed by the plant for resistance. Those organisms which are capable of oxidizing, as well as reducing, Mn may increase disease severity under conditions in which they immobilize Mn in the rhizosphere (decreasing Mn availability for the plant), or they may decrease disease under conditions that thermodynamically favor their Mn-reducing activity so they can increase the availability of Mn (and plant resistance). Increased yield of take-all-infected wheat following bacterization with various strains of *Bacillus* was correlated with their Mn-reducing activity (Scott and Wilken, 1995).

USING NUTRITION FOR DISEASE CONTROL

Disease control through nutrient manipulation can be achieved by either modifying nutrient availability or modifying nutrient uptake. Besides the use of fertilizer, changing the environment through pH modification, tillage, seedbed firmness, moisture control (irrigation or drainage), and specific crop sequences can have striking effects on nutrient availability. Inhibiting nitrification can increase the efficiency and availability of N in high leaching or denitrifying conditions. Microbial enrichment with mycorrhizae or plant-growth-promoting organisms also increase nutrient uptake—especially P, Zn, and/or Mn; microbes in the rhizosphere influence minor element availability through their oxidation-reduction reactions or siderophore production (Leong, 1986; Huber and McCay-Buis, 1993). Biological control of root pathogens that immobilize nutrients in the rhizosphere, and removing weed and pest competition, can increase nutrient availability.

The application of Mn is not always effective because of its rapid oxidation by microorganisms to nonavailable forms in soil. Application to foliage can relieve above-ground deficiency symptoms, but Mn is not well translocated in phloem, so that root tissues in contact with the pathogen may remain Mn deficient (Huber and McCay-Buis, 1993). Inhibiting nitrification of NH_4^+ fertilizers has suppressed Mn oxidation, as well as nitrification, and increased the availability of Mn, P, and Zn for plant uptake (Huber, Warren, and Tsai, 1986). A common characteristic of climax ecosystem plants is their inhibition of nitrification, which increases the efficiency of N in the environment (Rice, 1984).

Nutrient uptake can be modified by genetically improving a plant's efficiency in root absorption, translocation, and metabolic efficiency or by increasing tolerance to otherwise excess amounts of a specific nutrient (Graham, 1988; Huber and Wilhelm, 1988; Graham and Webb, 1991; Rengel, Pedler, and Graham, 1994; Huber, Tsai, and Stromberger, 1995). Selection of wheat seed with a higher Mn content produced plants with less take-all, compared with the same cultivar with a lower concentration of Mn in the seed (Huber and McCay-Buis, 1993; Pedler, 1994). Breeding for increased nutrient content is being actively pursued as a means of improving human nutrition and may concurrently increase plant resistance to a variety of diseases (Graham, 1984, 1988; Graham and Welch, 1996). The selection of cotton cultivars with multidisease resistance was based on the composition of their seed exudates (Bird, 1974; Bird et al., 1979; Bush, 1980).

EXAMPLES OF NUTRIENT-DISEASE INTERACTIONS

The following examples of Streptomyces scab of potato, Verticillium wilt, take-all of wheat, stalk rot of corn, clubroot of crucifers, Fusarium wilt, and tissue-macerating diseases are presented to characterize the interacting relationships between the plant, the pathogen, and the environment as observed in production agriculture. Other than fertilization, crop rotation and adjustment of soil pH have been the most common means of nutrient manipulation relative to disease control because of their effect on nutrient availability and microbial activity, which influences the availability of nutrients.

Streptomyces Scab of Potato

Streptomyces sp. are strong Mn oxidizers in soil (Bromfield, 1979; Graham and Webb, 1991). Cultural and soil conditions, such as crop rotation, soil amendment with specific crop residues, N fertilizers, soil acidification and irrigation (Keinath and Loria, 1989), all of which increase the availability of Mn, have been developed over the years to control common scab. Conversely, conditions that lower Mn availability increase this disease. Experimental applications of Mn to soil have decreased scab when the population of Mn-oxidizing organisms was low. Scab-suppressive soils have characteristically high total Mn or high activity of Mn-reducing organisms, while scab-conducive soils have low total Mn. Under acid conditions, when soil Mn would normally be solubilized, scab only occurs when total Mn in the soil is inadequate to meet plant needs, regardless of soil reaction. Soil applications of Mn to control scab have not become commercial practice because fertilizer Mn is quickly immobilized (Huber and Wilhelm, 1988), and environmental conditions need to be modified to increase or stabilize Mn availability.

Verticillium Wilt

Vascular wilts caused by *Verticillium albo-atrum* and *V. dahliae* are some of the most damaging diseases of vegetables, ornamentals, fruits, herbs, and field, forage, and forest crops in cooler climates. Many weed species also are hosts of *Verticillium* and may maintain a high population of the pathogen in soil (Pennypacker, 1989). Recommendations for wilt control of cultivated crops include selection of cultivars with genetic resistance, specific crop sequence, sanitation, soil fumigation, and nutritional sufficiency (Huber, 1989; Pennypacker, 1989). Verticillium wilt is more severe in alkaline than acid soils, and adequate levels of N, P, and K reduce disease. Soil fumigation and nitrification inhibitors maintain NH_4^+ in the soil and increase uptake of Mn, Cu, and Zn that are important in resistance (Graham and Webb, 1991). Although direct amendment with Mn has decreased the severity of Verticillium wilt, it has not provided consistent disease control because of its transient availability in soil. "Flood fallowing," the plowing down of a green manure crop and flooding the soil to maintain moisture saturation, has provided control of Verticillium wilt of potatoes and tomatoes by reducing the inoculum potential of the pathogen under the anaerobic conditions. This practice also increases the availability of Mn and other nutrients involved in resistance.

Take-All of Wheat

Take-all is the most serious disease of wheat and occurs wherever it is grown. Few diseases respond as dramatically to nutrition as take-all; twelve of the fourteen principal mineral nutrients required for plant growth affect take-all either individually or collectively. Although the application of either form of N will reduce losses from take-all, NH_4^+ is most effective because it stimulates plant resistance and increases the availability of Mn, Zn, and Fe. The pathogen is a strong oxidizer of Mn in the infection court (Wilhelm, Graham, and Rovira, 1990; Pedler et al., 1996), and severity of the disease is correlated with conditions that increase NH_4^+ and Mn availability to the plant. Crop rotation is recommended; however, long-term monocropping of wheat provides a natural biological control of this disease called take-all decline. Oat is the most effective precrop for wheat because its root exudates modify the soil microflora to increase manganese availability for the subsequent wheat crop. Balanced fertility, to include P sufficiency and inhibiting nitrification, along with crop rotation, has provided effective control of take-all in most areas. Reported biological control organisms for this disease are all manganese reducers (Huber and McCay-Buis, 1993).

Eyespot and Sharp Eyespot Diseases of Wheat

Unlike take-all, eyespot caused by *Pseudocercosporella herpotrichoides* and sharp eyespot of winter wheat caused by *Rhizoctonia cerealis* are increased by spring-applied N. Top-dressing N when wheat is dormant, and cool, wet conditions are prevalent, increases severity of both eyespot and sharp eyespot by stimulating the extracellular macerating enzymes involved in pathogenesis by these pathogens. Although NH_4^+ increases severity of these diseases more than NO_3^-, the effect of N is decreased by applying all of the N in the fall as NH_4^+, with a nitrification inhibitor to prevent leaching and denitrification losses over the winter. This eliminates the need for early spring applications of N that increase virulence of these pathogens.

Fusarium Wilt

Fusarium oxysporum causes vascular wilt or yellows disease of many vegetable, fruit, fiber, and ornamental crops. This disease had threatened to eliminate commercial vegetable production under glasshouse conditions in major production areas until effective cultural controls could be combined with genetic resistance (Jones, Engelhard, and Woltz, 1989). In contrast to Verticillium wilt, diseases caused by *Fusarium oxysporum* are generally more severe in warmer, low pH soils. Ammonium increases wilt while NO_3^- decreases wilt severity. Although there are many interactions of the various nutrients with the Fusarium wilt/yellows disease, effective control has been accomplished through adequate liming and the use of a NO_3^- source of N (Jones, Engelhard, and Woltz, 1989). The beneficial effects of the lime and NO_3^- application appear to be through decreased availability of Mn and Fe created by the high soil pH. Growth and virulence of the pathogen are restricted when these micronutrients and P become limiting at high pH. The application of lignosulfonate complexes of these minor elements, which are available for plant uptake at high pH, nullified the beneficial effects of lime and Ca-nitrate applications.

Clubroot of Crucifers

Nutrient manipulation has been practiced for clubroot control for more than 200 years (Campbell and Greathead, 1989). Similar to Fusarium wilt, better control of clubroot is achieved if Ca-nitrate is used as the N source after liming to bring the soil pH above 6.8. Both high Ca and high pH reduce infection and the subsequent plant-pathogen interaction to reduce clubbing. Little is known of the actual mechanism of clubroot control by liming, probably because of the effective control obtained in most areas. The requirements of high Ca and alkaline soil pH make rotation of crucifers with potatoes inadvisable because of the potential for increased scab on the potatoes.

Damping-Off, Root and Hypocotyl Rots, and Fruit Rots

Calcium and NO_3^- are important considerations in reducing the severity of damping-off and other macerating diseases (soft rots, cortical rots, fruit rots). The source of Ca, its particle size, time of application, and soil type affect the degree of disease control achieved. The application of NO_3^- or animal manures, and certain crop sequences that enhance nitrification, improve the control of root and hypocotyl rots caused by *Fusarium solani* and *Rhizoctonia solani* (Huber and Watson, 1970). Late-season Ca sprays and infiltration of fruit with Ca solutions increase the resistance of these tissues to maceration by soft-rotting organisms during postharvest storage. The mechanism of control by Ca is considered to be increased plant tissue resistance to macerating pectolytic enzymes of the various pathogens (Kelman, McGuire, and Tzeng, 1989); however, the involvement of other soil microorganisms in disease-suppressive soils was required along with high levels of Ca for effective control of damping-off and root diseases caused by *Pythium* spp. (Ko and Kao, 1989).

Corn Stalk Rot

Stalk rot describes a symptom of one of the most destructive diseases of corn. It is generally associated with nutrient, or other environmental stress, or senescence. Many high-yielding corn hybrids are never released because of their susceptibility. Nutrient deficiency and imbalances, and drought and high temperatures that reduce the metabolic ability of the plant to use NO_3^-, predispose plants to stalk rot. Although N sufficiency throughout the growth of the plant is important to reduce stalk rot, K may buffer the effects of excess N and help reduce disease severity. A full sufficiency of N prevents degradation of functional and structural plant components as N sources for developing grain.

SUMMARY/CONCLUSIONS

Diseases are a major factor affecting the efficient use of fertilizers by reducing crop yield, quality, and aesthetic value. Although disease resistance is genetically controlled, it is mediated through physiological and biochemical processes interrelated with the nutritional status of the plant or pathogen. Nutrition of a plant may determine its resistance or susceptibility to disease, its histological or morphological structure or properties, the function of tissues to hasten or slow pathogenesis, and the apparent virulence and ability of pathogens to survive. The complex problem of mineral nutrition of plants can no longer be considered in isolation of minerals' effects on disease. Highly resistant plants are less affected by alterations in nutrition than plants tolerant to disease, and highly susceptible plants may remain susceptible to nutrient conditions that greatly increase the resistance of intermediate or tolerant plants.

Most nutrients influence disease potential more than inoculum potential, and some nutrients may decrease disease even though the population of a pathogen is increased. The intricate relationship of plant pathogens with other microorganisms, environmental factors, and the plant is dynamic and complex. Although disease cannot be totally eliminated by any specific nutrient element, the severity of most diseases can be greatly decreased by proper nutrition. Knowledge of the relationship of plant nutrition to disease provides a basis for modifying current agricultural practices to reduce disease severity in integrated crop production systems.

REFERENCES

Alten, F. and H. Orth (1941). Untersuchungen uber den Aminosäurengehalt und die Anfalligkeit der Kartoffel gegen die Kraut-und Knollenfaule (*Phytophthora infestans* de By.). *Phytopathologische Zeitschrift* 13: 243-247.

Bateman, D.F. and H.G. Basham (1976). Degradation of plant cell walls and membranes by microbial enzymes. In *Encyclopedia of Plant Physiology, Physiological Plant Pathology.* Volume 4, Eds. R. Heitefuss and P.H. Williams. New York: Springer Verlag, pp. 316-355.

Beckman, C.H. (1980). Defenses triggered by the invader: Physical defenses. In *Plant Disease, An Advanced Treatise,* Volume 5, Eds. J.G. Horsfall and E.B. Cowling. New York: Academic Press, pp. 225-246.

Bird, L.S. (1974). The dynamics of cotton seedling disease control. In *The Relation of Soil Microorganisms to Soilborne Plant Pathogens.* Virginia Agricultural Experiment Station Southern Cooperative Service Bulletin No. 183, pp. 75-79.

Bird, L.S., C. Liverman, R.G. Percy, and D.L. Bush (1979). The mechanism of multi-adversity resistance in cotton. *Proceedings of the Beltwide Cotton Producers Research Conference, Cotton Disease Council* 39: 226-228.

Brain, P. J. and W.J. Whittington (1981). The influence of soil pH on the severity of swede powdery mildew, *Erysiphe cruciferarum*, infection. *Plant Pathology* 30: 105.

Bromfield, S.M. (1979). Manganous ion oxidation at pH values below 5.0 by cell-free substances from *Streptomyces* sp. cultures. *Soil Biology and Biochemistry* 11: 115-118.

Bruck, R.I. and P.D. Manion (1980). Interacting environmental factors associated with the incidence of Hypoxylon canker on trembling aspen, *Populus tremuloides. Canadian Journal of Forest Research* 10: 17.

Burr, T.J. and A. Ceasar (1984). Beneficial plant bacteria. *CRC Critical Reviews in Plant Sciences* 2: 1-20.

Bush, D.L. (1980). Variation in Root Leachate and Rhizosphere-Rhizoplane Microflora Among Cultivars Representing Different Levels of Multi-Adversity Resistance in Cotton. Doctoral thesis, Texas A&M University, College Station, Texas.

Butler, F.C. (1961). *Root and Foot Rot Diseases of Wheat.* Science Bulletin No. 77. New South Wales Department of Agriculture, Australia.

Campbell, R.N. and A.S. Greathead (1989). Control of clubroot of crucifers by liming. In *Soilborne Plant Pathogens: Management of Diseases with Macro- and Microelements*, Ed. A.W. Engelhard. St. Paul, MN: APS Press, pp. 90-101.

Chaudhry, F.M. and J.F. Loneragan (1970). Effects of nitrogen, copper, and zinc fertilizers on the copper and zinc nutrition of wheat plants. *Australian Journal of Agricultural Research* 21: 865-879.

Choong-Hoe, K. (1986). Effect of Water Management on the Etiology and Epidemiology of Rice Blast Caused by *Pyricularia oryzae* Cov. Doctoral thesis, Louisiana State University, Baton Rouge, LA.

Cooper, K.V., R.D. Graham, and N.E. Longnecker (1988). Triticale: A cereal for manganese-deficient soils. In *Manganese in Soils and Plants: Contributed Papers*, Eds. M.J. Webb, R.O. Nable, R.D. Graham, and R.J. Hannam. Adelaide, Australia: Manganese Symposium 1988, Inc., pp. 113-115.

Csinos, A.S. and D.K. Bell (1989). Pathology and nutrition in the peanut pod rot complex. In *Soilborne Plant Pathogens: Management of Diseases with Macro- and Microelements*, Ed. A.W. Engelhard. St. Paul, MN: APS Press, pp. 124-136.

Dastur, R.H. and J.G. Bhatt (1964). Relation of potassium to Fusarium wilt of flax. *Nature* 20: 1243.

Datnoff, L.E., R.N. Raid, G. Snyder, and D.B. Jones (1991). Effect of calcium silicate on blast and brown spot intensities and yields of rice. *Plant Disease* 75: 729-732.

Easton, G.D. (1964). The results of fumigating *Verticillium* and *Rhizoctonia* infested potato soils in Washington. *American Potato Journal* 41: 296.

Engelhard, A.W., Ed. (1989). *Soilborne Plant Pathogens: Management of Diseases with Macro- and Microelements.* St. Paul, MN: APS Press.

Falcon, M.F., R.L. Fox, and E.E. Trujillo (1984). Interactions of soil pH, nutrients, and moisture on Phytophthora root rot of avocado. *Plant and Soil* 81: 165-176.

Garas, N.A. and J. Kuc (1981). Potato lectin lyses zoospores of *Phytophthora infestans* and precipitates elicitors of terpenoid accumulation produced by the fungus. *Physiological Plant Pathology* 18: 227-238.

Graham, R.D. (1983). Effect of nutrient stress on susceptibility of plants to disease with particular reference to the trace elements. *Advances in Botanical Research* 10: 221-276.

Graham, R.D. (1984). Breeding for nutritional characteristics in cereals. *Advances in Plant Nutrition* 1: 57-102.

Graham, R.D. (1988). Genotypic differences in tolerance to manganese deficiency. In *Manganese in Soils and Plants*, Eds. R.D. Graham, R.J. Hannam, and N.C. Uren. Dordrecht, Netherlands: Kluwer Academic Publishers, pp. 261-276.

Graham, R.D. and A.D. Rovira (1984). A role for manganese in the resistance of wheat plants to take-all. *Plant and Soil* 78: 441-444.

Graham, R.D. and M.J. Webb (1991). Micronutrients and disease resistance and tolerance in plants. In *Micronutrients in Agriculture*, Second Edition, Eds. J.J. Mortvedt, F.R. Cox, L.M. Shuman, and R.M. Welch. Madison, WI: Soil Science Society of America, pp. 329-370.

Graham, R.D. and R.M. Welch (1996). Breeding for staple-food crops with high micronutrient density. International Workshop on Food Policy and Agricultural Technology to Improve Diet Quality and Nutrition, Annapolis, Maryland, January 10-12, 1994. (Agricultural Strategies for Micronutrients, Working Paper No. 3. International Food Policy Research Institute, Washington, DC).

Grewal, H.S., R.D. Graham, and Z. Rengel (1996). Genotypic variation in zinc efficiency and resistance to crown rot disease (*Fusarium graminearum* Schw. Group 1) in wheat. *Plant and Soil* 186: 219-226.

Haag, H.P., E. Balmer, and A. De Carvalho (1967). Influencia da "Murcha do algodoerio" No composicao mineral do algodoeiro. *Anais Escola Superior de Agricultura, "Luiz de Queiroz," University of Sao Paulo, Brazil* 24: 333-342.

Huber, D.M. (1963). Investigations on root rot of beans caused by *Fusarium solani f. phaseoli*. *Dissertation Abstracts* 25: 17.

Huber, D.M. (1968). Effects of cultural practices on the incidence and severity of foot rot of winter wheat in North Idaho. Idaho Agricultural Experiment Station Research Progress Report No. 118.

Huber, D.M. (1978). Disturbed mineral nutrition. In *Plant Disease, An Advanced Treatise,* Volume 3, *How Plants Suffer from Disease*, Eds. J.G. Horsfall and E.B. Cowling. New York: Academic Press, pp. 163-181.

Huber, D.M. (1980). The role of mineral nutrition in defense. In *Plant Disease, An Advanced Treatise,* Volume 5, *How Plants Defend Themselves*, Eds. J.G. Horsfall and E.B. Cowling. New York: Academic Press, pp. 381-406.

Huber, D.M. (1989). Introduction. In *Soilborne Plant Pathogens: Management of Diseases with Macro- and Microelements*, Ed. A.W. Engelhard. St. Paul, MN: APS Press, pp. 1-8.

Huber, D.M. (1991). The use of fertilizers and organic amendments in the control of plant disease. In *Handbook of Pest Management in Agriculture*, Volume 1, Second Edition, Ed. D. Pimentel. Boca Raton, FL: CRC Press, pp. 405-494.

Huber, D.M. (1994). The influence of mineral nutrition on vegetable diseases. *Horticultura Brasileira* 12: 206-214.

Huber, D.M. and D.C. Arny (1985). Interactions of potassium with plant disease. In *Potassium in Agriculture*, Ed. R.D. Munson. Madison, WI: American Society of Agronomy, pp. 467-488.

Huber, D.M., H. El-Nasshar, L.W. Moore, D.E. Mathre, and J.E. Wagner (1986). Interactions of a peat carrier and potential biological control agents. *Phytopathology* 76: 1104-1105.

Huber, D.M. and R.D. Graham (1992). Techniques for studying nutrient-disease interactions. In *Methods for Research on Soilborne Phytopathogenic Fungi*, Eds. L.L. Singleton, J.D. Michail, and C.M. Rush. St. Paul, MN: APS Press, pp. 204-214.

Huber, D.M. and R.R. Keeler (1977). Alteration of wheat peptidase activity after infection with powdery mildew. *Proceedings of the American Phytopathological Society* 4: 163.

Huber, D.M. and D.N. Mburu (1984). The relationship of rhizosphere bacteria to disease tolerance, the form of N, and amelioration of take-all with manganese. In *Proceedings of the 4th International Congress of Plant Pathology, Melbourne, Australia*. St. Paul, MN: APS Press.

Huber, D. M. and T.S. McCay-Buis (1993). A multiple component analysis of the take-all disease of cereals. *Plant Disease* 77: 437-447.

Huber, D.M., T.S. McCay-Buis, C. Riegel, R.D. Graham, and N. Robinson (1993). Correlation of zinc sufficiency with resistance of wheat to Rhizoctonia winter-kill. In *6th International Congress of Plant Pathology, Montreal, Canada*, Abstract 115.

Huber, D.M., C.Y. Tsai, and J.A. Stromberger (1995). Interaction of K with N and their influence on growth and yield potential of maize. In *Proceedings of the 49th Annual Corn and Sorghum Industry Research Conference*. Washington DC: American Seed Trade Association, pp. 165-175.

Huber, D.M., H.L. Warren, and C.Y. Tsai (1986). The curse of stalk rot. *Solutions Magazine* January (1): 26-30.

Huber, D.M. and R.D. Watson (1970). Effect of organic amendment on soilborne plant pathogens. *Phytopathology* 60: 20-26.

Huber, D.M. and R.D. Watson (1974). Nitrogen form and plant disease. *Annual Review of Phytopathology* 12: 139-165.

Huber, D.M., R.D. Watson, and G.W. Steiner (1965). Crop residues, nitrogen, and plant disease. *Soil Science* 100: 302-308.

Huber, D.M. and N.S. Wilhelm (1988). The role of manganese in resistance to plant diseases. In *Manganese in Soils and Plants*, Eds. R.D. Graham, R.J. Hannam, and N.C. Uren. Dordrecht, Netherlands: Kluwer Academic Publishers, pp. 155-173.

Hutchinson, W.G. (1935). Resistance of *Pinus sylvesteri* to a gall-forming *Peridermium*. *Phytopathology* 25: 819.

Jones, J.P., A.W. Engelhard, and S.S. Woltz (1989). Management of Fusarium wilt of vegetables and ornamentals by macro- and microelement nutrition. In *Soilborne Plant Pathogens: Management of Diseases with Macro- and Microelements*, Ed. A.W. Engelhard. St. Paul, MN: APS Press, pp. 18-32.

Keinath, A.P. and R. Loria (1989). Management of common scab of potato with plant nutrients. In *Soilborne Plant Pathogens: Management of Diseases with Macro- and Microelements*, Ed. A.W. Engelhard. St. Paul, MN: APS Press, pp. 152-166.

Kelman, A., R.G. McGuire, and K-C. Tzeng (1989). Reducing the severity of bacterial soft rot by increasing the concentration of calcium in potato tubers. In *Soilborne Plant Pathogens: Management of Diseases with Macro- and Microelements*. Ed. A.W. Engelhard. St. Paul, MN: APS Press, pp. 102-123.

Kiraly, Z. (1976). Plant disease resistance as influenced by biochemical effects of nutrients and fertilizers. In *Fertilizer Use and Plant Health*. Proceedings of Colloquium 12. Atlanta, GA: International Potash Institute, pp. 33-46.

Klein, E.K. (1957). Uber den Einfluss der Mineral salzernährung auf den Gehalt des Blattes an freien Aminosauren und Monosacchariden und seine Bedeutung für die Empfanglichkeit der Pflanze gegenuber parasitaren Pilzen. *Soils and Fertility* 20: 303.

Kliejunas, J.T. and W.H. Ko (1975). Continuous versus limited growth of fungi. *Mycologia* 67: 362-366.

Ko, W-H. and C-W.C. Kao (1989). Evidence for the role of calcium in reducing root disease incited by *Pythium* spp. In *Soilborne Plant Pathogens: Management of Diseases with Macro- and Microelements*, Ed. A.W. Engelhard. St. Paul, MN: APS Press, pp. 205-217.

Kunoh, H., K. Kuno, and H. Ishizaki (1985). Cytological studies of the early stages of powdery mildew of barley, *Hordeum vulgare*, and wheat. XI. Autofluorescence and haloes at penetration sites of appressoria of *Erysiphe graminis hordei* and *Erysiphe pisi* on barley coleoptiles. *Canadian Journal of Botany* 63: 1535-1539.

Last, F.T. (1962). Effects of nutrition on the incidence of barley powdery mildew. *Plant Pathology* 11: 133-135.

Leong, S.A. (1986). Siderophores: Their biochemistry and possible role in the biocontrol of plant pathogens. *Annual Review of Phytopathology* 24: 187-209.

Marschner, H. (1995). *Mineral Nutrition of Higher Plants*, Second Edition. London, UK: Academic Press.

McGregor, A.J. and G.C.S. Wilson (1964). The effect of applications of manganese sulfate to a neutral soil on the yield of tubers and the incidence of common scab in potatoes. *Plant and Soil* 20: 59-64.

McGregor, A.J. and G.C.S. Wilson (1966). The influence of manganese on the development of potato scab. *Plant and Soil* 25: 3-16.

McNew, G.L. and E.L. Spencer (1939). Effect of nitrogen supply of sweet corn on the wilt bacterium. *Phytopathology* 29: 1051.

Mortvedt, J.J., K.C. Berger, and H.M. Darling (1963). Effect of manganese and copper on the growth of *Streptomyces scabies* and the incidence of potato scab. *American Potato Journal* 40: 96-102.

Nelson, D.W. (1963). The Relationship Between Soil Fertility and the Incidence of Diplodia Stalk Rot and Northern Leaf Blight in *Zea mays*. Master of science thesis, University of Illinois, Urbana, IL.

Pearson, C.J. and B.C. Jacobs (1986). Elongation and retarded growth of rice during short-term submergence at three stages of development. *Field Crops Research* 13: 331-344.

Pedler, J.F. (1994). Resistance to Take-all Disease by Manganese-Efficient Wheat Cultivars. Doctoral thesis, University of Adelaide, Adelaide, Australia.

Pedler, J.F., M.J. Webb, S.C. Buchhorn, and R.D. Graham (1996). Manganese oxidizing ability of isolates of the take-all fungus correlates to virulence. *Biology and Fertility of Soils* 22: 272-278.

Pennypacker, B.W. (1989). The role of mineral nutrition in the control of Verticillium wilt. In *Soilborne Plant Pathogens: Management of Diseases with Macro- and Microelements*, Ed. A.W. Engelhard. St. Paul, MN: APS Press, pp. 33-45.

Reis, E.M., R.J. Cook, and B.L. McNeal (1982). Effect of mineral nutrition on take-all of wheat. *Phytopathology* 72: 224-229.

Reis, E.M., R.J. Cook, and B.L. McNeal (1983). Elevated pH and associated reduced trace-nutrient availability as factors contributing to take-all of wheat upon soil liming. *Phytopathology* 73: 411-413.

Rengel, Z. and R.D. Graham (1995). Importance of seed Zn content for wheat growth on Zn-deficient soil. II. Grain yield. *Plant and Soil* 173: 267-274.

Rengel, Z., J.F. Pedler, and R.D. Graham (1994). Control of Mn status in plants and rhizosphere: Genetic aspects of host and pathogen effects in the wheat take-all interaction. In *Biochemistry of Metal Micronutrients in the Rhizosphere*, Eds. J.A. Manthey, D.E. Crowley, and D.G. Luster. Boca Raton, FL: Lewis Publishers/CRC Press, pp. 125-145.

Rice, E.L. (1984). *Allelopathy*. Orlando, FL: Academic Press.

Rogers, P.F. (1969). Organic manuring for potato scab control and its relation to soil manganese. *Annals of Applied Biology* 63: 371-378.

Rothrock, C.S. and B.M. Cunfer (1991). Influence of small grain rotations on take-all in a subsequent wheat crop. *Plant Disease* 75: 1050-1052.

Savant, N.K., G.H. Snyder, and L.E. Datnoff (1997). Silicon management and sustainable rice production. *Advances in Agronomy* 58: 151-199.

Schulze, D.G., D.M. Huber, A.K.M. Shahjahan, M. Levy, S. Bajt, and B. Illman (1996). Manganese oxidation: A characteristic of susceptibility in the *Pyricularia*—rice disease (rice blast) system. Annual Report of the National Synchrotron Light Source, Brookhaven National Laboratory, Upton, NY.

Scott, D.B. and K. Wilken (1995). Biological control of root diseases in wheat by means of seed treatment with manganese-reducing bacteria. South Africa Agricultural Research Council SGI Technology Report, 36 pp.

Sharvelle, E.G. (1936). The nature of resistance of flax to *Melampsora lini*. *Journal of Agricultural Research* 53: 81-127.

Skou, P. (1981). Morphology and cytology of the infection process. In *Biology and Control of Take-All*, Eds. M.J.C. Asher and P.J. Shipton. London, UK: Academic Press, pp. 175-198.

Smiley, R.W. (1975). Forms of nitrogen and the pH in the root zone, and their importance to root infections. In *Biology and Control of Soilborne Plant Pathogens*, Ed. G.W. Bruehl. St. Paul, MN: APS Press, pp. 55-62.

Smiley, R.W. and R.J. Cook (1973). Relationship between take-all of wheat and rhizosphere pH in soils fertilized with ammonium and nitrate nitrogen. *Phytopathology* 63: 822-890.

Sparrow, D.H. and R.D. Graham (1988). Susceptibility of zinc-deficient wheat plants to colonization by *Fusarium graminearum* Schw. Group 1. *Plant and Soil* 112: 261-266.

Stetter, S. (1971). Influence of artificial fertilizers on *Ophiobolus graminis* and *Cercosporella herpotrichoides* in continuous cereal growing. *Tidsskrift for Planteavl* 75: 274-277.

Sutton, A.L., D.M. Huber, B.C. Joern, and D.D. Jones (1995). Management of nitrogen in swine manure to enhance crop production and minimize pollution. In *Proceedings of the 7th International Symposium on Agricultural Food Processing Wastes*. Chicago, IL: American Society of Agricultural Engineers, pp. 532-540.

Thongbai, P., R.D. Graham, S.M. Neate, and M.J. Webb (1993). Interaction between zinc nutritional status of cereals and *Rhizoctonia* root rot. II. Effect of zinc on disease severity of wheat under controlled conditions. *Plant and Soil* 153: 215-222.

Timonin, M.E. (1965). Interaction of higher plants and soil microorganisms. In *Microbiology and Soil Fertility*, Eds. C.M. Gilmore and O.N. Allen. Corvallis, OR: Oregon State University Press, pp. 135-138.

Tsai, C.Y., D.M. Huber, H.L. Warren, and C.L. Tsai (1986). Sink regulation of source activity by nitrogen utilization. In *Regulation of Carbon and Nitrogen Reduction and Utilization in Maize*, Eds. J.C. Shannon, C.D. Boyer, and D.P. Knievel. Rockville, MD: American Society of Plant Physiology, pp. 247-259.

Wagner, F. (1940). Die Bedeutung der Kieselsäure für das Wachstum einiger Kulturpflanzen, ihren Nahrstoffhaushalt und ihre Anfalligkeit gegen echte Mehltaupilze. *Phytopathologische Zeitschrift* 12: 427.

Warren, H.L., D.M. Huber, C.Y. Tsai, and D.W. Nelson (1980). Effect of nitrapyrin and N fertilizer on yield and mineral composition of corn. *Agronomy Journal* 72: 729-732.

Wilhelm, N.S. (1991). Investigations into *Gaeumannomyces graminis* var *tritici* Infection of Manganese-Deficient Wheat. Doctoral thesis, University of Adelaide, Adelaide, Australia.

Wilhelm, N.S., R.D. Graham, and A.D. Rovira (1988). Application of different sources of manganese sulphate decreases take-all (*Gaeumannomyces graminis* var *tritici*) of wheat grown in manganese deficient soil. *Australian Journal of Agricultural Research* 39: 1-10.

Wilhelm, N.S., R.D. Graham, and A.D. Rovira (1990). Control of Mn status and infection rate by genotype of both host and pathogen in the wheat take-all interaction. *Plant and Soil* 123: 267-275.

Chapter 8

Importance of Seed Mineral Nutrient Reserves in Crop Growth and Development

Ross M. Welch

Key Words: macronutrients, micronutrients, phloem mobility, seedling establishment, seed quality, seed viability, seed vigor.

INTRODUCTION

Seed vigor and viability are important components influencing seedling establishment, crop growth, and productivity (Association of Official Seed Analysis, 1983; Welch, 1986; TeKrony and Egli, 1991; McDonald and Copeland, 1997). Any factor (biotic and/or environmental) that negatively affects seed vigor and viability during seed development will have adverse consequences for crop production, especially when the seeds are sown under environmentally stressful conditions (Fenner, 1992; Welch, 1995). Because significant amounts of seed nutrient reserves can be acquired from vegetative tissues, both size and number of seeds produced by maternal plants are most likely determined by their nutritional status at the time of flowering and bud initiation. Additionally, the timing of nutrient supplies to the maternal plant is critical to seed size, with earlier applications of nutrients having greater effects than later applications. Furthermore, the most important single determinant of mineral nutrient reserves in seeds is the mineral nutrient availability to the maternal plant during reproductive development, with increasing supplies of a particular mineral nutrient enhancing the nutrient concentration in the mature seed (Fenner, 1992).

Many environmental stresses (e.g., water deficits, temperature extremes, pathogens, nutrient deficiencies, mineral toxicities, salinity, soil acidity, anaerobiosis, etc.) that directly affect the growth and nourishment of the maternal plant can indirectly influence seed development, seed nutrient reserves, and ultimately seed quality (including vigor and viability) (Welch, 1986).

Soil-nutrient availability and adequacy during maternal plant development are important factors that can have dramatic effects on seed viability and vigor. Unfortunately, little is known about the importance of most mineral nutrient seed reserves on seed quality (i.e., vigor and viability) and the seed's ability to resist various soil-imposed stresses during germination and seedling establishment. This chapter focuses on the importance of mineral nutrient reserves in seeds, as components of seed vigor and seed viability that can affect seedling establishment and, thereby, crop productivity, especially when the seeds are sown to soils of low fertility. This chapter is not intended to encompass all the literature on this subject, but strives to present examples that demonstrate the importance of seed nutrient reserves on resultant crop performance. For earlier reviews concerning seed vigor and viability, refer to TeKrony and Egli (1991) and Welch (1986). The topics of seed coatings and seed treatments and their effects on seedling establishment, although important, are not presented here. Readers are referred to the review by Scott (1989) for a discussion of these topics.

GENERAL CONSIDERATIONS

An earlier review documented the general effects of environmental stresses on maternal plant reproductive development and on seed vigor and viability (Welch, 1986). Little research has been conducted concerning this topic since that earlier review was published. The following is a brief summary of the conclusions reached in that previous review.

All environmental factors and their interactions that influence maternal plant growth can potentially influence the complicated and dynamic processes that control seed initiation, development, and seed nutrient reserves. Therefore, any one or any combination of these factors can modify the ultimate vigor and viability of seeds. Generally, the effects of mineral nutrient deficiency stress on seed development are indirect, resulting from their direct affects on maternal plant development (Welch, 1986) and subsequent seed quality. This is especially true for the phloem-mobile macronutrients (i.e., N [nitrogen], K [potassium], S [sulfur], P [phosphorus], and Mg [magnesium]). However, for the nutrients with variable (see Table 8.1) or conditional phloem mobilities, this generalization is not always true (Welch, 1986). When nutrient deficiencies affect maternal plant development, thereby decreasing seed vigor, the potential for survival of subsequent generations is impaired. A corollary to this generalization is that the greater the

TABLE 8.1. Categories of Phloem Mobility of Mineral Nutrient Elements

Mobile	Variable mobility	Conditional mobility
N	Fe	Ca
K	Zn	B
P	Cu	Mn
S	Mo	
Mg	Ni	
Cl	Co	

Source: Adapted from Welch (1986).

nutrient reserves (i.e., starch, protein, oil, and mineral nutrients) in a seed, the greater its vigor, and therefore the greater its potential for survival when grown under environmentally stressful conditions. Further, any stored nutrient in a seed that controls seedling development rate, under any set of circumstances, is a potential factor in seedling survival and crop performance. All factors affecting nutrient reserves in seeds potentially can change seed vigor, viability, seedling establishment, and seedling growth rate.

Some nutrient deficiencies, when incurred by maternal plants during reproductive growth, not only affect nutrient stores in developing seeds but also other important chemical constituents of seeds. Thus, nutrient deficiencies can thereby indirectly affect seed vigor and viability through their effects on important seed constituents. These include adverse nutrient deficiency effects on seed coat structures, seed filling rates, and seed hormone concentrations (Gray and Thomas, 1982). Maternal plant nutrient deficiencies of most phloem-mobile macronutrients (e.g., N, K, S, and Mg) primarily affect seed production by reducing the number of seeds per plant (Austin, 1972). A summary of the general effects of nutrient deficiencies on seed development and seed vigor gleaned from the literature (Austin, 1972; Pollock and Roos, 1972; Gray and Thomas, 1982; Welch, 1986) follows:

- Macronutrient deficiencies (e.g., N, K, P, S, Mg) predominantly affect the number of seeds produced, while the effects on nutrient reserves are minimal unless the deficiency is severe.
- Seeds with low nutrient reserves are disadvantaged unless sown to soils with adequate available nutrients for seedling establishment and growth after germination.

- A deficient supply of one nutrient to a maternal plant affects other nutrient stores in the developing seed; therefore, attributing effects on seedling performance to the maternal plant's deficient nutrient at a later date is difficult.
- Maternal plant nutrient deficiency effects on seed vigor, seed viability, and seedling performance are variable, depending on environmental conditions existing during seedling development and maturation of the progeny.
- Limited water transpiration rates and low phloem mobility of certain nutrients (e.g., Ca [calcium], Mn [manganese], and B [boron]) are important factors affecting seed nutrient stores and seed development.

As mentioned previously, evidence is accumulating that nutrient reserves (both organic and inorganic) within seeds affect the ability of seedlings to withstand various environmental stresses. For example, seeds containing low reserves of Cu (copper), Mn, and Zn (zinc) have decreased resistance to root pathogens, water stress, cold temperatures, and soil acidity conditions. Recent examples of these generalizations will be discussed subsequently for selected nutrients in various plant species. For earlier literature on this subject, see Welch (1986).

IMPORTANCE OF PHLOEM TRANSPORT AND TRANSFER CELLS

The movement of mineral nutrients from maternal plant source tissues (mature leaves, stems, and roots) into reproductive sink tissues adjoining the seed coat or testa is primarily controlled by their movement within vascular elements comprised of phloem vessels and not by their movement within xylem vascular elements into these tissues. This condition exists because in most plants, no mature xylem vessels extend completely into the fruiting body containing the developing seed tissues (Welch, 1986; 1995). Additionally, most developing seed tissues transpire relatively little water because of the humid microenvironment inside the enclosed seed pod walls, thus precluding significant xylem transport of mineral nutrients into tissues surrounding developing seeds. Consequently, the control of nutrient element transport to developing seed tissues is primarily dependent on those processes that control source-tissue phloem loading, phloem sap transport from source tissues to reproductive tissues, unloading of phloem sap nutrients within reproductive organs, and xylem-phloem mineral exchange processes. Once mineral nutrients are unloaded from phloem sap into apoplasmic spaces adjoining the seed coat, they must be absorbed by seed coat cells and moved within the symplasm of cells associated with the seed coat. These cells then deliver the nutrients to all cells developing within the maturing ovule, many of which are characteristically transfer cell types with extensive cell wall ingrowths and greatly increased plasma membrane surface areas (for a more complete discussion on this topic see Welch, 1986).

Unfortunately, very little is known about the mechanisms that control seed-loading processes for mineral nutrients. Recently, Laszlo (1994) studied the changes in transport of Ca and K in soybean fruit [*Glycine max* (L.) Merr.] during seed development. He reported that data obtained from Sr^{2+} (used as an analog for Ca^{2+} influx) transport into the ovule from the fruit suggested that the pod and seed coat impose a transport barrier that could account for the relative decline in embryo Ca^{2+} influx during seed development. The rubidium ion, Rb^{+}, (used as an analog for K^{+}) uptake into the embryo increased with seed fresh weight, in contrast to Sr^{2+} uptake, which did not. He summarized that soybean seed coats displayed mineral accumulation patterns (i.e., kinetics) that were cultivar and mineral specific, but that were different from the embryo accumulation patterns, suggesting that the seed coat is involved in regulating mineral uptake by the embryo.

Because of the importance of phloem transport in mineral nutrient accumulation in developing seeds, the following discussion has been separated into phloem-mobile, variable phloem-mobile, or conditionally phloem-mobile nutrient elements.

PHLOEM-MOBILE NUTRIENT ELEMENTS

Most of the research performed to determine the effects of phloem-mobile macronutrient elements (i.e., N, P, K, Mg, and S) on seed vigor and viability has focused on N, K, and P, with little attention given to the other phloem-mobile macronutrients. Examples of some typical findings are discussed next.

Nitrogen, Potassium, and Phosphorus

Increasing the supply of N to maternal plants has been reported to increase seed yield primarily by increasing the number of reproductive structures on the maternal plant rather than by increasing the number of seeds within each fruit. Usually, mean seed weight is not greatly influenced by N supply to the maternal plant (Welch, 1986). Some research has demonstrated that increasing the supply of N to maternal plants can affect seed vigor and viability, seedling growth, and final crop yields. For example, Schweizer and Ries (1969) reported that the amount of an unknown endogenous protein, or of a proteinaceous moiety, may be an important factor in subsequent yield of major agronomic crops, and that this moiety was increased in seeds developed from maternal plants receiving increased soil N supplies, as compared to controls not receiving N treatments. They came to this conclusion from their results showing that oat (*Avena sativa* L.) grain with a higher protein content (from maternal plants previously grown on N-enriched soil) resulted in 21 to 42 percent greater grain production. Further, wheat grain, containing higher protein levels, obtained from various locations within the United States, developed into larger seedlings. Ries (1971) reported that increasing the N levels to maternal bean (*Phaseolus vulgaris* L.) plants from 0 to 100 kg ha^{-1} resulted in a positive correlation between N supplied and seed weights and seed protein content. Greenhouse studies were performed on seeds differing in weight and protein content. Results showed that both seed size and seed protein content were positively correlated to the size and protein content of the seedlings produced from the seeds. Further research with these seeds led Ries to conclude that seedling size was more highly correlated to seed protein content than to seed size and that the N status of the maternal plant affected the growth and yield of the progeny. Ries also suggested that this may not be simply a relationship between N supply and seed protein content alone, but possibly, an unknown seed factor is increased by increasing N supply to the maternal plant, which improves seed yields and seed vigor.

Sawan, Maddah El Din, and Gregg (1989) reported that increasing the supply of N, from 108 to 216 kg N/ha, to maternal cotton (*Gossypium barbadense* L.) plants resulted in increased seed viability (i.e., germination velocity, second germination count, and total germination capacity) and seedling vigor (i.e., length of hypocotyl, radicle, and entire seedling, and in seedling fresh and dry weights). Additionally, Saraswathy and Dharmalingam (1992) studied the effects of maternal plant nutrition on seed quality of mustard (*Brassica juncea* L.) plants grown on a sandy loam under irrigation. Seed vigor index (calculated from percent germination and total seedling length) and the protein content of the seeds were significantly increased in those seeds obtained from maternal plants receiving the high N and K fertilization compared to controls. Other reports have confirmed that various forms of N fertilizer can promote seed vigor in various crops (see references in Welch, 1986). To my knowledge, even today, the identity of these seed vigor-stimulating nitrogenous factors are unknown.

In 1982, Gray and Thomas speculated that these factors may be phytohormones, such as cytokinins, but evidence is still lacking.

Potassium nutrition of maternal plants can also affect seed vigor and the ability of the progeny to resist K-deficient soil conditions. Ozanne and Asher (1965) reported that withholding K from small-seeded plant species grown on a K-deficient sandy soil delayed seedling emergence and reduced the proportion of seedlings that eventually emerged. Further study revealed a fairly close relationship between the maximum depth of root penetration and the seed K content. They concluded that seeds of some small seeded species may contain insufficient K to allow them to emerge from a K-poor soil.

Harrington (1960) presented results of research with pepper plants (*Capsicum frutescens* L.) grown under K-deficient conditions. Here, premature sprouting of seeds in the pepper fruits resulted if the K deficiency of the maternal plant occurred after a period of K adequacy, indicating that K may play a role in the biosynthesis of germination inhibitors. Pepper seeds produced on the maternal K-deficient plants deteriorated more rapidly than did seeds from K-adequate pepper plants. Harrington concluded that K deficiency during maternal plant reproductive development reduces seed longevity and, therefore, seed vigor.

The effect of K fertilizers on various physiological parameters and yield of cotton grown on a K-deficient soil was studied by Sun and colleagues (1989). They reported that increasing K supply to maternal cotton plants increased 1000-seed weight, seed viability and germination rate, crude fat content, and percentage of linolenic acid in the total fatty acids.

Although culturing maternal plants on media inadequate in P can produce seeds with reduced vigor, the degree to which lower seed vigor affects progeny growth can depend on the postgermination conditions. For example, Austin (1966) reported that watercress [*Rorippa nasturtium aquaticum* (L.) Hayek] grown in sand culture and supplied deficient levels of P produced fewer, but larger, seeds. The mineral composition of the seeds was also altered, with increasing P supply reducing the concentration of N in seed, but increasing that of K and P. There was no effect of P supply on seed germination, but if the seeds were propagated on P-deficient media, the size of the progeny produced was directly related to the P supplied to the maternal plant. However, if P-enriched seeds were grown on P-adequate media, there was no effect of P supply to the maternal plants on subsequent seedling growth. Therefore, increased P stores in seeds can have advantageous effects on seedling establishment if the seeds are propagated on soils with inadequate P availability.

The effects of varying P reserves (e.g., 0.21, 0.26, and 0.43 percent P) in uniformly sized seeds of lupin (*Lupinus angustifolius* L.) on early shoot, root, and nodule growth, and the growth response of the seedlings to increased P supply, have been studied (Thomson, Bell, and Bolland, 1992). Four days after germination, seedling fresh weights were depressed by low seed P stores when compared to those seeds with medium or high P stores. This effect of low seed P reserves on whole plant fresh weight persisted to final harvest at day thirty-two even if seedlings were supplied adequate or luxury levels of P in nutrient solutions during development. However, by day six after germination, seedlings propagated from low-P seeds recovered relative growth rates to match those of seedlings produced from medium- or high-P seeds when the P supplied to the seedlings was adequate. Low-P seeds strongly depressed nodule number and mass at all levels of P supplied to seedlings in nutrient solution. Apparently, seed P may be specifically involved in nodule development at an early stage of nodule formation in lupin roots. Similar results were also found for lupin seeds varying in P stores that were grown in both a P-deficient loam soil and a P-adequate sandy soil. Further, they suggested that low P concentrations in lupin seeds may limit successful seedling establishment in the field, especially if the seeds are sown in a soil with inadequate supplies of available P or when the seedlings are subjected to early soil stresses.

Root mycorrhizal (*Glomus intraradices* Schenck & Smith) infection of maternal wild oat (*Avena fatua* L.) plants has been reported to enhance seed and seedling vigor (Lu and Koide, 1991; Koide and Lu, 1992). Apparently, mycorrhizal root infection of oats resulted in increased P stores in the seeds developed from those plants. Additionally, the progeny showed improved early leaf expansion rates and greater leaf areas, shoot and root nutrient content, and root/shoot ratios compared to progeny from seeds of nonmycorrhizal maternal plants. Mycorrhizal infection of the maternal oats increased P concentrations and the number of seeds per spikelet produced by the progeny, but significantly reduced the weight of the seeds produced. The authors suggested that mycorrhizal infection influenced plant vitality by increasing progeny vigor and reproductive success in addition to increased maternal fecundity.

The effects of P supply to white mustard (*Sinapis alba* L.) and velvetleaf (*Abutilon theophrasti* Medic.) on seed vigor and progeny development has been reported (Lewis and Koide, 1990). Seed weight, seed N content, and seed P were influenced by maternal mycorrhizal infection and by P supply. However, these factors did not significantly affect seedling emergence rates of the progeny. Maternal P supply did not greatly change growth or root/shoot ratios within the progeny in either plant species.

Fungal infection of velvetleaf maternal plant roots significantly influenced progeny vigor as measured by shoot weight, root weight, height and leaf area. About 25 percent of the variation in these variables could be accounted for by total maternal seed P content. Mycorrhizal infection may have affected progeny vigor via its effect on changing the P fractions in seeds produced from the maternal plants. The authors concluded that mycorrhizal infection may influence progeny fitness through its effect on progeny vigor.

Sulfur and Magnesium

To my knowledge, no new information is available concerning the effects of S and Mg on seed vigor and viability since the last review of this subject (Welch, 1986). The following discussion is a brief synopsis of the findings from the previous review. Because S supply to maternal plants affects the accumulation of seed storage reserves (e.g., protein, oil, and Mo [molybdenum]), it has the potential to affect seed vigor and seedling establishment. Magnesium supply can affect seed vigor also. For example, in experiments with barley (*Hordeum vulgare* L.), Mg-depleted seedlings (germinated in sand) had inadequate Mg supply from grain of maternal plants and contained less chlorophyll in their leaves than seedlings from Mg-sufficient control grain. Possibly, plants require a continuous supply of adequate amounts of Mg for normal growth and maximum seed vigor.

Summary of General Effects of Phloem-Mobile Nutrients

- Maternal plants deficient in N, K, or P can rapidly remobilize these nutrients from storage tissues to developing seeds.
- When grown under moderate deficiency stress of N, K, or P, plants will minimize the number of seeds produced, thereby allowing the development of the highest seed vigor possible under prevailing environmental conditions.
- Seed development is influenced only when severe nutrient deficiencies of N, K, or P are encountered and after their stores in other plant organs are exhausted.
- Both S and Mg are released slowly from storage pools in vegetative organs under certain circumstances, leading to limitations of phloem supply of these nutrients to developing seeds. Thus, excessive stores of these nutrients cannot guarantee adequate nutrition for developing seeds, and a continuous supply of these nutrients to the maternal plant must be maintained to ensure maximum seed vigor and seed yield.

VARIABLE PHLOEM-MOBILE NUTRIENT ELEMENTS

For nutrients with variable phloem mobility (see Table 8.1), the consequences of maternal plant deficiencies of these elements on seed development, vigor, and viability are complex. The efficiency of remobilization of these nutrients via the phloem from storage sites in vegetative organs to sinks in developing reproductive organs is an important factor in determining ultimate seed vigor and seedling performance. This is especially true when the maternal plants are supplied inadequate levels of these nutrients. Unfortunately, the degree of phloem mobility of these nutrient elements is highly variable between plant species, and even between genotypes within the same species. Additionally, their mobility is dependent not only on genetic variability but also on environmental conditions during development and on plant growth stage. Under certain circumstances, these nutrients can be highly mobile in the phloem, but under others, they appear to be immobile.

Copper, Iron, and Zinc

Some examples of differences in phloem mobility for Cu, Fe (iron), and Zn were discussed in a previous review (Welch, 1986). Copper accumulation in seeds is strongly influenced by Cu and N supply to the maternal plant, with increasing supplies of these nutrients increasing the amount of Cu remobilized from vegetative organs to the developing seed. Apparently, protein catabolites are responsible for the movement of Cu from vegetative phloem sources into phloem sap for transport to phloem sinks within rapidly developing seed tissues.

Iron has been reported to be loaded into developing ovules of peas (*Pisum sativum* L.) via the phloem in an unknown Fe-chelate complex (Grusak, 1994). Control of the biosynthesis of this unknown Fe-chelator may be an important factor in regulating Fe movement in the phloem to reproductive organs. Certain nonprotein amino acids, for example, the phytometallophores such as nicotianamine, may be the chelating molecules responsible for the phloem transport of Fe into reproductive organs (Welch, 1995). These types of physiological studies should be expanded to include other micronutrients and other plant species to ascertain the pathway and regulatory mechanisms of phloem transport to, and entry of these essential nutrients into, developing seeds and grains.

Zinc mobility in plants is also highly dependent on the adequacy of Zn available to plants from their growth media. When Zn is provided in excess of that required for optimum yield, it can be readily transported out of vegetative tissues and into reproductive tissues and seeds (Welch, House, and Allaway, 1974; Welch, 1986). However, when Zn supply is inadequate, little is transported out of vegetative tissues and into developing seeds. This is true even when maternal plants suffer from severe Zn deficiency. Why this is so is not known, but it may be the result of Zn binding tightly to Zn-containing essential metabolites in vegetative tissues, such as in Zn-activated enzymes (e.g., carbonic anhydrase) or to essential Zn-binding sites at the surface of cellular membranes (Welch, 1995).

The effect of increasing grain Zn reserves in wheat subsequently grown on Zn-deficient soils in Turkey has been studied by Yilmaz and colleagues (1997). They reported that increasing the Zn content of bread wheat grain from an average of 355 ng to 1,465 ng Zn per grain resulted in increased yields of progeny if no Zn fertilizer was applied to the Zn-deficient soil, especially under rain-fed conditions. However, even the highest grain Zn levels could not overcome Zn deficiency without Zn applications to the soil.

Rengel and Graham (1995) studied the effects of grain Zn content on early wheat seedling establishment on soils with low Zn fertility. They reported that, for both the Zn-efficient genotype Excalibur and the Zn-inefficient genotype Gatcher, increasing the seed Zn content from an average of 250 ng Zn to 700 ng Zn per grain resulted in greater root and shoot seedling growth three weeks after germination on a Zn-deficient soil. After six weeks of growth, greater root and shoot growth of the progeny from high-Zn grain, compared to those from low-Zn grain, was apparent for those seedlings not supplied additional Zn in the soil. Fertilization with 0.2 mg Zn per kg soil was required to achieve 90 percent optimum yield for plants derived from high-Zn grain, but 0.8 mg Zn per kg soil was required for maximum yield in plants grown from low-Zn grain. They concluded that higher grain Zn content acted similarly to a "starter" fertilizer by improving early vegetative growth and by dissipating differences between wheat genotypes that differ in Zn efficiency.

Nickel

Nickel (Ni), another variable phloem-mobile element, has been reported to be an essential nutrient for all higher plants (Brown, Welch, and Cary, 1987; Brown et al., 1987; Brown, Welch, and Madison, 1990). Interestingly, Ni has been shown to perform an essential function during seed germination. The effect of Ni supply to maternal barley plants, grown in purified nutrient solutions, on grain vigor and viability is shown in Table 8.2. As depicted, without adequate Ni supply, maternal barley plants developed nonviable grain that did not germinate upon imbibition (Brown, Welch, and Cary, 1987). Figure 8.1 illustrates the relationship between maternal barley plant Ni supply and the percent germination of grain produced. As shown, the critical level of grain Ni required for optimum germination of barley appears to be about 100 ng Ni per g dry weight of grain.

TABLE 8.2. Effect of Ni Supplied to Maternal Barley Plant on Vigor and Viability of the Grain, Estimated by a Triphenyl-Tetrazolium-Chloride (TTC) Staining Method As Percent of All Grain

Viability and vigor	Ni supplied to plant during growth	
	0 μM %	1.0 μM %
Not viable	23 ± 5[a]	0[b]
Low vigor	70 ± 12[c]	40 ± 15[d]
High vigor	7 ± 7[e]	60 ± 2[f]

Source: Adapted from Brown (1988).

Note: Means ± SD; means followed by different letters differ significantly ($P \leq 0.05$).

The exact metabolic role of Ni in seed germination is not known. Possibly, it may be related to the utilization of N stored in compounds in the seed that require the cycling of N through urea during their catabolism upon germination because Ni is an essential component of the enzyme urease (Welch, 1981). The importance of Ni in cereal seed viability suggests that more research should be performed to delineate its function(s) in seed viability and to extend the findings beyond cereals to include other agriculturally important species.

Molybdenum

The total seed content of micronutrient metals with variable phloem mobility, such as Mo, can have significant effects on seed vigor and viability, especially when sown to micronutrient-poor soils. For example, in some large legume seeds, the normal seed Mo content can supply ten times the amount of Mo required by plants to reach maturity (Welch, 1986). Further, Gurley and Giddens (1969) reported that increasing the Mo content in soybean seeds through Mo fertilization of maternal plants supplied the Mo requirements of the progeny for their entire life cycle. Brodrick, Sakala, and Giller (1992) studied the effects of Mo seed reserves on various bean (*Phaseolus vulgaris* L.) genotypes, varying in seed size and grown in nutrient solutions. They reported that neither high (1.6 μg Mo per seed) nor low (0.1 μg Mo per seed) seed reserves in a small-seeded genotype (BAT 1297) could prevent Mo deficiency (i.e., reduced shoot,

FIGURE 8.1. Effect of Changing Ni Concentrations in the Grain of Barley on Grain Viability (Percent Germination)

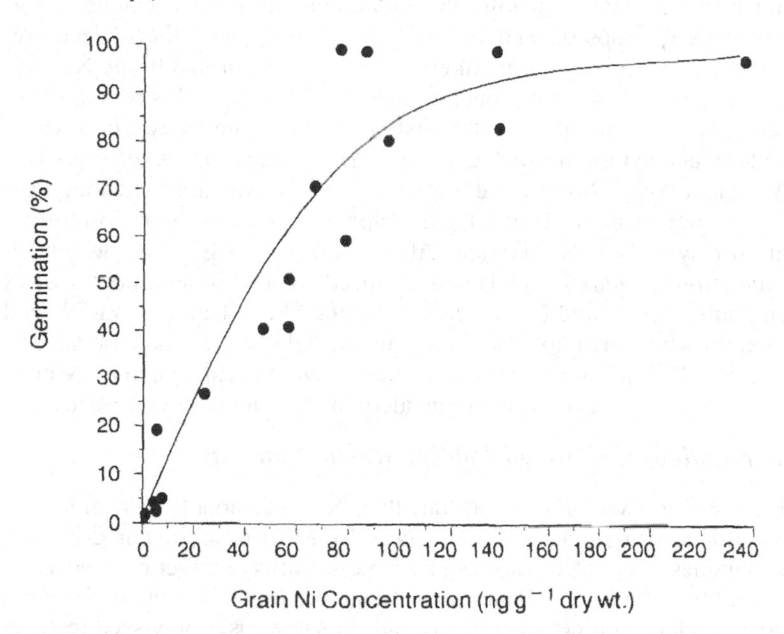

Source: Adapted from Brown (1988).

Note: Maternal plants were supplied 0, 0.6, or 1.0 μM Ni in nutrient solution during growth to produce grain with varying Ni concentrations.

root, and nodule dry weight, N_2 fixation, and seed production) if Mo was not supplied adequately. However, in one large-seeded genotype (Canadian Wonder), both low- (1.4 μg Mo per seed) and high-Mo (3.6 μg Mo per seed) seed stores provided enough Mo to prevent Mo deficiency, even if no Mo was supplied in the growth media. Producing soybean seeds from maternal plants derived from low-Mo seed and not supplied with additional Mo in their growth media reduced seed yields by as much as 38 percent. Both small and large seed soybean genotypes having low-Mo seed reserves (i.e., <1.4 μg Mo per seed) produced plants with both lower nodule weight and lower N content, which ultimately resulted in these plants producing less seed. Furthermore, the viability of low-Mo seeds compared to high-Mo seeds was lower, with percent germination reduced by as much as 50 percent in the low-Mo seeds.

Cobalt

In Co (cobalt)-deficient soils, the cobalt content of seeds can be important to legume crops dependent on N_2 fixation to meet their N requirements because Co is an essential element required for symbiotic N_2 fixation by nodules of legume roots (Marschner, 1995). For this reason, there is a close relationship between the Co content of legume seeds, crop yield, N content, and symptoms of N deficiency in legume crops when grown on Co-deficient soils (Robson and Snowball, 1987). Apparently, enough Co can be stored in large-seeded legume species to meet their Co requirements for symbiotic N_2 fixation. About 200 ng Co g^{-1} dry weight in *Lupinus angustifolius* L. seeds was required for maximum shoot growth when planted in N- and Co-deficient soil, and about 100 ng Co g^{-1} seed dry weight when sown to a Co-deficient soil (Gladstones, Loneragan, and Goodchild, 1977). Thus, legume seed quality when propagated on N-deficient, Co-poor soils is dependent on adequate Co stores in their seeds.

Generalizations for Variable Phloem-Mobile Nutrients

The previous examples demonstrate that, because variable phloem-mobile micronutrients, for example, Cu, Zn, Ni, Mo, and Co, are required in such low quantities, their seed reserves can have significant effect on seed vigor and viability. Adequate seed stores can delay the onset of acute deficiencies in crops produced on deficient media and, in some cases, low seed reserves of these nutrients can significantly decrease final crop yield even if these nutrients are provided to the plants at adequate levels after germination.

In summary, severe deficiencies of variable phloem-mobile nutrients can result in complete failure of a plant to develop seeds. Thus, these elements must be provided at adequate levels to ensure maximum seed yield and vigor. If plants, suffering from moderate deficiencies of these elements, cannot remobilize enough of these elements from storage sites in vegetative tissues to reproductive organs during development, then seed yield will suffer and seed vigor and viability will diminish. In contrast to phloem-mobile nutrients, the mobility of variable phloem-mobile nutrients is highest at superoptimal supplies and lowest at deficient supplies. Maximizing seed stores of these nutrients will ensure optimum seed vigor and viability. This can be accomplished by providing these nutrients at levels in excess of those required for optimum vegetative yields, especially during reproductive growth.

CONDITIONALLY PHLOEM-MOBILE NUTRIENT ELEMENTS

Nutrients with conditional phloem mobilities include Ca, B, and Mn, although the evidence for the mobility of Mn in the phloem is conflicting

(Welch, 1986). Apparently, Mn phloem mobility is related to Mn status of the plant and can differ between plant species or even within the same species. Manganese is included in this section because plants that do not mobilize enough Mn via the phloem to reproductive tissues develop seed disorders characteristic of those produced by Ca and B deficiencies.

Importance of Xylem Transport

The xylem sap rather than the phloem sap is the primary source of conditionally phloem-mobile nutrient elements entering developing fruit. Because of this, a complex interaction exists between the movement of these elements to reproductive organs and the water relations of plants. Rapidly growing reproductive organs act as if they are temporary water reserves, filling with water during nocturnal periods of low transpirational water loss, followed by depletion of water during periods of high transpirational water loss during the day. Because of this diurnal change in water flux in the xylem, there are long periods during the day when water either ceases to move via the xylem to developing fruits, or water is lost from the organ via the xylem to other tissues. This generates problems of insufficient supply of conditionally phloem-mobile nutrients to reproductive organs because they can only enter the organ in sufficient quantities during periods of low transpirational water loss. This can lead to nutritional disorders in rapidly developing tissues within the reproductive organ, without affecting the growth of other vegetative organs such as leaves and roots. If plant genotypes that do not efficiently remobilize Ca, B, and Mn in their xylem sap, become deficient in these nutrients, especially during reproductive development, they will develop damaged seeds with low vigor and viability. If the deficiency is severe enough, supplying adequate levels of these nutrients in the growth media at germination will not prevent low vigor in the seedlings produced from the damaged seed. Therefore, it is imperative that these nutrients be provided at adequate levels continuously during growth to ensure maximum seed vigor and viability. Possibly, these nutrients should be applied foliarly to sensitive crops during reproductive growth to ensure that adequate amounts of these nutrients are available to seeds during this critical period of development.

Calcium

Adams and Hartzog (1991) studied the effects of gypsum ($CaSO_4$) applications and soil Ca levels on seed quality in runner peanuts (*Arachis hypogea* L.). They reported significant effects of gypsum treatment on Ca concentrations in the seeds, with three out of eight sites producing higher Ca concentrations in the seeds from maternal plants receiving the gypsum applications. Additionally, at one field site, a highly significant increase in seed germination and seedling vigor resulted from gypsum applications to the soil, and at two other sites, gypsum application caused a significant increase in seedling survival, without simultaneous increase in germination. The minimum Ca concentration in seed needed for maximum germination was 282 mg Ca kg^{-1}, while the minimum Ca concentration in seed for maximum seedling survival was 260 mg kg^{-1}. This suggests that a higher level of available soil Ca is required for maximum seed quality when compared to the maximum Ca required for optimum yield or for the development of sound mature kernels.

Keiser and Mullen (1997) studied the effects of Ca supply (in nutrient solution) and relative humidity (RH) on soybean (*Glycine max* L.) seed Ca concentrations and, subsequently, seed quality. Changes in relative humidity did not significantly affect seed Ca or seed germination, but increasing Ca supply from 0.6 to 2.5 mM Ca significantly increased Ca in the seeds from about 1.7 to 2.4 mg Ca g^{-1} dry weight. They suggested that RH did not affect seed Ca nor seed germination because Ca may have been significantly remobilized from other tissues to seeds via the phloem, or an adequate supply of Ca may have been delivered from the xylem via root pressure flow. There was a decrease in germination from 97 to 42 percent, with decreasing seed Ca concentrations from 2.4 to 0.9 mg g^{-1}, respectively. Decreased germination was correlated with an increase in the percentage of abnormal seedlings. Therefore, if conditions exist that change the balance of Ca and other nutrients in the growth media or limit seed Ca concentrations below a critical level (2.0 mg Ca g^{-1} seed for achieving 90 percent germination), seed quality can be adversely affected.

Manganese

Field trials were performed by Longnecker, Graham, and Marcar (1988) on the effects of low- and high-Mn grain concentrations on barley (*Hordeum vulgare* L.) growth. Plots sown with grain containing high Mn levels produced higher plant densities than those plots sown with grain containing lower Mn concentrations. Furthermore, grain yield was significantly affected by seed Mn level in two different years and for five different barley genotypes.

Seeds of narrow-leafed lupin (*Lupinus angustifolius* L.) have low seedling establishment percentages when grown on Mn-poor soils. Crosbie and colleagues (1993) and Longnecker and colleagues (1996) studied the effects of varying seed Mn concentrations on narrow-leafed lupin emergence.

Three different seed lots differing in seed Mn concentrations (i.e., 7, 15, and 35 mg Mn kg^{-1} seed dry weight) were used. There were no visible differences in seed quality or size among seed lots. Emergence of seedlings varied between 35 to 52 percent in pots sown with low-Mn seeds, while, in comparison, emergence from high-Mn seeds varied between 88 to 100 percent. The detrimental effect of low seed Mn concentrations on seedling emergence could not be reversed by applying Mn to the soil. All (100 percent) of the high-Mn seeds were viable, while the percentage was only 66 percent for the low-Mn seeds. Furthermore, plants that emerged from seeds containing 7 mg Mn kg^{-1} seed had lower shoot dry weight compared to seeds containing higher Mn concentrations. They concluded that inadequate Mn supply during seed-filling periods of narrow-leafed lupin reproductive growth can result in the production of nonviable seeds that cannot be visually distinguished from viable seeds.

Foliar application of Mn (25 mg Mn l^{-1} as MnEDTA) to cotton (*Gossypium barbadense* L.) grown on Mn-poor soils increased seed yield, seed weight, seed viability (germination), and seedling vigor (length of hypocotyl, radicle, and entire seedling, and seedling fresh and dry weight) (Sawan, Maddah El Din, and Gregg, 1993).

The effect of seed Mn content on take-all [root, crown, and foot rot disease of cereals caused by the soilborne fungus *Gaeumannomyces graminis* (Sacc.)] was studied by McCay-Buis and colleagues (1995). Grain of five genotypes of soft red winter wheat (*Triticum aestivum* L.), differing in Mn levels from low (38 mg Mn kg^{-1} seed) to high (68 mg Mn kg^{-1} seed), were sown to fields at three locations in Indiana representing different soil types, management practices, and environmental conditions. Soils at all sites were known to have moderate to severe take-all disease problems as the primary root disease. Under these conditions, plants from high-Mn seeds were generally more vigorous, averaging 11 percent less take-all (white heads), and yielding a mean of 165 kg ha^{-1} more grain when compared to plants produced from low-Mn seeds. They concluded that the limited ability of roots to remobilize Mn basipetally has a serious detrimental effect for both root development and plant disease resistance. Higher levels of Mn in seeds may overcome this limitation and constitute a means for disease avoidance through enhanced early root growth, or through stimulation of more root lignin and phenolic compound biosynthesis as a direct disease defense mechanism.

Moussavi-Nik and colleagues (1997) studied the influence of grain Mn concentrations on early growth and tiller number of six wheat (*Triticum aestivum* L. and *Triticum durum* Desf.) genotypes grown in a growth chamber in pots containing a Mn-deficient calcareous sand. Genotypes grown from high-Mn grain produced more tillers and had greater shoot weight in comparison to low-Mn grain, even if adequate Mn was supplied as a fertilizer during growth. Therefore, high seed Mn concentration is more efficient at increasing shoot and root production in wheat grown on Mn-poor soils than is applying Mn fertilizer to the soil.

Boron

Boron deficiency in maternal plants results in seeds that are defective and with characteristic "hollow-heart," or knurled, hollow, off-colored areas on cotyledons. Severe B deficiency can result in poorly developed, off-colored (yellow or tan), and pointed plumules. A large proportion of cells in B-deficient seeds are empty and collapsed. Apparently, B deficiency primarily affects the development of cotyledons (Welch, 1986). Since B plays a critical role in pollen tube growth as well as seed and fruit set, an unusually high demand for B exists during seed and grain development when compared to B demand during vegetative growth (Marschner, 1995).

Seed produced from marginally B-deficient maternal plants can result in lower seed viability, even if maternal plant yields are not depressed. In black gram (*Vigna mungo* L.), B concentration of 6 mg kg^{-1} dry weight of seed has been reported as a critical level for normal seed viability (Bell et al., 1989). Thus, a continuous supply of adequate levels of B must be maintained within the maternal plant during reproductive growth if maximum seed vigor and viability are to be achieved. Because B has conditional phloem mobility, any environmental condition that reduces the flux of B via xylem into reproductive tissues during seed development would greatly decrease seed vigor and viability.

Generalizations for Mineral Elements with Conditional Phloem Mobility

The effects of deficiencies of conditionally phloem-mobile nutrients (Ca, Mn, and B) on seed production, vigor, and viability are summarized here:

- Deficient plants produce seeds with reduced vigor.
- Germination of seeds produced from maternal plants deficient in these nutrients will be inferior, and resulting seedlings will be abnormal and will grow poorly, even if these nutrients are supplied in sufficient amounts during the germination.
- Permanent damage to the embryo usually occurs during seed development if the maternal plant is deficient in these nutrients, which may account for the resulting low vigor of the seeds produced.

- If the deficiency is severe, seeds will not be produced by the maternal plant because rapidly growing reproductive organs are affected first.

GENERAL CONCLUSIONS

Clearly, inadequate seed stores of any mineral nutrient can have adverse effects on seed vigor and viability, and on seedling performance, when the seeds are germinated on nutrient-poor soils. Maternal plants deficient in nutrients with limited phloem mobility during reproductive development will produce seeds irreversibly damaged, even though the vegetative growth of the maternal plant is not visually affected. Supplying adequate nutrients to seedlings produced from these damaged seeds after germination will not overcome the negative effects on subsequent progeny growth. Adequate supplies of these nutrients must be continuously supplied to maternal plants, possibly in superoptimal amounts, to ensure optimum seed vigor and viability.

Unfortunately, since the last general review of this subject was published over a decade ago (Welch, 1986), little new knowledge has been forthcoming concerning the regulation of mineral nutrient movement to, and deposit in, developing seeds and grains. Without question, more basic research is needed to expand our understanding of these processes and the roles that mineral nutrients play in seed vigor and viability, especially the parts played by micronutrients. Most of the information available is empirical, obtained without a clear understanding of the physiology and biochemistry of the processes involved. Without this basic knowledge, it is impossible to take meaningful advantage of modern molecular biological techniques to genetically modify plants in ways that will improve seed vigor and viability, and that could lead to improved productivity and crop quality. Lastly, this type of information is greatly needed to develop crops with enhanced nutritional and health-promoting qualities, goals that are becoming increasingly important to the agriculture, human nutrition, and health communities globally (Welch, Combs Jr., and Duxbury, 1997).

REFERENCES

Adams, J.F. and D.L. Hartzog (1991). Seed quality of runner peanuts as affected by gypsum and soil calcium. *Journal of Plant Nutrition* 14: 841-851.

Association of Official Seed Analysis (1983). Seed vigor testing. In *Handbook 32*. Springfield, IL: Association of Official Seed Analysis.

Austin, R.B. (1966). The growth of watercress [*Rorippa nasturtium aquaticum* (L.) Hayek] from seed as affected by the phosphorus nutrition of the parent plant. *Plant and Soil* 24: 113-120.

Austin, R.B. (1972). Effects of environment before harvesting on viability. In *Viability of Seeds*, Ed. E.H. Roberts. London, UK: Chapman and Hall, pp. 114-149.

Bell, R.W., L. McLay, D. Plaskett, B. Dell, and J.F. Loneragan (1989). Germination and vigor of black gram [*Vigna mungo* (L.) Hepper] seed from plants grown with and without boron. *Australian Journal of Agricultural Research* 40: 273-279.

Brodrick, S.J., M.K. Sakala, and K.E. Giller (1992). Molybdenum reserves of seed, and growth and N_2 fixation by *Phaseolus vulgaris* L. *Biology and Fertility of Soils* 13: 39-44.

Brown, P.H. (1988). Nickel: An Essential Element for Higher Plants. Doctoral thesis, Cornell University, Ithaca, New York.

Brown, P.H., R.M. Welch, and E.E. Cary (1987). Nickel: A micronutrient essential for all higher plants. *Plant Physiology* 84: 801-803.

Brown, P.H., R.M. Welch, E.E. Cary, and R.T. Checkai (1987). Beneficial effects of nickel on plant growth. *Journal of Plant Nutrition* 10: 2125-2135.

Brown, P.H., R.M. Welch, and J.T. Madison (1990). Effect of nickel deficiency on soluble anion, amino acid, and nitrogen levels in barley. *Plant and Soil* 125: 19-27.

Crosbie, J., N. Longnecker, F. Davies, and A. Robson (1993). Effects of seed manganese concentration on lupin emergence. In *Plant Nutrition—From Genetic Engineering to Field Practice*, Ed. N.J. Barrow. Dordrecht, Netherlands: Kluwer Academic Publishers, pp. 665-668.

Fenner, M. (1992). Environmental influences on seed size and composition. *Horticultural Reviews* 13: 183-213.

Gladstones, J.S., J.F. Loneragan, and N.A. Goodchild (1977). Field responses to cobalt and molybdenum by different legume species, with inferences on the role of cobalt in legume growth. *Australian Journal of Agricultural Research* 28: 619-628.

Gray, D. and T.H. Thomas (1982). Seed germination and seedling emergence as influenced by the position of development of the seed on, and chemical application to, the parent plant. In *The Physiology and Biochemistry of Seed Development, Dormancy, and Germination*, Ed. A.A. Khan. New York: Elsevier, pp. 81-110.

Grusak, M.A. (1994). Iron transport to developing ovules of *Pisum sativum* L. Seed import characteristics and phloem iron-loading capacity of source regions. *Plant Physiology* 104: 649-655.

Gurley, W.H. and J. Giddens (1969). Factors affecting uptake, yield response, and carryover of molybdenum in soybean seed. *Agronomy Journal* 61: 7-9.

Harrington, J.F. (1960). Germination of seeds from carrot, lettuce, and pepper plants grown under severe nutrient deficiencies. *Hilgardia* 30: 219-235.

Keiser, J.R. and R.E. Mullen (1997). Calcium and relative humidity effects on soybean seed nutrition and seed quality. *Crop Science* 33: 1345-1349.

Koide, R.T. and X. Lu (1992). Mycorrhizal infection of wild oats: Maternal effects on offspring growth and reproduction. *Oecologia* 90: 218-226.

Laszlo, J.A. (1994). Changes in soybean fruit Ca^{2+} (Sr^{2+}) and K^+ (Rb^+) transport ability during development. *Plant Physiology* 104: 937-944.

Lewis, J.D. and R.T. Koide (1990). Phosphorus supply, mycorrhizal infection and plant offspring vigor. *Functional Ecology* 4: 695-702.

Longnecker, N., J. Crosbie, F. Davies, and A. Robson (1996). Low seed manganese concentration and decreased emergence of *Lupinus angustifolius*. *Crop Science* 36: 355-361.

Longnecker, N.E., R.D. Graham, and N.E. Marcar (1988). The effect of seed manganese on barley (*Hordeum vulgare* L.) growth and yield. In *International Symposium on Manganese in Soils and Plants: Contributed Papers*, Eds. M.J. Webb, R.O. Nable, R.D. Graham, R.J. Hannam. Adelaide, Australia: Manganese Symposium 1988, Inc., pp. 47-49.

Lu, X. and T. Koide (1991). *Avena fatua* L. seed and seedling nutrient dynamics as influenced by mycorrhizal infection of the maternal generation. *Plant, Cell and Environment* 14: 931-938.

Marschner, H. (1995). *Mineral Nutrition of Higher Plants*, Second Edition. London, UK: Academic Press.

McCay-Buis, T.S., D.M. Huber, R.D. Graham, J.D. Phillips, and K.E. Miskin (1995). Manganese seed content and take-all of cereals. *Journal of Plant Nutrition* 18: 1711-1721.

McDonald, M.B. and L. Copeland (1997). *Seed Production Principles and Practices*. New York: Chapman and Hall.

Moussavi-Nik, M., Z. Rengel, G.J. Hollamby, and J.S. Ascher (1997). Seed manganese (Mn) content is more important than Mn fertilization for wheat growth under Mn deficient conditions. In *Plant Nutrition—For Sustainable Food Production and Environment*, Eds. T. Ando, K. Fujita, T. Mae, H. Matsumoto, S. Mori, and J. Sekiya. Dordrecht, Netherlands: Kluwer Academic Publishers, pp. 267-268.

Ozanne, P.G. and C.J. Asher (1965). The effect of seed potassium on emergence and root development of seedlings in potassium-deficient sand. *Australian Journal of Agricultural Research* 16: 773-784.

Pollock, B.M. and E.E. Roos (1972). Seed and seedling vigor. In *Seed Biology*. Volume 1. *Importance, Development and Germination*, Ed. T.T. Kozlowski. New York: Academic Press, pp. 313-387.

Rengel, Z. and R.D. Graham (1995). Importance of seed Zn content for wheat growth on Zn-deficient soil. I. Vegetative growth. *Plant and Soil* 173: 259-266.

Ries, S.K. (1971). The relationship of protein content and size of bean seed with growth and yield. *Journal of the American Society for Horticultural Science* 96: 557-560.

Robson, A.D. and K. Snowball (1987). Response of narrow-leafed lupins to cobalt application in relation to Co concentration in seed. *Australian Journal of Experimental Agriculture* 27: 657-660.

Saraswathy, S. and C. Dharmalingam (1992). Mother crop nutrition influencing seed quality of mustard (*Brassica juncea*) grown in the western tract of Tamil Nadu. *Seed Research* 20: 88-91.

Sawan, Z.M., M.S. Maddah El Din, and B.R. Gregg (1989). Effect of nitrogen fertilisation and foliar application of calcium and micro-elements on cotton seed yield, viability and seedling vigor. *Seed Science and Technology* 17: 421-431.

Sawan, Z.M., M.S. Maddah El Din, and B.R. Gregg (1993). Cotton seed yield, viability and seedling vigor as affected by plant density, growth retardants, copper and manganese. *Seed Science and Technolology* 21: 417-431.

Schweizer, C.J. and S.K. Ries (1969). Protein content of seed: Increase improves growth and yield. *Science* 165: 73-75.

Scott, J.M. (1989). Seed coatings and treatments and their effects on plant establishment. *Advances in Agronomy* 42: 43-83.

Sun, X., L.H. Rao, Y.S. Zhang, Q.H. Ying, C.X. Tang, and L.X. Qin (1989). Effect of potassium fertilizer application on physiological parameters and yield of cotton grown on a potassium deficient soil. *Zeitschrift für Pflanzenernährung und Bodenkunde* 152: 269-272.

TeKrony, D.M. and D.B. Egli (1991). Relationship of seed vigor to crop yield: A review. *Crop Science* 31: 816-822.

Thomson, B.D., R.W. Bell, and M.D.A. Bolland (1992). Low seed phosphorus concentration depresses early growth and nodulation of narrow-leafed lupin (*Lupinus angustifolius* cv. Gungurru). *Journal of Plant Nutrition* 15: 1193-1214.

Welch, R.M. (1981). The biological significance of nickel. *Journal of Plant Nutrition* 3: 345-356.

Welch, R.M. (1986). Effects of nutrient deficiencies on seed production and quality. *Advances in Plant Nutrition* 2: 205-247.

Welch, R.M. (1995). Micronutrient nutrition of plants. *CRC Critical Reviews in Plant Science* 14: 49-82.

Welch, R.M., G.F. Combs Jr., and J.M. Duxbury (1997). Toward a "Greener" revolution. *Issues of Science and Technology* 14: 50-58.

Welch, R.M., W.A. House, and W.H. Allaway (1974). Availability of zinc from pea seeds to rats. *Journal of Nutrition* 104: 733-740.

Yilmaz, A., H. Ekiz, I. Gülekin, B. Torun, S. Karanlik, and I. Cakmak (1997). Effect of seed zinc content on grain yield and zinc concentration of wheat grown in zinc-deficient calcareous soils. In *Plant Nutrition for Sustainable Food Production and Environment*. Eds. T. Ando, K. Fujita, T. Mae, H. Matsumoto, S. Mori, and J. Sekiya. Dordrecht, Netherlands: Kluwer Academic Publishers, pp. 283-284.

Chapter 9

Physiological Mechanisms Underlying Differential Nutrient Efficiency of Crop Genotypes

Zdenko Rengel

Key Words: copper, cultivar, genotypic difference, germplasm, iron, manganese, nutrient availability, nutrient uptake, phosphorus, rhizosphere, root growth, zinc.

INTRODUCTION

Nutrient deficiencies in plants are widespread throughout agricultural regions of the world and can result in reduced growth or declining yields of many crop species (Welch et al., 1991; Cakmak, Sari, Marschner, Kalayci, et al., 1996; Cakmak, Yilmag, et al., 1996; Pearson and Rengel, 1997). However, some plant species and genotypes have a capacity to grow and yield well on soils of low fertility (e.g., Graham, Ascher, and Hynes, 1992; Haynes, 1992; Graham and Rengel, 1993; Gourley, Allan, and Russelle, 1994; Cakmak, Sari, Marschner, Kalayci, et al., 1996; Cakmak, Yilmaz, et al., 1996; Cakmak, Derici, et al., 1997; Cakmak, Ekiz, et al., 1997; Pearson and Rengel, 1997); these species and genotypes are considered tolerant to nutrient deficiency (= nutrient efficient). Efficient genotypes grow and yield well on nutrient-deficient soils by employing specific physiological mechanisms that allow them to gain access to sufficient quantities of nutrients (uptake efficiency) and/or to more effectively utilize nutrients taken up (utilization efficiency) (Sattelmacher, Horst, and Becker, 1994).

The subject of plant responses to Zn (zinc), Fe (iron), Mn (manganese), and P (phosphorus) deficiency has been covered to various extents in a number of recent reviews (Batten, 1992; Blair, 1993; Graham and Rengel, 1993; Marschner and Dell, 1994; Rengel, Pedler, and Graham, 1994; Randall, 1995; Welch, 1995; Jolley et al., 1996; Pearson and Rengel, 1997; Crowley and Rengel, 1999; Cianzio, 1999). Therefore, this chapter will only emphasize genotypic differences in physiological mechanisms behind a differential capacity of crop genotypes to take up and utilize Zn, Fe, Mn, and P.

NUTRIENT EFFICIENCY

Considerable work has been undertaken to elucidate genotypic differences in uptake and utilization efficiency of a number of nutrients (N [nitrogen], P, K [potassium], Fe, Zn, Mn, and Cu [copper]). In the case of N, NO_3^- may be present in the soil solution in the plant-available form; an increased capacity of genotypes to capture NO_3^- may prevent or minimize its leaching losses. In contrast, the other nutrients listed are usually present in relatively large total amounts in soil, but the available fraction and the concentration in the soil solution may be insufficient to satisfy plant needs. In such a case, desirable characteristics of efficient genotypes would be a capacity to increase an available nutrient fraction and take up relatively larger amounts of a nutrient (uptake efficiency), even though nutrient utilization efficiency may also be important.

Uptake efficiency may consist of increased capacity to solubilize nonavailable nutrient forms into plant-available ones and/or increased capacity to transport nutrients across the plasma membrane. However, it appears likely that the increased capacity to convert nonavailable into available nutrient forms is of greater importance for efficient uptake, especially for nutrients that are transported to roots by diffusion, which is a relatively slow process that frequently limits nutrient uptake. For diffusion-supplied nutrients, larger amounts are transported toward roots if a larger concentration gradient between the root surface and the bulk soil can be maintained by vigorous nutrient uptake at the root surface. However, when a capacity of root cells to take up nutrients is considerably greater than the rate of nutrient replenishment at the root surface, the uptake rate will be governed by the nutrient supply rather than by the nutrient uptake capacity of the plant (Rengel, 1993). Therefore, an increased capacity of root cells to take up nutrients, due to an increased expression of high-affinity nutrient uptake systems in the plasma membranes under nutrient stress (a greater number of transporters, a greater nutrient affinity, or a greater rate of turnover of transporters), is expected to be of secondary importance as an efficiency mechanism for diffusion-supplied nutrients (e.g., P, K, Zn, Mn, and Cu).

A mechanistic explanation of differential efficiency among genotypes of crop plants is still lacking for most nutrients studied. It may, however, be assumed that (1) efficiency mechanisms vary among crop species and genotypes, (2) more than one mechanism is often responsible for the level of nutrient efficiency in a particular genotype, and (3) increased efficiency of one genotype in comparison to the other is due to involvement of additional mechanisms not present (expressed) in the less-efficient genotype. An additional complexity stems from the mechanisms of nutrient efficiency likely operating at various levels of plant organization (molecular, physiological, structural, or developmental). It is expected that efficiency mechanisms (or regulation of their expression) will serve as a basis for developing a well-targeted screening procedure for use in breeding programs aimed at producing genotypes of superior nutrient efficiency for cropping soils of low nutrient availability in a more sustainable way.

Elucidation of mechanisms of nutrient efficiency is hampered by inability to distinguish between causes and effects of nutrient deficiency. The time between withdrawal of a nutrient (strictly speaking, that would be the time of imposition of deficiency, while in reality, plants do not become deficient immediately, but only when internal nutrient stores required for normal functioning of the biochemical processes have been exhausted) and the time when measurable changes can be observed is relatively long (hours at best for macronutrients, and days most often for micronutrients). This is in contrast with ion toxicity, for which the time elapsed between imposition of the stress (an addition of the toxic concentration of an ion) and measurable response may be only seconds (e.g., membrane-bound Ca [calcium] is replaced by Al [aluminum] instantaneously; Rengel, 1992). Some preliminary results have shown that membrane leakiness, measurable after 24 h (hours), may be an early sign of Zn deficiency (Cakmak and Marschner, 1988a), but more recent research indicates that leakage of K after 24 h of Zn deficiency is quite variable and, therefore, longer times are required for reliable measurement of membrane leakiness and thus the extent of the Zn-deficiency stress (Cakmak, 1995).

In contrast to ion toxicity studies (e.g., see Delhaize and Ryan, 1995; Rengel, 1997a), few reports have been published so far that tested differential nutrient efficiency of isogenic lines, that is, lines with a uniform genetic background (more than 95 percent similarity), differing practically only in the trait of nutrient efficiency under examination. If genotypes are not isogenic, they differ in a number of traits that may interactively bring about the overall apparent difference in the efficiency. In such a situation, it is difficult to decipher physiological mechanisms specifically associated with increased efficiency of acquisition and/or utilization of a particular nutrient. A comparison between mutants and the wild-type parental genotypes offer advantages similar to those arising from using isogenic lines. Mutants were used in studying features of P (a corn mutant *rootless*; Sattelmacher, Horst, and Becker, 1994) and Fe efficiency (e.g., several Fe-hyperaccumulating mutants of pea [Grusak, Welch, and Kochian, 1990; Welch and La Rue, 1990; Grusak and Pezeshgi, 1996; Romera et al., 1996] and tomato [*chloronerva*; Stephan and Grün, 1989; Pich et al., 1997] as well as Fe-inefficient mutants [*fefe* muskmelon; Welkie, 1996; and *ys1* corn; Von Wiren, Marschner, and Römheld, 1995]). In addition, in Cu deficiency research, wheat genotypes containing a long arm of the rye chromosome 5 were compared with the ones without that translocation (Graham et al., 1987).

MECHANISMS OF ZINC EFFICIENCY

Zinc-efficient plants appear to employ a plethora of physiological mechanisms that allow them to withstand the Zn deficiency stress better than the Zn-inefficient plants. These Zn efficiency mechanisms include:

1. a greater proportion and longer length of fine roots with diameters ≤ 0.2 mm (Dong, Rengel, and Graham, 1995; Rengel and Wheal, 1997a);
2. differential changes in the rhizosphere chemistry and biology, including the release of greater amounts of Zn-chelating phytosiderophores (Cakmak et al., 1994; Walte et al., 1994; Cakmak, Sari, Marschner, Ekiz, et al., 1996; Cakmak, Sari, Marschner, Kalayci, et al., 1996; Cakmak, Yilmaz et al., 1996; Rengel, Römheld, and Marschner, 1998);
3. an increased maximum uptake rate (I_{max}) resulting in increased net Zn accumulation (Cakmak, Sari, Marschner, Kalayci, et al., 1996; Rengel and Graham, 1996; Rengel and Wheal, 1997b; Rengel, Römheld, and Marschner, 1998); and
4. more efficient utilization and compartmentalization of Zn within cells, tissues, and organs (Graham and Rengel, 1993), including a greater activity of carbonic anhydrase (Rengel, 1995a) and antioxidative enzymes (e.g., superoxide dismutase; Cakmak, Öztürk, et al., 1997), maintaining sulfhydryl groups in the root-cell plasma membranes in a reduced state (Rengel, 1995b), and a differential pattern of biosynthesis of the root-cell plasma membrane polypeptides (Rengel and Hawkesford, 1997).

The previous list of Zn efficiency mechanisms is fairly long, mainly because of inherent problems in trying to distinguish a cause and the effects of Zn efficiency, as discussed earlier. Some mechanisms of Zn efficiency (such as differential root geometry) may operate only in soil environments, while others (differential uptake kinetics, compartmentalization, transport, retranslocation and utilization of Zn, differences in production of Zn-mobilizing phytosiderophores, etc.) may be operational in both soil and nutrient solution environments. In addition, the Zn-P and Zn-Fe interactions may be expressed to a different extent in various genotypes and thus may variously influence the level of Zn efficiency.

The Zn-efficient wheat genotypes had greater fertilizer efficiency and a greater harvest index compared to the Zn-inefficient genotypes when grown in soils of low Zn availability (Rengel and Graham, 1995d). The greater fertilizer efficiency of the Zn-efficient wheat genotypes compared to the Zn-inefficient ones offers some promise in managing severely Zn-deficient soils by a combination of (1) growing Zn-efficient genotypes and (2) Zn fertilization at low rates. Given that Zn deficiency is typically patchy in a single field (Kubota and Allaway, 1972), growing Zn-efficient genotypes can overcome spatial (horizontal and down the profile; see also Nable and Webb, 1993) as well as temporal variation in Zn availability to plants.

Breeding for greater Zn efficiency, at least in wheat, should be possible by transferring rye genes controlling the Zn efficiency trait. Individual genes have not been deciphered yet, but transferring rye chromosomes 1R and 7R into wheat has increased its Zn efficiency (Cakmak, Derici, et al., 1997).

Root Growth

The root morphology varies among genotypes as well as among species (Itoh and Barber, 1983; Barber and Silberbush, 1984; Fitter, 1991; Schwarz, Léon, and Geisler, 1991). Different root development may not only affect nutrient uptake but also plays an important role in differential Zn efficiency among various genotypes (Dong, Rengel, and Graham, 1995; Rengel and Wheal, 1997a).

Root growth in a number of plant species remains unchanged or slightly increases under Zn deficiency (Chapin, 1988; Cakmak and Marschner, 1988a; Rengel, 1995a, b; Rengel and Graham, 1995a, 1996; Cakmak, Öztürk, et al., 1996; Cakmak, Sari, Marschner, Kalayci, et al., 1996; Cakmak, Yilmaz, et al., 1996; Rengel and Hawkesford, 1997). However, such a response also occurs when other nutrients are limited (Haynes, Koide, and Elliott, 1991) and should be considered a general adaptive mechanism to deficiency stresses rather than a primary mechanism of Zn efficiency.

The Zn-efficient wheat genotypes developed longer and thinner roots (a greater proportion of fine roots with diameter ≤ 0.2 mm) than the Zn-inefficient ones when grown in the Zn-deficient soils (Dong, Rengel, and Graham, 1995; Rengel and Wheal, 1997a), thus resulting in exploration of a larger volume of soil. As Zn ions are transported toward roots by diffusion (Warncke and Barber, 1972), an increased root surface area is particularly important because it reduces the distance Zn ions have to travel in soil solution to the root absorption sites (Marschner, 1993), hence allowing more efficient scavenging of the small amounts of Zn ions.

Mycorrhizae can increase plant capacity to take up Zn from soils low in available Zn. The myccorrhizal roots of green gram (*Vigna radiata*) (Sharma and Srivastava, 1991), corn (Faber et al., 1990; Kothari, Marschner, and Römheld, 1991; Sharma et al., 1992), and pigeon pea (*Cajanus cajan*) (Wellings, Wearing, and Thompson, 1991) have higher tissue Zn concentrations than in nonmycorrhizal plants. However, in the case of wheat grown on Zn-deficient soils of the Australian cereal belt, it has been shown that (1) a natural level of the mycorrhizal inoculum is low (see Graham and Rengel, 1993), (2) mycorrhizal infection is nil when soil is fertilized with P (Baon et al., 1992), and (3) mycorrhizae do not seem to be involved in the expression of differential Zn efficiency of wheat genotypes (Pearson and Rengel, 1997). An apparently negligible role of mycorrhizae in Zn uptake by wheat plants grown on Zn-deficient soils further stresses the importance of developing longer and thinner roots as traits associated with the Zn-efficient wheat genotypes during the early stages of growth.

In lowland rice, bicarbonate is the major factor inducing Zn deficiency (Yang, Römheld, and Marschner, 1993). It inhibits root growth in the Zn-inefficient rice genotype, probably because of accumulation of organic acids; these bicarbonate-related effects do not occur in the Zn-efficient rice genotype (Yang, Römheld, and Marschner, 1994).

Chemistry and Biology of Rhizosphere

Exudation of Phytosiderophores

The differences in nutrient efficiency may be due to quantitative and/or qualitative differences in root exudation among species/genotypes. When subjected to Zn deficiency, Zn-efficient genotypes release greater amounts of phytosiderophores (PS) (Cakmak et al., 1994; Walter et al., 1994; Cakmak, Sari, Marschner, Ekiz, et al., 1996; Cakmak, Sari, Marschner, Kalayci, et al., 1996; Cakmak, Yilmaz, et al., 1996; Rengel, Römheld, and Marschner, 1998) and take up more Zn than Zn-inefficient genotypes (Rengel, Römheld, and Marschner, 1998).

Recent research has shown that the positive correlation between PS exudation and Zn efficiency of wheat genotypes may not always hold. An imperfect correlation was obtained between the amount of PS released under Zn deficiency by plants of seven wheat genotypes and the ranking of these genotypes for Zn efficiency (Erenoglu et al., 1996). Such a result indicates the complexity of the syndrome of Zn efficiency in wheat. Further work is necessary before a definite conclusion about the importance of PS release in the response to Zn deficiency of various wheat genotypes can be put forward.

Rhizosphere Microflora

Preliminary results show that wheat genotypes differing in Zn efficiency may differentially influence the microbial populations in the rhizosphere (Rengel et al., 1996; Rengel, 1997b; Rengel, Ross, and Hirsch, 1998). Zinc deficiency increased numbers of fluorescent pseudomonads in the rhizospheres of all wheat genotypes tested so far, but the effect was particularly obvious for Zn-efficient genotypes (Rengel et al., 1996; Rengel, 1997b). A possible causal relationship between increased populations of rhizosphere bacteria and an increased capacity of wheat genotypes to acquire Zn under deficient conditions warrants further research.

Zinc Uptake and Transport

Net Zinc Uptake

Differences in Zn uptake kinetics were noted among genotypes of wheat, rice, sorghum, and corn (for the review, see Graham and Rengel, 1993). However, the relevance of these studies is questionable because unphysiologically high concentrations of Zn were used (up to 250 μM) or the Zn efficiency ranking of genotypes was not provided.

Wheat genotypes differed in the rate of net Zn uptake from chelate-buffered nutrient solutions having various total Zn concentrations (Rengel and Graham, 1995b, 1996; Rengel and Hawkesford, 1997; Rengel and Wheal, 1997b; Rengel, Römheld, and Marschner, 1998). No significant correlation was found between the rate of net Zn uptake and dry matter production for the large number of genotypes tested under low-Zn conditions (Rengel and Graham, 1995b). However, when the most Zn-efficient and most Zn-inefficient wheat genotypes were grown at Zn^{2+} activities likely to resemble the activities in the soil solution, there was a strong positive correlation between Zn efficiency, the rate of Zn uptake, and dry matter production (Rengel and

Graham, 1996; Rengel and Hawkesford, 1997; Rengel and Wheal, 1997b; Rengel, Römheld, and Marschner, 1998). Zinc deficiency caused an increase in I_{max} (maximum net uptake rate) in the Zn-efficient wheat genotypes, but not in the Zn-inefficient ones (Rengel and Wheal, 1997b). A greater rate of Zn uptake in Zn-efficient than in Zn-inefficient genotypes was observed under deficient, but not under sufficient, Zn supply, indicating an inducible, rather than a constitutive, efficiency mechanism. However, an increase in the Zn uptake rate by Zn-efficient genotypes may not be measurable if the Zn deficiency stress is not severe enough (Wheal and Rengel, 1997). The short-term Zn uptake kinetics and the rate of transport of Zn to the plant tops during vegetative growth relate well to the expression of Zn efficiency at maturity because Zn-efficient genotypes accumulate more Zn in grain than inefficient ones (Graham, Ascher, and Hynes, 1992).

Recent experiments have demonstrated that Zn uptake in wheat genotypes is inversely related to symplasmic Zn concentration (following withdrawal of Zn supply, the ability of plants to take up Zn increases; Rengel and Graham, 1996; Rengel and Hawkesford, 1997; Rengel, Römheld, and Marschner, 1998). Similarly, a partial inhibition of the Zn uptake system by Zn is obvious from the fact that Zn-deficient plants take up Zn at a greater rate than Zn-sufficient ones.

Zinc Efflux

Zinc efflux may be a significant component of net Zn uptake (up to 85 percent of labeled Zn may be effluxed into external solution; Santa Maria and Cogliatti, 1988). Recent studies (Rengel, 1995b; Rengel and Graham, 1996) suggested, for the first time, that Zn efflux may be genotype-dependent (greater in the Zn-inefficient genotype), in addition to being dependent on the internal Zn status and external solution Zn^{2+} activities. The root-cell plasma membranes of the Zn-inefficient genotypes may have a greater requirement for external Zn than those of the Zn-efficient ones.

Zinc Transport from Roots to Shoots

Under Zn deficiency, the Zn-efficient wheat genotypes transported larger amounts of Zn from roots to shoots than did plants of Zn-inefficient genotypes (Cakmak, Sari, Marschner, Kalayci, et al., 1996; Rengel, Römheld, and Marschner, 1998). This greater Zn transport from roots to shoots in Zn-efficient genotypes occurred early in the growth period (Wheal and Rengel, 1997); such a characteristic may contribute to overall Zn efficiency, especially if Zn availability is decreased later.

Utilization of Zinc in Cells

Zinc Tissue Concentration

Total Zn concentration in roots and leaves is a poor estimate of the wheat Zn status (Rengel and Graham, 1995b) (also in other crops; Gibson and Leece, 1981; Cakmak and Marschner, 1987). No difference in total Zn concentration in roots and shoots could be found among wheat genotypes differing in Zn efficiency when grown under Zn-deficient conditions in soil (Dong, Rengel, and Graham, 1995) or in the nutrient solution (Rengel and Graham, 1995b); similar results were reported for bean (Jolley and Brown, 1991). Therefore, no correlation was found between Zn tissue concentration and Zn efficiency in various crops (see also Cakmak, Sari, Marschner, Kalayci, et al., 1996; Erenoglu et al., 1996; Cakmak, Derici, et al., 1997; Cakmak, Ekiz, et al., 1997; Cakmak, Öztürk, et al., 1997). However, since total Zn and the physiologically active Zn pools may not correspond in size (Cakmak and Marschner, 1987), it remains to be tested whether genotypes differing in Zn efficiency differ in concentration of physiologically active Zn in relevant cell compartments.

Although Zn-inefficient wheat genotypes had the same root and leaf concentrations of total Zn as Zn-efficient ones when grown under Zn deficiency (see Table 9.1), only Zn-inefficient genotypes showed typical visual symptoms of Zn deficiency (Cakmak et al., 1994; Rengel and Graham, 1995a; Cakmak, Derici, et al., 1997; Cakmak, Öztürk, et al., 1997; Fischer, Thimm, and Rengel, 1997; Rengel and Hawkesford, 1997). The Zn-efficient genotypes produced significantly greater amounts of shoot dry matter than the Zn-inefficient genotypes, without any difference between the two groups in the total Zn concentration in shoot (see Table 9.1). If one considers that physiological processes in both Zn-efficient and Zn-inefficient genotypes occurred at the same Zn concentration, a greater amount of dry matter accumulated in the Zn-efficient genotypes clearly testifies to greater utilization efficiency in the Zn-efficient genotypes compared to the Zn-inefficient ones.

Carbonic Anhydrase Activity

Under sufficient Zn supply, wheat genotypes differing in Zn efficiency had the same carbonic anhydrase (CA) activity; under Zn deficiency, however, a twofold higher CA activity was recorded in the Zn-efficient genotype compared to the Zn-inefficient one. For any given Zn concentration in leaf tissue, the Zn-efficient genotype showed greater CA activity than the Zn-inefficient one (Rengel, 1995a).

TABLE 9.1. Total Zn Concentration and Dry Matter Production in Shoots of Wheat Plants Grown in Chelate-Buffered Nutrient Solution Supplemented with Different Concentrations of Total Zn

Genotype[a]	Shoot Zn concentration mg kg^{-1} DM		Shoot dry matter mg $plant^{-1}$	
	0.1 μM total Zn	10 μM total Zn	0.1 μM total Zn	10 μM total Zn
Durati	8	15	11	37
Gatcher	9	15	13	37
Aroona	8	16	16	45
Warigal	8	18	22	38
Tukey's $HSD_{0.05}$	1[b]	2	2	5

Source: Adapted from Rengel and Graham (1995b).

[a] Durati and Gatcher are Zn-inefficient genotypes; Aroona and Warigal are Zn-efficient genotypes.

[b] Tukey's $HSD_{0.05}$ is Honestly Significant Difference at $\alpha = 0.05$ ($n = 3$) for the interaction genotype × Zn supply.

Wheat genotypes differing in Zn efficiency showed different gas exchange characteristics. The Zn-efficient genotype showed a higher net CO_2 fixation rate than the Zn-inefficient one, especially under conditions of low Zn supply and high light intensity (Fischer, Thimm, and Rengel, 1997). Since the difference between genotypes in the net CO_2 fixation rate was enhanced under high irradiance (where net CO_2 fixation may be limited by CO_2 availability), the higher net CO_2 fixation rate in the Zn-efficient genotype, as compared to the Zn-inefficient genotype grown under Zn deficiency (Fischer, Thimm, and Rengel, 1997), may be related to higher CO_2 availability due to higher carbonic anhydrase activity (Rengel, 1995a).

Greater activities of CA in the Zn-efficient wheat genotype compared to the Zn-inefficient one, when grown under Zn deficiency stress (Rengel, 1995a), are more likely to be a consequence, rather than a cause, of differential Zn efficiency because a differential decrease in growth under Zn deficiency (smaller in the Zn-efficient compared to the Zn-inefficient genotype) became apparent before significant differences in CA activity between the genotypes could be detected (Rengel, 1995a).

Given that plants may be exposed to Zn deficiency (or suboptimal Zn supply) for extended periods of time (up to the whole growth period), an ability to maintain relatively greater CA activities in the Zn-efficient genotypes (Rengel, 1995a) may prove beneficial in maintaining the photosynthetic rate at a greater level (Fischer, Thimm, and Rengel, 1997). This may be especially important in wheat because it has a basal level of CA activity that is significantly lower than most other plant species (see Makino et al., 1992), which contain CA in excess of the requirement for the normal photosynthetic rate and dry matter production.

Antioxidative Enzymes

Similar to a whole range of other environmental stresses (Foyer, Descourvieres, and Kunert, 1994), Zn deficiency increases production of free oxygen radicals (Cakmak and Marschner, 1988b; Wenzel and Mehlhorn, 1995), thus causing oxidative stress that may result in lipid peroxidation, protein denaturation, DNA mutation, photosynthesis inhibition, and so on. Such a response may at least partly be due to the fact that Zn deficiency decreases activity of antioxidative enzymes, particularly superoxide dismutase (SOD), which detoxifies free oxygen radicals (Cakmak and Marschner, 1987, 1988b). This decrease in SOD activity was especially strong in the Zn-inefficient wheat genotypes compared with the Zn-efficient wheats and rye (Cakmak, Öztürk, et al., 1997), indicating that the capacity of plants to resist oxidative damage through maintaining higher activity of SOD may be one of the mechanisms of Zn efficiency.

Sulfhydryl Groups in the Root-Cell Plasma Membrane

Smaller amounts of 5,5′-dithio-bis(2-nitrobenzoic acid)-reactive sulfhydryl groups were found in Zn-deficient than in Zn-sufficient barley (Welch and Norvell, 1993), as well as in wheat roots (Rengel, 1995b). In addition, the Zn-efficient wheat genotype had a greater amount of sulfhydryl groups than the Zn-inefficient one, regardless of Zn supply (Rengel, 1995b).

In wheat roots, a relatively small amount of Zn is required for preventing oxidation of sulfhydryl groups into disulfides (Rengel, 1995b). The increased amount of reactive sulfhydryl groups in the roots may be one of the mechanisms that, under conditions of Zn deficiency, allow better growth and productivity of Zn-efficient wheat genotypes in comparison to the Zn-inefficient ones.

Polypeptide Biosynthesis

Various Zn supplies differentially influenced biosynthesis of polypeptides in the root-cell plasma membranes of wheat genotypes differing in

Zn efficiency (Rengel and Hawkesford, 1997). Biosynthesis of the 34-kDa polypeptide increased under Zn deficiency in the root-cell plasma membranes of the Zn-efficient wheat genotype (Warigal), but not in the Zn-inefficient one (Durati), leading to a suggestion that the polypeptide might be connected with the Zn-efficiency mechanisms. Both Zn-efficient and Zn-inefficient wheat genotypes had a capacity to synthesize the 34-kDa polypeptide early in the growing period, but this capacity disappeared in the Zn-inefficient genotype with the onset of Zn deficiency stress; only de novo biosynthesis of the 34-kDa polypeptide was regulated by Zn deficiency (Rengel and Hawkesford, 1997).

Recent research failed to identify increased biosynthesis of the 34-kDa polypeptide in a number of other Zn-efficient wheat and barley genotypes grown under Zn deficiency (Z. Rengel and T. Grünewald, unpublished results). It is therefore unlikely that the 34-kDa polypeptide may represent a part of the widespread Zn efficiency mechanism, even though it appears to be linked to Zn efficiency of Warigal wheat. The 34-kDa polypeptide is the first reported root-cell plasma membrane polypeptide specifically induced under Zn deficiency; further studies are therefore warranted.

MODIFICATION OF PHENOTYPIC EXPRESSION OF ZINC EFFICIENCY

Ionic Interactions

Zinc-Phosphorus Interactions

Genotype Excalibur (Zn-efficient when soil grown; Graham, Ascher, and Hynes, 1992; Rengel and Graham, 1995c, d) ranks as a relatively inefficient genotype when grown in the Zn-deficient nutrient solution (Rengel and Graham, 1995a, b). However, changing the composition of the nutrient solution (daily additions of 5 to 10 μmoles of P per liter per day to maintain low P concentration) caused a significant drop in root and shoot P concentrations in Excalibur, thus dramatically improving its growth in the Zn-deficient nutrient solution (Rengel and Wheal, 1997a, b; Wheal and Rengel, 1997) and increasing its perceived Zn efficiency. The Zn-deficiency-induced P toxicity, a phenomenon that has been well described in the literature (Loneragan and Webb, 1993, and references therein), should therefore be taken into account when assessing relative Zn efficiency of crop genotypes.

Zinc-Iron Interactions

The Zn-inefficient bean genotype had enhanced exudation of reductants, increased reduction of Fe^{3+}, and increased accumulation of Fe in leaves (typical Fe deficiency responses in bean as the strategy I plant, see below) as compared to the Zn-efficient genotype when grown under Zn stress. It therefore appears that Zn deficiency stimulated the initiation of the Fe deficiency response in the Zn-inefficient genotype, thus increasing reduction and uptake of Fe, which in turn competitively inhibited Zn uptake and utilization (Kochian, 1993) and enhanced Zn deficiency (Jolley and Brown, 1991).

In wheat, shoot concentrations of Fe, Mn, and Cu were higher when plants were grown at deficient, rather than sufficient, Zn supply. The Zn-efficient wheat genotypes showed a greater rate of net Fe uptake (Rengel and Graham, 1995b; 1996; Rengel, Römheld, and Marschner, 1998) and released a greater amount of PS (Cakmak et al., 1994; Walter et al., 1994; Cakmak, Öztürk, et al., 1996; Cakmak, Sari, Marschner, Ekiz, et al., 1996; Cakmak, Sari, Marschner, Kalayci, et al., 1996; Cakmak, Öztürk, et al., 1997; Rengel, Römheld, and Marschner, 1998) compared to the Zn-inefficient genotypes when grown under Zn deficiency. Such an observation is consistent with reports on Zn-deficiency-induced Fe deficiency (Walter et al., 1994; Rengel and Graham, 1996), both being expressed to a greater extent (and therefore resulting in greater Fe as well as Zn uptake) in the Zn-efficient than in the Zn-inefficient wheat genotypes. Under Zn deficiency, translocation of Fe from roots to shoots is depressed in the Zn-efficient genotypes (Walter et al., 1994; Rengel and Graham, 1995b, 1996; Rengel, Römheld, and Marschner, 1998). As a consequence, hidden, physiological Fe deficiency may develop in leaves, causing increased PS release.

Zinc-Copper Interactions

Warigal wheat is Cu inefficient, while the translocation line Warigal-5RL/4B (sometimes designated as Warigal-5R), carrying a part of the long arm of the rye chromosome 5, is Cu efficient (Graham et al., 1987); these two genotypes, however, are of similar Zn efficiency (Rengel and Graham, 1995a, b). Warigal-5R (Cu-efficient) showed greater uptake of Cu than Warigal when grown at deficient Zn supply for up to 15 d (days) (Rengel and Graham, 1995b). Further research is needed to ascertain the significance of Zn-Cu interactions in Zn-efficient versus Cu-efficient wheat genotypes.

Seed Zinc Content

Larger seed size results in improved seedling vigor and early growth of cereals (Mian and Nafziger, 1992) and better crop stands that out-yield crops derived from the small seeds (Hampton, 1981). In addition to seed size, seed quality (higher quality expressed as higher seed Zn content) also influences vegetative growth and grain yield of cereals grown under Zn deficiency (Rengel and Graham, 1995c, d; Genc et al., 1998).

Different seed Zn content may differentially influence performance of different genotypes under Zn-deficient conditions. Wheat (Rengel and Graham, 1995c, d) and barley (Genc et al., 1998) genotypes with higher seed Zn content produce bigger roots and shoots during early growth before seed reserves of Zn are exhausted. Greater root mass would be beneficial in the later stages when plants have to rely on scavenging Zn from the environment. Although Excalibur wheat is more Zn efficient than Gatcher when grown from the low-Zn seed (around 250 ng Zn $seed^{-1}$), the difference in Zn efficiency is dissipated when seed Zn content is threefold greater (Rengel and Graham, 1995c, d). Therefore, the higher seed Zn content acts similarly to the starter fertilizer.

Plant species differ in their ability to load Zn into seeds (White, Robson, and Fisher, 1981; Longnecker and Robson, 1993). For plants grown under Zn deficiency, grain is a poor sink for Zn accumulation in comparison with other tissues (Rengel and Graham, 1995d). In addition, various parts of the grain have a differential ability to accumulate Zn (Pearson and Rengel, 1994; Pearson et al., 1995). It remains to be established whether various crop genotypes have different potentials for loading of nutrients into various seed tissues and/or remobilization of these nutrients to a young seedling. This is of particular interest for genotypes that differ in nutrient efficiency because a greater nutrient efficiency may start as a faster and more thorough remobilization of nutrients from various seed tissues. Preliminary results indeed showed that wheat genotypes have different abilities to load nutrients to different parts of seed (Moussavi-Nik et al., 1997). However, no correlation was found between the level of Zn efficiency of the genotypes studied and the early growth, while dependent on seed nutrient reserves.

Given the importance of a relatively higher seed Zn content for plant growth, it is to be expected that seed Zn content will be an important factor in any future screening of crop genotypes based on growth measurements. However, seed with similar seed Zn content for such testing may be, at best, impractical or, at worst, impossible to obtain, thus obviating a need for other (e.g., biochemical, molecular) means of testing for the Zn efficiency trait.

MECHANISMS OF IRON EFFICIENCY

Generally, plants employ two types of mechanisms under Fe deficiency (for reviews, see Römheld, 1991; Marschner and Römheld, 1994; Mori, 1994; Welch, 1995; Jolley et al., 1996; Pearson and Rengel, 1997; Crowley and Rengel, 1999). Strategy I (characteristic for dicots and nongraminaceous monocots) involves increased reduction of Fe^{3+} to Fe^{2+} at the root-cell plasma membrane, acidification of the rhizosphere (enhanced net exudation of protons), enhanced exudation of reducing and/or chelating compounds into the rhizosphere, and changes of root histology and morphology (root tip swelling, formation of rhizodermal transfer cells, induced root branching, more root hairs, etc.). Strategy II (employed by grasses) involves increased exudation of PS into the rhizosphere.

Within both Strategy I and Strategy II plants, various species and genotypes within species differ in Fe efficiency (Römheld and Marschner, 1990; Jolley and Brown, 1994; Pearson and Rengel, 1997). As a response to Fe deficiency, Fe-efficient species and genotypes induce strong metabolic and structural changes that allow them to grow and yield better under deficiency conditions in comparison to Fe-inefficient species and genotypes.

Strategy I

Relative importance of various Fe deficiency responses (enhancement of Fe^{3+} reduction and exudation of protons or reducing/chelating compounds into the rhizosphere, and histological/morphological changes in root growth) is poorly understood because relatively few studies actually provide comparative measurements of all these components of the Fe deficiency response. A recent report (Wei, Loeppert, and Ocumpaugh, 1997) showed that under Fe deficiency, Fe-efficient clover had exudation of protons threefold higher than the Fe-inefficient genotype; in contrast, there was no difference between genotypes in the Fe^{3+} reduction rate, while exudation of reductants was negligible in both genotypes. More studies of this type are required to gain an insight into relative importance of different Fe deficiency responses in various plant species.

Among woody plants with Strategy I, *Vitis berlandieri* and *V. vinifera* show greater rhizosphere acidification, extrude greater amounts of organic acids, and have a greater capacity to reduce Fe^{3+} chelates in comparison to Fe-inefficient *V. riparia* (Brancadoro et al., 1995). It is interesting to note that the level of Fe efficiency in various *Vitis* cultivars was not related to the total Fe concentration in leaves (Bavaresco, 1997). In contrast, the Fe-efficient groundnut genotypes had a greater concentration of active Fe in their leaves compared to the Fe-inefficient genotypes (Reddy, Ashalatha, and Venkaiah, 1993).

The Fe-efficient genotypes of apples (Zhang, Liu, and Mao, 1996) and soybeans (Longnecker and Welch, 1990) accumulate more apoplasmic Fe in roots than Fe-inefficient ones; the apoplasmic Fe pool is important in Fe nutrition of shoots. In addition, the Fe-efficient soybean genotypes had an increased root-cell reducing capacity and acidified the rhizosphere by up to 1.5 pH units under Fe deficiency, while the Fe-inefficient genotypes did not show such responses (Liu, Zhang, and Mao, 1996).

Mutants of tomato (Stephan and Grün, 1989) and pea (Welch and La Rue, 1990; Grusak, Welch, and Kochian, 1990; Grusak and Pezeshgi, 1996) show typical Fe-deficiency responses (increased rates of Fe^{3+} reduction, increased rhizosphere acidification, and/or increased exudation of reducing/chelating compounds into the rhizosphere), regardless of Fe availability, thus suffering from Fe toxicity if Fe levels in the medium are somewhat above deficient. Recent research has shown that overexpression of root Fe^{3+} reductase activity by *E107* mutant of pea and *chloronerva* mutant of tomato was inhibited by ethylene inhibitors, indicating that large Fe accumulation capacity of these two mutants may be due to a genetic defect in their ability to regulate root ethylene production (Romera et al., 1996), even though the effect might be via the IAA-related changes in ethylene biosynthesis and action (*cf.* Landsberg, 1996). In addition, the immunohistochemical localization of nicotianamine, low-molecular-weight Fe-chelator, in situ in wild-type tomato genotype showed that the mutant *chloronerva* does not contain any nicotianamine (Pich et al., 1997); this technique may prove valuable in deciphering molecular mechanisms involved in the Fe deficiency response and differential Fe efficiency.

Differential expression of the Fe deficiency response in roots of Fe-efficient and Fe-inefficient genotypes may be due to a transducible signal travelling from shoot to root and regulating Fe homeostasis in plants, as experiments with *dgl* pea mutant and its parental genotype showed (Grusak and Pezeshgi, 1996). Reductase studies using plants with reciprocal shoot/root grafts demonstrated that shoot expression of the *dgl* gene enhances Fe(III) reductase activity in roots.

The Fe-efficient wild-type muskmelon genotype had a greater proton exudation and Fe^{3+} reducing capacity than the Fe-inefficient mutant; interestingly, the Fe-efficient genotype also exuded several-fold greater amounts of riboflavin than the Fe-inefficient mutant (Welkie, 1996). Riboflavin may be involved in Fe^{3+} reduction as an intermediate in the plasma membrane NAD(P)H-Fe-chelate reductase, thus alleviating the Fe deficiency stress.

Root Fe^{3+} reducing activity of soybean genotypes was correlated with genotypic resistance to Fe chlorosis measured in field nurseries and was used as a method for identifying chlorosis-resistant (Fe-efficient) genotypes. Stevens and colleagues (1993) have developed such a screening method for Fe efficiency based on Fe^{3+}-reducing capacity of roots of soybean genotypes grown in $CaCO_3$-buffered, Fe-free nutrient solution. In contrast, Graham and colleagues (1992) based their screening method on the growth of callus in vitro: the callus from Fe-efficient soybean genotypes grew better than callus from Fe-inefficient genotypes in the bicarbonate-containing media, thus mimicking the ranking observed in the field trials (for the review, see Cianzio, 1999).

Strategy II

Graminaceous species acquire Fe by releasing PS and taking up ferrated PS complexes through a specific uptake system that is highly activated under Fe deficiency (Römheld, 1991; Von Wiren et al., 1994; Von Wiren, Marschner, and Römheld, 1995). The rate of PS release is different for different crops and also for different genotypes of a particular crop. The rate of release of PS is positively related to Fe efficiency of species; the relative effectiveness in PS release, and thus Fe efficiency, decreases in the following order: barley > corn > sorghum (Römheld and Marschner, 1990) or oats > corn (see Mori, 1994). The Fe-efficient genotypes of wheat (Hansen, Jolley, and Brown, 1995; Hansen et al., 1996) and oats (Jolley and Brown, 1989; Hansen and Jolley, 1995) exuded more PS than the Fe-inefficient genotypes, making a PS-exudation test suitable for screening genotypes in the breeding program. In contrast, there is no difference in the amount of PS released under Fe deficiency between the two corn genotypes differing in Fe efficiency (Von Wiren et al., 1994), but the Fe-inefficient one has almost tenfold lower maximum uptake capacity (I_{max}) of the saturable component of uptake, arising from either decreased activity or a lower number of membrane transporters in its root-cell plasma membrane in comparison to the Fe-efficient genotype (Von Wiren, Marschner, and Römheld, 1995).

Although all wheat genotypes tested to date (Mori and Nishizawa, 1987; Zhang, Römheld, and Marschner, 1989; 1991; Cakmak et al., 1994; Walter et al., 1994; Cakmak, Öztürk, et al., 1996; Cakmak, Sari, Marschner, Ekiz, et al., 1996; Cakmak, Sari, Marschner, Kalayci, et al., 1996; Rengel, Römheld, and Marschner, 1998), as well as rice (Mori and Nishizawa, 1987) and corn (Mori, 1994), release only deoxymugineic acid (DMA), different oat genotypes exude a wide variety of PS types (Mori, 1994). Rye, a species very tolerant to Fe deficiency in calcareous soils, exudes mugineic acid, hydroxymugineic acid (HMA), and DMA into the rhizosphere. It has been suggested that capacity of rye to exude HMA may be related to superior tolerance of rye to Fe, as well as Zn, deficiency when compared to wheat, which exudes only DMA (Cakmak, Öztürk, et al., 1996).

The release of PS and the subsequent uptake of the Fe-PS complex are under different genetic control (Römheld and Marschner, 1990). Indeed, it was found that the Fe-inefficient yellow-stripe mutant of corn (*ys1*) maintains the rate of PS release similar to that of Fe-efficient cultivars (e.g., Alice), but Fe inefficiency of *ys1* is the result of a defect in the uptake system for the Fe-PS complex (up to twentyfold lower uptake rate in *ys1* compared to Alice corn; Von Wiren et al., 1994). The reason for lower Fe uptake rate by Fe-inefficient *ys1* compared to Fe-efficient Alice was the mutation in *ys1* that affected a high-affinity uptake component, leading to a decrease in activity and/or number of Fe-PS transporters (decreased I_{max}, unchanged K_m) (Von Wiren, Marschner, and Römheld, 1995).

MECHANISMS OF MANGANESE EFFICIENCY

Yield of cereals grown on calcareous soils is frequently limited by Mn deficiency resulting from low Mn availability rather than low Mn content in the soil (Welch et al., 1991). Considerable differences in tolerance to Mn-deficient soils have been found among cereal genotypes (Bansal, Nayyar, and Takkar, 1991), but the mechanisms of Mn efficiency by which one genotype (Mn-efficient) may tolerate Mn-deficient soils better than the other genotype (Mn-inefficient) are poorly understood (Graham, 1988; Huang, Webb, and Graham, 1993, 1994, 1996; Rengel, Pedler, and Graham, 1994). Superior internal utilization of Mn, a lower physiological requirement, a faster specific rate of Mn absorption, and better root geometry are unlikely to be major contributors to Mn efficiency (Graham, 1988). The remaining two possible mechanisms of Mn efficiency are (1) better internal compartmentalization and remobilization of Mn (Huang, Webb, and Graham, 1993) and (2) excretion by roots of Mn-efficient genotypes of greater amounts of substances capable of mobilizing insoluble Mn (protons, reductants, Mn-binding ligands, and microbial stimulants) (Rengel et al., 1996; Rengel, 1997b). The latter mechanism is of particular importance because genotypic differences in Mn efficiency among barley cultivars were expressed only in the soil-based systems, not in the solution culture, indicating that different acquisition of Mn from soils with low Mn availability may be responsible for the different abilities of barley genotypes to tolerate Mn deficiency (Huang, Webb, and Graham, 1994).

Rhizosphere Acidification

Differential expression of Mn efficiency among barley (Tong, Rengel, and Graham, 1997) as well as wheat genotypes (Marcar, 1986) is not associated with differences in Mn availability expected to be produced by differential rhizosphere acidification as a response to different forms of N supply. The reason for such an observation may be the relatively high pH buffering capacity that calcareous soils frequently have (e.g., the rhizosphere pH changes may be smaller than 0.2 pH units in a calcareous soil of higher pH; see Marschner, 1995).

Rhizosphere Microflora

Different plant species (Lemanceau et al., 1995; Wiehe and Hoflich, 1995), as well as genotypes within a species (Timonin, 1946, 1965; Hornby and Ullstrup, 1967; Howie and Echandi, 1983; Liljeroth, Van Veen, and Miller, 1990), differentially influence the quantitative and qualitative composition of microbial populations in the rhizosphere. The nature and activity of root exudate components that might be involved in promoting or inhibiting growth of various bacterial populations in the rhizosphere is the subject of current research (Van Overbeek, Van Veen, and Van Elsas, 1997; P. Curnow and P. Hirsch, 1998).

It is possible that the genetic control of Mn efficiency is expressed through the composition of root exudates encouraging a more favorable balance of Mn reducers to Mn oxidizers in the rhizosphere. Timonin (1946) found greater numbers of Mn-oxidizing microbes in the rhizosphere of Mn-inefficient than in Mn-efficient oat cultivars and concluded that this was the basis of genotypic differences in Mn uptake. Later research confirmed these findings and linked greater Mn efficiency to production of root exudates that are toxic to Mn-oxidizing microorganisms in the rhizosphere (Timonin, 1965).

Rhizosphere of wheat genotypes contained an increased proportion of Mn reducers under Mn-deficient compared to Mn-sufficient conditions. When grown at the sufficient Mn supply, no difference in the ratio of Mn reducers and Mn oxidizers was found in the rhizosphere of genotypes differing in Mn efficiency. However, under Mn deficiency, the Mn-efficient Aroona had the ratio of Mn reducers to Mn oxidizers increased threefold in comparison to the Mn-inefficient genotype. In contrast, microflora in the rhizosphere of other Mn-efficient wheat genotypes (such as C8MM) did not show the same response as Aroona. It therefore appears that different mechanisms may underlie the expression of Mn efficiency in wheat genotypes (Rengel, 1997b). Further characterization of Mn reducers and oxidizers from the rhizosphere of wheat genotypes is in progress.

Considerably more fluorescent pseudomonads were found in the rhizosphere of some (e.g., Machete), but not other (e.g., Bodallin), Mn-efficient wheat genotypes (Rengel et al., 1996; Rengel, 1997b). Fluorescent pseudomonads are the most important bacterial group with the capacity to reduce Mn (Posta, Marschner, and Römheld, 1994). It is therefore expected that an increased number of fluorescent pseudomonads (i.e., Mn reducers) will contribute to increased Mn efficiency.

MECHANISMS OF PHOSPHORUS EFFICIENCY

Plant species, as well as genotypes within the species, differ in P efficiency when grown in soils with low P availability (e.g., Caradus, 1994, 1995; Randall, 1995; Mamo, Richter, and Hoppenstedt, 1996; Trolove et al., 1996a, b). The C_4 species appear to have similar P efficiency as C_3 species, but monocots are more P efficient than dicots because deficiency affects branching (dicots) more than tillering (monocots) (Halsted and Lynch, 1996). In contrast, P deficiency severely affected tillering in soil-grown wheat (Horst, Abdou, and Wiesler, 1996), a species that was not tested in the study by Halsted and Lynch (1996).

Generally, three broad categories of P efficiency mechanisms exist in plants to increase availability and uptake of P under P deficiency (Pearson and Rengel, 1997): (1) secretion or exudation of chemical compounds into the rhizosphere, (2) alteration of the geometry or architecture of the root system, and (3) association with microorganisms.

Differences among genotypes in utilization efficiency are sometimes found when the amount of shoot dry matter produced per unit of P acquired was considered (Hedley, Kirk, and Santos, 1994; Gorny, 1996; Stelling, Wang, and Römer, 1996; Rao et al., 1997; Subbarao, Ae, and Otani, 1997a), but there are numerous contrary examples (e.g., Buso and Bliss, 1988; Jones, Blair, and Jones, 1989; Gourley, Allan, and Russelle, 1994; Kemp and Blair, 1994). Attempts to screen genotypes for P efficiency based on an index of utilization efficiency, although sometimes apparently successful (greater utilization index associated with P-efficient genotypes; e.g., Trolove et al., 1996a), should be avoided (Jones, Blair and Jessop, 1989; Blair, 1993) because such indexes may not reflect efficiency mechanisms, but just a generally better physiological status of the P-efficient genotypes, in comparison to P-inefficient genotypes when grown under P deficiency.

Exudation of Organic Compounds into the Rhizosphere

Under P deficiency, plants exude a wide range of organic and inorganic compounds. Acidification of rhizosphere by extrusion of protons causes dissolution of plant-unavailable P forms, such as rock phosphate (Trolove et al., 1996b) and Ca-P complexes in calcareous soils (Yan, Lynch, and Beebe, 1996). Rhizosphere acidification is more prominent in P-efficient plant species (Trolove et al., 1996b), and genotypes within the species (Yan, Lynch, and Beebe, 1996), than in their P-inefficient counterparts. In contrast, no difference in the rhizosphere acidification was noted for a range of wheat and barley (Gahoonia and Nielsen, 1996) and rice genotypes (Hedley, Kirk, and Santos, 1994) differing in P efficiency.

Genotypes differing in P efficiency also differ in solubilizing activity of their root exudates, but differences in capacity to solubilize $FePO_4$ could not fully account for differences in P uptake from the $FePO_4$ (Subbarao, Ae, and Otani, 1997b), indicating a complex nature of P efficiency, at least in pigeon pea. The P-efficient bean (Helal, 1990) and barley genotypes (Asmar, Gahoonia, and Nielsen, 1995) had a greater activity of extracellular phosphatases in the rhizosphere soil in comparison to the P-inefficient genotypes. In contrast, such a response could not be ascertained for rice genotypes differing in P efficiency (Hedley, Kirk, and Santos, 1994).

Although genetic studies have been done on manipulating root size and morphology and phosphorus uptake (for the review, see Caradus, 1995), no studies have been reported so far on heritability of P efficiency mechanisms associated with root exudation and solubilization of P complexes considered plant-unavailable. There is, however, genetic variability in root exudation reported for several species (for acid phosphatase, see Caradus, 1995; for unspecified $FePO_4$ solubilizing activity of root exudates see Subbarao, Ae, and Otani, 1997b). Comparison of durum and flax genotypes of unknown P efficiency has shown large differences in the amounts of organic acids released into the rhizosphere (Cieslinski et al., 1997). More research into the genetic basis of qualitative and quantitative differences in root exudation is warranted.

Root Morphology

Plants growing in P-deficient soil allocate a greater proportion of assimilates to root growth and tend to have fine roots of small diameter and relatively large surface area (Schenk and Barber, 1979; Snapp, Koide, and Lynch, 1995). Not only fine roots, but especially root hairs (Itoh and Barber, 1983; Barber and Silberbush, 1984), are effective in scavenging P from the soil environment. The genotype of rye with long root hairs is more P-efficient than the one with short root hairs (Baon, Smith, and Alston, 1994). Selection of white clover for longer root hairs resulted in 14 percent longer hairs after just one cycle of selection (Caradus, 1995), indicating a significant potential of such a breeding approach in increasing P efficiency of plants.

Wheat genotypes differing in P efficiency had different P depletion profiles in the rhizosphere, indicating that differential root hair length, as well as root exudation, might have contributed to differential P efficiency (Gahoonia and Nielsen, 1996). Fine roots with a small diameter also contributed to effective P uptake and increased P efficiency in wheat (Horst, Abdou, and Wiesler, 1996); generally, a good correlation existed between root length and P uptake for a number of crops (Nielsen and Schjorring, 1983; Blair, 1993; Caradus, 1995; Otani and Ae, 1996), including wheat genotypes (Römer, Augustin, and Schilling, 1988). However, efficiency in P uptake cannot account for all the genotypic variation in P efficiency in pigeon pea (Subbarao, Ae, and Otani, 1997b). In addition, the pattern of P accumulation over the whole growth period may influence P efficiency of wheat genotypes grown in the field conditions; genotypes accumulating more P early in the growth period had a poor grain yield compared to genotypes with a large accumulation of P later in the growth period (Jones, Jessop, and Blair, 1992).

Mycorrhizae and Rhizosphere Microflora

It is well-known that association between roots and mycorrhizae allows plants to improve their access to soil P pools (e.g., Marschner and Dell, 1994; Pearson and Rengel, 1997); this subject is beyond the scope of this chapter. However, emphasis will be placed on genotypic differences in mycorrhizal colonization resulting in differential access to soil P pools.

The rye genotype with short root hairs was more dependent on mycorrhizal colonization for growth in the P-deficient soil than the genotype with long root hairs (Baon, Smith, and Alston, 1994), the mycorrhizae apparently compensating for the reduced root absorptive surface area in the genotype with short root hairs. Similar results were obtained with white clover genotypes differing in root hair length (see Caradus, 1995).

Barley genotypes differed in the mycorrhizal colonization rate, but P efficiency was negatively correlated with the colonization rate (Baon, Smith, and Alston, 1993). Wheat genotypes differed in the benefits arising from the mycorrhizal colonization: up to 20 percent increase in shoot yield, with a decrease in root growth, was recorded (Xavier and Germida, 1997). In contrast, differences in P efficiency among bean (*Phaseolus vulgaris*) genotypes were not related to the mycorrhizal colonization (Yan, Lynch, and Beebe, 1995).

Although it is clear that rhizosphere microflora influence P nutrition of plants grown in soil (for the review, see Crowley and Rengel, 1999), qualitative and quantitative differences among crop genotypes in rhizosphere microflora, as related to the P uptake and P efficiency, have not yet been unequivocally shown.

BREEDING FOR NUTRIENT EFFICIENCY

The role of nutrient efficient genotypes in modern agriculture has been outlined by Lynch (1998). The agronomic significance of nutrient efficiency in modern crop cultivars is an important adjunct to the use of fertilizers on soils in which the effectiveness of fertilizers may be limited by chemical and biological reactions, topsoil drying, subsoil constraints, or disease interactions (e.g., Graham and Rengel, 1993; Rengel, Pedler, and Graham, 1994). Growing nutrient-efficient genotypes of crop plants on soils of low nutrient availability represents an environmentally friendly approach that would reduce land degradation by reducing the use of machinery (e.g., Thongbai et al., 1993) and by minimizing application of chemicals (fertilizers) on agricultural land. The danger of exhaustion of soil nutrient resources ("land mining") would be negligible, at least for micronutrients and P (Graham, 1984), because the total supply of micronutrients and P in soils is sufficient for hundreds of years of sustainable cropping by new, efficient genotypes that are able to gain access to the micronutrient pools generally considered plant-unavailable.

Growing nutrient-efficient plants on soils that are low in plant-available nutrients would represent the strategy of "tailoring the plant to fit the soil," in contrast to the older strategy of "tailoring the soil to fit the plant," (terminology according to Foy, 1983). The significance of such an approach should be assessed bearing in mind that, out of all agricultural innovations, farmers most readily accept new cultivars because these can achieve improvement in yield without necessitating much change in agricultural practice (see Little, 1988). However, the relatively slow progress in deciphering the genetics, physiology, and biochemistry behind the mechanisms of nutrient efficiency has hampered development of genotypes of superior nutrient efficiency through conscious breeding efforts geared specifically toward that purpose.

The root morphology is both genetically controlled and environmentally induced (Barber and Silberbush, 1984; Fitter, 1991); incorporation of the traits controlling root morphology into a given genotype through the conventional breeding programs appears to be difficult. However, as the genetic basis of the factors controlling root morphology is being deciphered (Sharma and Lafever, 1992), the variation in root morphology may be utilized in the future in the breeding programs geared specifically for the production of nutrient-efficient genotypes (e.g., Zn-efficient; Dong, Rengel, and Graham, 1995; Rengel and Wheal, 1997a; and P-efficient; Schenk and Barber, 1979).

Breeding of beans in the past has produced (in most cases inadvertently) modern genotypes that, in comparison to wild relatives, partition a relatively larger proportion of assimilates to root growth (a valuable trait in P-deficient soils) and therefore have higher efficiency of P utilization (Araujo, Teixeira, and De Almeida, 1997). Similarly, modern durum and tef [*Eragrostis tef* (Zucc.) Trotter] genotypes have better P efficiency than older ones (Mamo, Richter, and Hoppenstedt, 1996).

A century of breeding wheat did not produce genetic improvement in performance under low-fertility conditions (Shroyer and Cox, 1993). However, where conscious breeding efforts to select wheat genotypes on low-P soils were employed, selected genotypes were relatively more P efficient than those produced in high-input systems on soils containing sufficient P (Sanchez and Salinas, 1981). Similarly in white clover, the natural selection pressure caused genotypes and wild accessions evolved on P-deficient soils to be P efficient (Caradus, 1994).

Genotypes of wheat (Soon, 1992), as well as white clover (Crush and Caradus, 1993), that are tolerant to Al toxicity and acid soils are also P efficient. In contrast, tolerance to Zn toxicity does not confer tolerance to toxicity of Cu, Fe, or Mn (Bradshaw, McNeilly, and Gregory, 1965; Walley, Khan, and Bradshaw, 1974). Such specificity has also been demonstrated in nutrient efficiency. Studies using addition lines showed that Cu, Zn, and Mn efficiency in rye were independent traits carried on different chromosomes (Graham, 1984; Graham et al., 1987). For a particular nutrient, the various mechanisms of efficiency are likely to be additive (as shown for Zn efficiency in rice; Majumder, Rakshit, and Borthakur, 1990), putting great emphasis on a breeding program that would be based on stepwise compounding of genetic information (see Yeo and Flowers, 1986) for a number of nutrient efficiency mechanisms into one locally adapted crop cultivar. In such a breeding program, genotypes having genes that control a particular mechanism of efficiency may be important, even though these genotypes themselves may not show phenotypically high overall nutrient efficiency. In addition, using genotypes from the local crop germplasm would be advantageous because development of cultivars with improved nutrient efficiency may be achieved without disrupting the previously selected adaptation to the local environmental conditions.

In general, nutrient efficiencies are simply inherited, with a relatively small number of genes involved and efficiency being dominant (Graham, Ascher, and Hynes, 1992; Cianzio, 1999). The furthest advances in breeding crops for nutrient-deficient soils have been made with wheat for Cu efficiency (transferred from rye; Graham et al., 1987) and soybeans for Fe (Cianzio, 1999) and Mn efficiency (Graham, Nickell, and Hoeft, 1995). Breeding soybeans for Fe efficiency relies on recurrent selection and field screening. Field screening was used to evaluate performance of soybean genotypes on Mn-deficient soil as well; good narrow sense heritability for Mn efficiency was found when F_2-derived F_3 families were used (Graham, Nickell, and Hoeft, 1995).

Screening for Nutrient Efficiency

Simple, fast, and inexpensive techniques are needed in breeding programs to permit the assessment of a large number of genotypes in segregating populations, without relying on expensive field testing (Graham and Rengel, 1993). The better nutrient efficiency mechanisms are understood, the more successful the screening technique that is developed. The general complexity of the nutrient deficiency syndrome is reflected in a relatively large number of criteria for defining efficiency in screening programs (e.g., Graham, 1984; Blair, 1993; Gourley, Allan, and Russelle, 1994).

One of the major problems with nutrient deficiency, stemming from the long time required for the expression of deficiency stress, is that the level of stress is frequently assessed based on plant growth. However, growth is a very complex characteristic, influenced by a myriad of interrelated processes. As such, growth is clearly not suitable as a parameter to distinguish causes and effects of nutrient deficiency stress. If breeding for nutrient efficiency is to be based on a quick and reliable screening technique, processes causing (or preventing) nutrient deficiency, rather than those appearing as a consequence of it, have to be specifically targeted for selection.

Screening in pots is common (e.g., Yan, Lynch, and Beebe, 1995; Gahoonia and Nielsen, 1996), being less expensive and faster than field work, and the soil can be made quite uniform within the treatment (Batten, 1992). However, the environment is less realistic, and root binding in small pots may be a limiting factor of its own, thus masking the full expression of nutrient deficiency stress and, consequently, mechanisms of efficiency (Graham and Rengel, 1993). Moreover, if the efficiency mechanism relies on a large volume of soil being explored by roots (Huang,

Webb, and Graham, 1996), pot testing obviously will not be suitable for genotype selection.

The correlation between nutrient efficiency ranking in the controlled environments and in the field occurred only to a limited extent for Zn (Rengel and Graham, 1995a), Fe (Graham et al., 1992; Stevens et al., 1993; Hansen and Jolley, 1995), Mn (Rengel, Pedler, and Graham, 1994), and P (Caradus, 1994). In general, good correlations between field and laboratory testing of a number of genotypes to various deficiency (Stelling, Wang, and Römer, 1996) and toxicity stresses (Rengel and Jurkic, 1992) are notoriously hard to achieve. There are, however, examples to the contrary (Hansen et al., 1996), when exudation of phytosiderophores from wheat genotypes subjected to Fe deficiency stress in nutrient solution correlated well with Fe efficiency in the field.

The assessment of nutrient efficiency in comparison with a standard cultivar can be done in a field experiment in which the soil is deficient enough in a nutrient to significantly limit the yield of some genotypes (for the review, see Graham and Rengel, 1993). When a large number of genotypes are to be screened, tests are conducted in soils not amended with the nutrient in question. Such one-level screening at the high degree of stress imposed is a good indicator of efficiency for native plant species (MacNair, 1993) because it mimics the natural course of evolution (as a selection of genotypes most able to sustain growth in a limiting environment). In contrast, for crop genotypes, a two-level screening (limiting and nonlimiting environment) is preferred to avoid selecting genotypes of superior performance in limiting environments but of relatively poor agronomic performance in nonlimiting environments. Adequate fertilization will define the yield potential and allow the calculation of the nutrient efficiency index based on the relative growth [(yield at −nutrient) / (yield at +nutrient)]. This normalization removes much of the background genetic effect but does not exclude nutrient × genotype × environment effects that can occasionally be important (Graham and Rengel, 1993; Gourley, Allan, and Russelle, 1994).

The expression of different nutrient efficiency mechanisms may be related to the intensity of the nutrient deficiency stress or to other environmental conditions (e.g., see Graham et al., 1992; Wei, Ocumpaugh, and Loeppert, 1994; Rengel and Graham, 1996). Therefore, screening for differential expression of individual nutrient efficiency mechanisms may only be achieved under controlled conditions: in contrast, field testing (if done over a number of years and in a number of relevant locations and soil types) may give an estimate of the overall level of nutrient efficiency as a net result of interactions among various efficiency mechanisms and the genotype × environment interaction.

It is to be expected that in the near future, molecular methods will dominate selection for nutrient efficiency. The molecular markers will be identified for regions of the genome segregating with the trait of a particular nutrient efficiency, leading to rapid and efficient selection. However, the molecular markers may be population dependent (e.g., markers for Fe efficiency in soybeans; Cianzio, 1999), making them unsuitable for use in breeding programs.

Even in screening based on molecular methods, seed nutrient content may need to be taken into consideration (differential seed nutrient content frequently presents a confounding factor when the screening methods are based on growth; e.g., Zn [Rengel and Graham, 1995a, c, d; Genc et al., 1998] and P [Kemp and Blair, 1994; Hedley, Kirk, and Santos, 1994; Yan, Lynch, and Beebe, 1995; 1996]) because expression of two genes associated with Fe efficiency in barley (Ids1 and Ids2) was greater after three, compared to two, weeks of growth, possibly indicating exhaustion of seed Fe reserves after prolonged growth (Huang et al., 1996).

CONCLUDING REMARKS

Nutrient-efficient genotypes can be effectively used in modern agriculture in conjunction with fertilization, especially in areas and for nutrients for which effectiveness of fertilization is low. Mechanisms conferring efficiency to particular nutrients are still not well understood, except in the case of Fe efficiency. This lack of understanding of the physiological basis of efficiency hampers efforts to design an effective screening method required for a successful breeding program. More collaboration among scientists of various disciplines is needed to tackle the apparent complexity of the nutrient efficiency syndrome in crop plants.

REFERENCES

Araujo, A.P., M.G. Teixeira, and D.L. De Almeida (1997). Phosphorus efficiency of wild and cultivated genotypes of common bean (*Phaseolus vulgaris* L.) under biological nitrogen fixation. *Soil Biology and Biochemistry* 29: 951-957.

Asmar, F., T.S. Gahoonia, and N.E. Nielsen (1995). Barley genotypes differ in activity of soluble extracellular phosphatase and depletion of organic phosphorus in the rhizosphere soil. *Plant and Soil* 172: 117-122.

Bansal, R.L., V.L. Nayyar, and P.N. Takkar (1991). Field screening of wheat cultivars for manganese efficiency. *Field Crops Research* 29: 107-112.

Baon, J.B., S.E. Smith, and A.M. Alston (1993). Mycorrhizal responses of barley cultivars differing in P efficiency. *Plant and Soil* 157: 97-105.

Baon, J.B., S.E. Smith, and A.M. Alston (1994). Growth response and phosphorus uptake of rye with long and short root hairs: Interactions with mycorrhizal infection. *Plant and Soil* 167: 247-254.

Baon, J.B., S.E. Smith, A.M. Alston, and R.D. Wheeler (1992). Phosphorus efficiency of three cereals as related to indigenous mycorrhizal infection. *Australian Journal of Agricultural Research* 43: 479-491.

Barber, S.A. and M. Silberbush (1984). Plant root morphology and nutrient uptake. In *Roots, Nutrient and Water Influx, and Plant Growth*, Eds. S.A. Barber and D.R. Bouldin. Madison, WI: Science Society of America, Crop Science Society of America, American Society of Agronomy, pp. 65-87.

Batten, G.D. (1992). A review of phosphorus efficiency in wheat. *Plant and Soil* 146: 163-168.

Bavaresco, L. (1997). Relationship between chlorosis occurrence and mineral composition of grapevine leaves and berries. *Communications in Soil Science and Plant Analysis* 28: 13-21.

Blair, G. (1993). Nutrient efficiency—What do we really mean? In *Genetic Aspects of Plant Mineral Nutrition*, Eds. P.J. Randall, E. Delhaize, R.A. Richards, and R. Munns. Dordrecht, Netherlands: Kluwer Academic Publishers, pp. 205-213.

Bradshaw, A.D., T.S. McNeilly, and R.P.G. Gregory (1965). Industrialization, evolution and the development of heavy metal tolerance in plants. In *Ecology and the Industrial Society*, Eds. G.T. Goodman, R.W. Edwards, and J.M. Lambert. Oxford, UK: Blackwell Scientific, pp. 327-343.

Brancadoro, L., G. Rabotti, A. Scienza, and G. Zocchi (1995). Mechanisms of Fe efficiency in roots of *Vitis* spp. in response to iron deficiency stress. *Plant and Soil* 171: 229-234.

Buso, G.S.C. and F.A. Bliss (1988). Variability among lettuce cultivars grown at two levels of available phosphorus. *Plant and Soil* 111: 67-74.

Cakmak, I. (1995). Personal communication. University of Cukurova, Adana, Turkey.

Cakmak, I., R. Derici, B. Torun, I. Tolay, H.J. Braun, and R. Schlegel (1997). Role of rye chromosomes in improvement of zinc efficiency in wheat and triticale. *Plant and Soil* 196: 249-253.

Cakmak, I., H. Ekiz, A. Yilmaz, B. Torun, N. Köleli, I. Gültekin, A. Alkan, and S. Eker (1997). Differential response of rye, triticale, bread and durum wheats to zinc deficiency in calcareous soils. *Plant and Soil* 188: 1-10.

Cakmak, I., K.Y. Gülüt, H. Marschner, and R.D. Graham (1994). Effect of zinc and iron deficiency on phytosiderophore release in wheat genotypes differing in zinc efficiency. *Journal of Plant Nutrition* 17: 1-17.

Cakmak, I. and H. Marschner (1987). Mechanism of phosphorus-induced zinc deficiency in cotton. III. Changes in physiological availability of zinc in plants. *Physiologia Plantarum* 70: 13-20.

Cakmak, I. and H. Marschner (1988a). Increase in membrane permeability and exudation in roots of zinc deficient plants. *Journal of Plant Physiology* 132: 356-361.

Cakmak, I. and H. Marschner (1988b). Enhanced superoxide radical production in roots of zinc-deficient plants. *Journal of Experimental Botany* 39: 1449-1460.

Cakmak, I., L. Öztürk, S. Eker, B. Torun, H.I. Kalfa, and A. Yilmaz (1997). Concentration of zinc and activity of copper/zinc superoxide dismutase in leaves of rye and wheat cultivars differing in sensitivity to zinc deficiency. *Journal of Plant Physiology* 151: 91-95.

Cakmak, I., L. Öztürk, S. Karanlik, H. Marschner, and H. Ekiz (1996). Zinc-efficient wild grasses enhance release of phytosiderophores under zinc deficiency. *Journal of Plant Nutrition* 19: 551-563.

Cakmak, I., N. Sari, H. Marschner, H. Ekiz, M. Kalayci, A. Yilmaz, and H.J. Braun (1996). Phytosiderophore release in bread and durum wheat genotypes differing in zinc efficiency. *Plant and Soil* 180: 183-189.

Cakmak, I., N. Sari, H. Marschner, M. Kalayci, A. Yilmaz, S. Eker, and K.Y. Gülüt (1996). Dry matter production and distribution of zinc in bread and durum wheat genotypes differing in zinc efficiency. *Plant and Soil* 180: 173-181.

Cakmak, I., A. Yilmaz, M. Kalayci, H. Ekiz, B. Torun, B. Erenoglu, and H.J. Braun (1996). Zinc deficiency as a critical problem in wheat production in Central Anatolia. *Plant and Soil* 180: 165-172.

Caradus, J.R. (1994). Selection for improved adaptation of white clover to low phosphorus and acid soils. *Euphytica* 77: 243-250.

Caradus, J.R. (1995). Genetic control of phosphorus uptake and phosphorus status in plants. In *Genetic Manipulation of Crop Plants to Enhance Integrated Nutrient Management in Cropping Systems. 1. Phosphorus.* Proceedings of an FAO/ICRISAT Expert Consultancy Workshop, March 15-18, 1994, Eds. C. Johansen, K.K. Lee, K.K. Sharma, G.V. Subbarao, and E.A. Kueneman. Patancheru, Andhra Pradesh, India: International Crops Research Institute for the Semi-Arid Tropics, pp. 55-74.

Chapin, F.S. (1988). Ecological aspects of plant mineral nutrition. *Advances in Plant Nutrition* 3: 161-191.

Cianzio, S. (1999). Breeding crops for improved nutrient efficiency: Soybean and wheat as case studies. In *Mineral Nutrition of Crops: Fundamental Mechanisms and Implications*, Ed. Z. Rengel. Binghamton, NY: The Haworth Press, Inc., pp. 267-287

Cieslinski, G., K.C.J. Van Rees, A.M. Szmigielska, and P.M. Huang (1997). Low molecular weight organic acids released from roots of durum wheat and flax into sterile nutrient solutions. *Journal of Plant Nutrition* 20: 753-764.

Crowley, D.E. and Z. Rengel (1999). Biology and chemistry of nutrient availability in the rhizosphere. In *Mineral Nutrition of Crops: Fundamental Mechanisms and Implications*, Ed. Z. Rengel. Binghamton, NY: The Haworth Press, Inc., pp. 1-40.

Crush, J.R. and J.R. Caradus (1993). Effect of different aluminium and phosphorus levels on aluminium-tolerant and aluminium-susceptible genotypes of white clover *Trifolium repens* L. *New Zealand Journal of Agricultural Research* 36: 99-107.

Curnow, P. and P. Hirsch (1998). Personal communication. Rothamsted Experimental Station, Harpenden, UK.

Delhaize, E. and P.R. Ryan (1995). Aluminum toxicity and tolerance in plants. *Plant Physiology* 107: 315-321.

Dong, B., Z. Rengel, and R.D. Graham (1995). Characters of root geometry of wheat genotypes differing in Zn efficiency. *Journal of Plant Nutrition* 18: 2761-2773.

Erenoglu, B., I. Cakmak, H. Marschner, V. Römheld, S. Eker, H. Daghan, M. Kalayci, and H. Ekiz (1996). Phytosiderophore release does not relate well with Zn efficiency in different bread wheat genotypes. *Journal of Plant Nutrition* 19: 1569-1580.

Faber, B.A., R.J. Zasoski, R.G. Burau, and K. Uriu (1990). Zinc uptake by corn as affected by vesicular-arbuscular mycorrhizae. *Plant and Soil* 129: 121-130.

Fischer, E.S., O. Thimm, and Z. Rengel (1997). Zinc nutrition influences gas exchange in wheat. *Photosynthetica* 33: 505-508.

Fitter, A.H. (1991). The ecological significance of root system architecture: An economic approach. In *Plant Root Growth*, Ed. D. Atkinson. Oxford, UK: Blackwell Scientific Publications, pp. 229-243.

Foy, C.D. (1983). Plant adaptation to mineral stress in problem soils. *Iowa State Journal of Research* 57: 355-391.

Foyer, C.H., P. Descourvieres, and K.J. Kunert (1994). Protection against oxygen radicals: An important defence mechanism studied in transgenic plants. *Plant, Cell and Environment* 17: 507-523.

Gahoonia, T.S. and N.E. Nielsen (1996). Variation in acquisition of soil phosphorus among wheat and barley genotypes. *Plant and Soil* 178: 223-230.

Genc, Y., G.K. McDonald, R.D. Graham, Z. Rengel, and I. Cakmak (1998). Screening for zinc efficiency in early stage of growth of barley (*Hordeum vulgare*). In preparation for submission.

Gibson, T.S. and D.R. Leece (1981). Estimation of physiologically active zinc in maize by biochemical assay. *Plant and Soil* 63: 395-406.

Gorny, A.G. (1996). Genetic variation in the response of winter wheat to low level of nitrogen and phosphorus supply. *Plant Breeding and Seed Science* 40: 125-130.

Gourley, C.J.P., D.L. Allan, and M.P. Russelle (1994). Plant nutrient efficiency: A comparison of definitions and suggested improvement. *Plant and Soil* 158: 29-37.

Graham, M.J., C.D. Nickell, and R.G. Hoeft (1995). Inheritance of tolerance to manganese deficiency in soybean. *Crop Science* 35: 1007-1010.

Graham, M.J., P.A. Stephens, J.M. Widholm, and C.D. Nickell (1992). Soybean genotype evaluation for iron deficiency chlorosis using sodium bicarbonate and tissue culture. In preparation for submission.

Graham, R.D. (1984). Breeding for nutritional characteristics in cereals. *Advances in Plant Nutrition* 1: 57-102.

Graham, R.D. (1988). Genotypic differences in tolerance to manganese deficiency. In *Manganese in Soils and Plants*, Eds. R.D. Graham, R.J. Hannam, and N.C. Uren. Dordrecht, Netherlands: Kluwer Academic Publishers, pp. 261-276.

Graham, R.D., J.S. Ascher, P.A.E. Ellis, and K.W. Shepherd (1987). Transfer to wheat of the copper efficiency factor carried on rye chromosome arm 5RL. *Plant and Soil* 99: 107-114.

Graham, R.D., J.S. Ascher, and S.C. Hynes (1992). Selecting zinc-efficient cereal genotypes for soils of low zinc status. *Plant and Soil* 146: 241-250.

Graham, R.D. and Z. Rengel (1993). Genotypic variation in zinc uptake and utilization. In *Zinc in Soils and Plants*, Ed. A.D. Robson. Dordrecht, Netherlands: Kluwer Academic Publishers, pp. 107-118.

Grusak, M.A. and S. Pezeshgi (1996). Shoot-to-root signal transmission regulates root Fe(III) reductase activity in the dgl mutant of pea. *Plant Physiology* 110: 329-334.

Grusak, M.A., R.M. Welch, and L.V. Kochian (1990). Physiological characterization of a single-gene mutant of *Pisum sativum* exhibiting excess iron accumulation. I. Root iron reduction and iron uptake. *Plant Physiology* 93: 976-981.

Halsted, M. and J. Lynch (1996). Phosphorus responses of C-3 and C-4 species. *Journal of Experimental Botany* 47: 497-505.

Hampton, J.G. (1981). The extent and significance of seed size variation in New Zealand wheats. *New Zealand Journal of Experimental Agriculture* 9: 179-183.

Hansen, N.C. and V.D. Jolley (1995). Phytosiderophore release as a criterion for genotypic evaluation of iron efficiency in oat. *Journal of Plant Nutrition* 18: 455-465.

Hansen, N.C., V.D. Jolley, W.A. Berg, M.E. Hodges, and E.G. Krenzer (1996). Phytosiderophore release related to susceptibility of wheat to iron deficiency. *Crop Science* 36: 1473-1476.

Hansen, N.C., V.D. Jolley, and J.C. Brown (1995). Clipping foliage differentially affects phytosiderophore release by two wheat cultivars. *Agronomy Journal* 87: 1060-1063.

Haynes, B., R.T. Koide, and G. Elliott (1991). Phosphorus uptake and utilization in wild and cultivated oats (*Avena* ssp.). *Journal of Plant Nutrition* 14: 1105-1118.

Haynes, R.J. (1992). Relative ability of a range of crop species to use phosphate rock and monocalcium phosphate as phosphorus sources when grown in soil. *Journal of the Science of Food and Agriculture* 60: 205-211.

Hedley, M.J., G.J.D. Kirk, and M.B. Santos (1994). Phosphorus efficiency and the forms of soil phosphorus utilized by upland rice cultivars. *Plant and Soil* 158: 53-62.

Helal, H.M. (1990). Varietal differences in root phosphatase activity as related to utilization of organic phosphates. *Plant and Soil* 123: 161-163.

Hornby, D. and A.J. Ullstrup (1967). Fungal populations associated with maize roots. Composition and comparison of mycofloras from genotypes differing in root rot resistance. *Phytopathology* 57: 869-875.

Horst, W.J., M. Abdou, and F. Wiesler (1996). Differences between wheat cultivars in acquisition and utilization of phosphorus. *Zeitschrift für Pflanzenernährung und Bodenkunde* 159: 155-161.

Howie, W.J. and E. Echandi (1983). Rhizobacteria: Influence of cultivar and soil type in plant growth and yield of potato. *Soil Biology and Biochemistry* 15: 127-132.

Huang, C., R.D. Graham, S.J. Barker, and S. Mori (1996). Differential expression of iron deficiency-induced genes in barley genotypes with differing manganese efficiency. *Journal of Plant Nutrition* 19: 407-420.

Huang, C., M.J. Webb, and R.D. Graham (1993). Effect of pH on Mn absorption by barley genotypes in a chelate-buffered nutrient solution. *Plant and Soil* 155/156: 437-440.

Huang, C., M.J. Webb, and R.D. Graham (1994). Manganese efficiency is expressed in barley growing in soil system but not in a solution culture. *Journal of Plant Nutrition* 17: 83-95.

Huang, C., M.J. Webb, and R.D. Graham (1996). Pot size affects expression of Mn efficiency in barley. *Plant and Soil* 178: 205-208.

Itoh, S. and S.A. Barber (1983). Phosphorus uptake by six plant species as related to root hairs. *Agronomy Journal* 75: 457-461.

Jolley, V.D. and J.C. Brown (1989). Iron inefficient and efficient oat cultivars. II. Characterization of phytosiderophore released in response to iron deficiency stress. *Journal of Plant Nutrition* 12(7): 923-937.

Jolley, V.D. and J.C. Brown (1991). Factors in iron-stress response mechanism enhanced by zinc-deficiency stress in Sanilac, but not Saginaw navy bean. *Journal of Plant Nutrition* 14: 257-266.

Jolley, V.D. and J.C. Brown (1994). Genetically controlled uptake and use of iron by plants. In *Biochemistry of Metal Micronutrients in the Rhizosphere*, Eds. J.A. Manthey, D.E. Crowley, and D.G. Luster. Boca Raton, FL: Lewis Publishers/CRC Press, pp. 251-266.

Jolley, V.D., K.A. Cook, N.C. Hansen, and W.B. Stevens (1996). Plant physiological responses for genotypic evaluation of iron efficiency in strategy I and strategy II plants. A review. *Journal of Plant Nutrition* 19: 1241-1255.

Jones, G.P.D., G.J. Blair, and R.S. Jessop (1989). Phosphorus efficiency in wheat—A useful selection criterion? *Field Crops Research* 21: 257-264.

Jones, G.P.D., R.S. Jessop, and G.J. Blair (1992). Alternative methods for the selection of phosphorus efficiency in wheat. *Field Crops Research* 30: 29-40.

Kemp, P.D. and G.J. Blair (1994). Phosphorus efficiency in pasture species. VIII. Ontogeny, growth, P acquisition and P utilization of Italian ryegrass and phalaris under P deficient and P sufficient conditions. *Australian Journal of Agricultural Research* 45: 669-688.

Kochian, L.V. (1993). Zinc absorption from hydroponic solutions by plant roots. In *Zinc in Soils and Plants*, Ed. A.D. Robson. Dordrecht, Netherlands: Kluwer Academic Publishers, pp. 45-57.

Kothari, S.K., H. Marschner, and V. Römheld (1991). Contribution of VA mycorrhizal hyphae in acquisition of phosphorus and zinc by maize grown in a calcareous soil. *Plant and Soil* 131: 177-185.

Kubota, J. and W.H. Allaway (1972). Geographic distribution of trace element problems. In *Micronutrients in Agriculture*, Eds. J.J. Mortvedt, P.M. Giordano, and W.L. Lindsay. Madison, WI: Soil Science Society of America, pp. 525-554.

Landsberg, E.C. (1996). Hormonal regulation of iron-stress response in sunflower roots: A morphological and cytological investigation. *Protoplasma* 194: 69-80.

Lemanceau, P., T. Corberand, L. Gardan, X. Latour, G. Laguerre, J.M. Boeufgras, and C. Alabouvette (1995). Effect of two plant species, flax (*Linum usitatissimum* L) and tomato (*Lycopersicon esculentum* Mill), on the diversity of soilborne populations of fluorescent pseudomonads. *Applied and Environmental Microbiology* 61: 1004-1012.

Liljeroth, E., J.A. Van Veen, and H.J. Miller (1990). Assimilate translocation to the rhizosphere of two wheat lines and subsequent utilization by rhizosphere microorganisms at two soil nitrogen concentrations. *Soil Biology and Biochemistry* 22: 1015-1021.

Little, R. (1988). Plant-soil interactions at low pH. Problem-solving genetic approach. *Communications in Soil Science and Plant Analysis* 19: 1239-1257.

Liu, S.J., F.S. Zhang, and D.R. Mao (1996). Genotypic difference in responses to iron deficiency of *Glycine max*. *Acta Phytophysiologica Sinica* 22: 1-5.

Loneragan, J. and M.J. Webb (1993). Interactions between zinc and other nutrients affecting the growth of plants. In *Zinc in Soils and Plants*, Ed. A.D. Robson. Dordrecht, Netherlands: Kluwer Academic Publishers, pp. 119-134.

Longnecker, N.E. and A.D. Robson (1993). Distribution and transport of zinc in plants. In *Zinc in Soils and Plants*, Ed. A.D. Robson. Dordrecht, Netherlands: Kluwer Academic Publishers, pp. 79-91.

Longnecker, N. and R.M. Welch (1990). Accumulation of apoplastic iron in plant roots: A factor in the resistance of soybeans to iron deficiency induced chlorosis? *Plant Physiology* 92: 17-22.

Lynch, J. (1998). The role of nutrient-efficient crops in modern agriculture. In *Nutrient Use in Crop Production*, Ed. Z. Rengel. Binghamton, NY: The Haworth Press, Inc., pp. 241-264.

MacNair, M.R. (1993). The genetics of metal tolerance in vascular plants. *New Phytologist* 124: 541-559.

Majumder, N.D., S.C. Rakshit, and D.N. Borthakur (1990). Genetic effects on uptake of selected nutrients in some rice (*Oryza sativa* L.) varieties in phosphorus deficient soil. *Plant and Soil* 123: 117-120.

Makino, A., H. Sahashita, J. Hidema, T. Mae, K. Ojima, and B. Osmond (1992). Distinctive responses of ribulose-1,5-bisphosphate carboxylase and carbonic anhydrase in wheat leaves to nitrogen nutrition and their possible relationships to the CO_2-transfer resistance. *Plant Physiology* 100: 1737-1743.

Mamo, T., C. Richter, and A. Hoppenstedt (1996). Phosphorus response studies on some varieties of durum wheat (*Triticum durum* Desf.) and tef (*Eragrostis tef* [Zucc.] Trotter) grown in sand culture. *Journal of Agronomy and Crop Science* 176: 189-197.

Marcar, N.E. (1986). Genotypic Variation for Manganese Efficiency in Cereals. Doctoral thesis, University of Adelaide, Adelaide, Australia.

Marschner, H. (1993). Zinc uptake from soils. In *Zinc in Soils and Plants*, Ed. A.D. Robson. Dordrecht, Netherlands: Kluwer Academic Publishers, pp. 59-77.

Marschner, H. (1995). Rhizosphere pH effects on phosphorus nutrition. In *Genetic Manipulation of Crop Plants to Enhance Integrated Nutrient Management in Cropping Systems. 1. Phosphorus.* Proceedings of an FAO/ICRISAT Expert Consultancy Workshop, March 15-18, 1994, Eds. C. Johansen, K.K. Lee, K.K. Sharma, G.V. Subbarao, and E.A. Kueneman. Patancheru, Andhra Pradesh, India: International Crops Research Institute for the Semi-Arid Tropics, pp. 107-115.

Marschner, H. and B. Dell (1994). Nutrient uptake in mycorrhizal symbiosis. *Plant and Soil* 159: 89-102.

Marschner H. and V. Römheld (1994). Strategies of plants for acquisition of iron. *Plant and Soil* 165: 261-274.

Mian, A.R. and E.D. Nafziger (1992). Seed size effects on emergence, head number, and grain yield of winter wheat. *Journal of Production Agriculture* 5: 265-268.

Mori, S. (1994). Mechanisms of iron acquisition by graminaceous (strategy II) plants. In *Biochemistry of Metal Micronutrients in the Rhizosphere*, Eds. J.A. Manthey, D.E. Crowley, and D.G. Luster. Boca Raton, FL: Lewis Publishers/ CRC Press, pp. 225-249.

Mori, S. and N. Nishizawa (1987). Methionine as a dominant precursor of phytosiderophores in *Graminaceae* plants. *Plant and Cell Physiology* 28: 1081-1092.

Moussavi-Nik, M., Z. Rengel, N.J. Pearson, and G. Hollamby (1997). Dynamics of nutrient remobilisation from seed of wheat genotypes during imbibition, germination and early seedling growth. *Plant and Soil* 197: 271-280.

Nable, R.O. and M.J. Webb (1993). Further evidence that zinc is required throughout the root zone for optimal growth and development. *Plant and Soil* 150: 247-253.

Nielsen, N.E. and J.K. Schjorring (1983). Efficiency and kinetics of phosphorus uptake from soil by various barley genotypes. *Plant and Soil* 72: 225-230.

Otani, T. and N. Ae (1996). Sensitivity of phosphorus uptake to changes in root length and soil volume. *Agronomy Journal* 88: 371-375.

Pearson, N.J. and Z. Rengel (1994). Distribution and remobilization of Zn and Mn during grain development in wheat. *Journal of Experimental Botany* 45: 1829-1835.

Pearson, N.J. and Z. Rengel (1997). Mechanisms of plant resistance to nutrient deficiency stresses. In *Mechanisms of Environmental Stress Resistance in Plants*, Eds. A.S. Basra and R.K. Basra. Amsterdam, Netherlands: Harwood Academic Publishers, pp. 213-240.

Pearson, J.N., Z. Rengel, C.F. Jenner, and R.D. Graham (1995). Transport of zinc and manganese to developing wheat grains. *Physiologia Plantarum* 95: 449-455.

Pich, A., S. Hillmer, R. Manteuffel, and G. Scholz (1997). First immunohistochemical localization of the endogenous Fe^{2+}-chelator nicotianamine. *Journal of Experimental Botany* 48: 759-767.

Posta, K., H. Marschner, and V. Römheld (1994). Manganese reduction in the rhizosphere of mycorrhizal and nonmycorrhizal maize. *Mycorrhiza* 5: 119-124.

Randall, P.J. (1995). Genotypic differences in phosphate uptake. In *Genetic Manipulation of Crop Plants to Enhance Integrated Nutrient Management in Cropping*

Systems. 1. Phosphorus. Proceedings of an FAO/ICRISAT Expert Consultancy Workshop, March 15-18, 1994, Eds. C. Johansen, K.K. Lee, K.K. Sharma, G.V. Subbarao, and E.A. Kueneman. Patancheru, Andhra Pradesh, India: International Crops Research Institute for the Semi-Arid Tropics, pp. 31-47.

Rao, I.M., V. Borrero, J. Ricaurte, R. Garcia, and M.A. Ayarza (1997). Adaptive attributes of tropical forage species to acid soils. III. Differences in phosphorus acquisition and utilization as influenced by varying phosphorus supply and soil type. *Journal of Plant Nutrition* 20: 155-180.

Reddy, K.B., M. Ashalatha, and K. Venkaiah (1993). Differential response of groundnut genotypes to iron-deficiency stress. *Journal of Plant Nutrition* 16: 523-531.

Rengel, Z. (1992). Role of calcium in aluminium toxicity. *New Phytologist* 121: 499-513.

Rengel, Z. (1993). Mechanistic simulation models of nutrient uptake: A review. *Plant and Soil* 152: 161-173.

Rengel, Z. (1995a). Carbonic anhydrase activity in leaves of wheat genotypes differing in Zn efficiency. *Journal of Plant Physiology* 147: 251-256.

Rengel, Z. (1995b). Sulfhydryl groups in root-cell plasma membranes of wheat genotypes differing in Zn efficiency. *Physiologia Plantarum* 95: 604-612.

Rengel, Z. (1997a). Mechanisms of plant resistance to aluminium and heavy metals. In *Mechanisms of Environmental Stress Resistance in Plants*, Eds. A.S. Basra and R.K. Basra. Amsterdam, Netherlands: Harwood Academic Publishers, pp. 241-276.

Rengel, Z. (1997b). Root exudation and microflora populations in rhizosphere of crop genotypes differing in tolerance to micronutrient deficiency. *Plant and Soil* 196: 255-260.

Rengel, Z. and R.D. Graham (1995a). Wheat genotypes differ in Zn efficiency when grown in chelate-buffered nutrient solution. I. Growth. *Plant and Soil* 176: 307-316.

Rengel, Z. and R.D. Graham (1995b). Wheat genotypes differ in Zn efficiency when grown in chelate-buffered nutrient solution. II. Nutrient uptake. *Plant and Soil* 176: 317-324.

Rengel, Z. and R.D. Graham (1995c). Importance of seed Zn content for wheat growth on Zn-deficient soil. I. Vegetative growth. *Plant and Soil* 173: 259-266.

Rengel, Z. and R.D. Graham (1995d). Importance of seed Zn content for wheat growth on Zn-deficient soil. II. Grain yield. *Plant and Soil* 173: 267-274.

Rengel, Z. and R.D. Graham (1996). Uptake of zinc from chelate-buffered nutrient solutions by wheat genotypes differing in Zn efficiency. *Journal of Experimental Botany* 47: 217-226.

Rengel, Z., R. Guterridge, P. Hirsch. and D. Hornby (1996). Plant genotype, micronutrient fertilization and take-all infection influence bacterial populations in the rhizosphere of wheat. *Plant and Soil* 183: 269-277.

Rengel, Z. and M.J. Hawkesford (1997). Biosynthesis of a 34-kDa polypeptide in the root-cell plasma membrane of Zn-efficient wheat genotype increases upon Zn starvation. *Australian Journal of Plant Physiology* 24: 307-315.

Rengel, Z. and V. Jurkic (1992). Genotypic differences in wheat Al tolerance. *Euphytica* 62: 111-117.

Rengel, Z., J.F. Pedler, and R.D. Graham (1994). Control of Mn status in plants and rhizosphere: Genetic aspects of host and pathogen effects in the wheat take-all interaction. In *Biochemistry of Metal Micronutrients in the Rhizosphere*, Eds. J.A. Manthey, D.E. Crowley, and D.G. Luster. Boca Raton, FL: Lewis Publishers/CRC Press, pp. 125-145.

Rengel, Z., V. Römheld, and H. Marschner (1998). Uptake of zinc and iron by wheat genotypes differing in zinc efficiency. *Journal of Plant Physiology* 152: 433-438.

Rengel, Z., G. Ross, and P. Hirsch (1998). Plant genotype and micronutrient status influence colonization of wheat roots by soil bacteria. *Journal of Plant Nutrition* 21: 99-113.

Rengel, Z. and M.S. Wheal (1997a). Herbicide chlorsulfuron decreases growth of fine roots and micronutrient uptake in wheat genotypes. *Journal of Experimental Botany* 48: 927-934.

Rengel, Z. and M.S. Wheal (1997b). Kinetics of Zn uptake by wheat is affected by herbicide chlorsulfuron. *Journal of Experimental Botany* 48: 935-941.

Römer, W., J. Augustin, and G. Schilling (1988). The relationship between phosphate absorption and root length in nine wheat cultivars. *Plant and Soil* 111: 199-201.

Romera, F.J., R.M. Welch, W.A. Norvell, S.C. Schaefer, and L.V. Kochian (1996). Ethylene involvement in the over-expression of Fe(III)-chelate reductase by roots of E107 pea (*Pisum sativum* L. (brz, brz) and chloronerva tomato (*Lycopersicon esculentum* L.) mutant genotypes. *Biometals* 9: 38-44.

Römheld, V. (1991). The role of phytosiderophores in acquisition of iron and other micronutrients in graminaceous species: An ecological approach. *Plant and Soil* 130: 127-134.

Römheld, V. and H. Marschner (1990). Genotypic differences among gramineaceous species in release of phytosiderophores and uptake of iron phytosiderophores. *Plant and Soil* 123: 147-153.

Sanchez, P.A. and J.G. Salinas (1981). Low-input technology for managing Oxisol and Ultisol in tropical America. *Advances in Agronomy* 34: 279-406.

Santa Maria, G.E. and D.H. Cogliatti (1988). Bidirectional Zn-fluxes and compartmentation in wheat seedling roots. *Journal of Plant Physiology* 132: 312-315.

Sattelmacher, B., W.J. Horst, and H.C. Becker (1994). Factors that contribute to genetic variation for nutrient efficiency of crop plants. *Zeitschrift für Pflanzenernährung und Bodenkunde* 157: 215-224.

Schenk, M.K. and S.A. Barber (1979). Root characteristics of corn genotypes as related to P uptake. *Agronomy Journal* 71: 921-924.

Schwarz, K.-U., J. Léon, and G. Geisler (1991). Genetic variability in root development of spring barley (*H. vulgare* L.) and oats (*A. sativa* L.). In *Plant Roots and Their Environment*, Eds. B.L. McMichael and H. Persson. Amsterdam, Netherlands: Elsevier Science Publishers, pp. 639-646.

Sharma, R.C. and H.N. Lafever (1992). Variation for root traits and their genetic control in spring wheat. *Euphytica* 59: 1-8.

Sharma, A.K. and P.C. Srivastava (1991). Effect of vesicular-arbuscular mycorrhizae and zinc application on dry matter and zinc uptake of green gram *Vigna radiata* L. Wilczek. *Biology and Fertility of Soils* 11: 52-56.

Sharma, A.K., P.C. Srivastava, B.N. Johri, and V.S. Rathore (1992). Kinetics of zinc uptake by mycorrhizal (VAM) and non-mycorrhizal corn (*Zea mays* L.) roots. *Biology and Fertility of Soils* 13: 206-210.

Shroyer, J.P. and T.S. Cox (1993). Productivity and adaptive capacity of winter wheat landraces and modern cultivars grown under low-fertility conditions. *Euphytica* 70: 27-33.

Snapp, S., R. Koide, and J. Lynch (1995). Exploitation of localized phosphorus-patches by common bean roots. *Plant and Soil* 177: 211-218.

Soon, Y.K. (1992). Differential response of wheat genotypes to phosphorus in acid soils. *Journal of Plant Nutrition* 15: 513-526.

Stelling, D., S.H. Wang, and W. Römer (1996). Efficiency in the use of phosphorus, nitrogen and potassium in topless faba beans (*Vicia faba* L.). Variability and inheritance. *Plant Breeding* 115: 361-366.

Stephan, U.W. and M. Grün (1989). Physiological disorders of the nicotianamine-auxotroph tomato mutant *chloronerva* at different levels of iron nutrition. II. Iron deficiency response and heavy metal metabolism. *Biochemie und Physiologie der Pflanzen* 185: 189-200.

Stevens, W.B., V.D. Jolley, N.C. Hansen, and D.J. Fairbanks (1993). Modified procedures for commercial adaptation of root iron-reducing capacity as a screening technique. *Journal of Plant Nutrition* 16: 2507-2519.

Subbarao, G.V., N. Ae, and T. Otani (1997a). Genetic variation in acquisition and utilization of phosphorus from iron-bound phosphorus in pigeonpea. *Soil Science and Plant Nutrition* 43: 511-519.

Subbarao, G.V., N. Ae, and T. Otani (1997b). Genotypic variation in iron- and aluminum-phosphate solubilizing activity of pigeonpea root exudates under P deficient conditions. *Soil Science and Plant Nutrition* 43: 295-305.

Thongbai, P., R.J. Hannam, R.D. Graham, and M.J. Webb (1993). Interaction between zinc nutritional status of cereals and *Rhizoctonia* root rot severity. I. Field observation. *Plant and Soil* 153: 207-214.

Timonin, M.I. (1946). Microflora of the rhizosphere in relation to the manganese-deficiency disease of oats. *Soil Science Society of America Proceedings* 11: 284-292.

Timonin, M.I. (1965). Interaction of higher plants and soil microorganisms. In *Microbiology and Soil Fertility*, Eds. C.M. Gilmore and O.N. Allen. Corvallis. OR: Oregon State University Press, pp. 135-138.

Tong, Y., Z. Rengel, and R.D. Graham (1997). Interactions between nitrogen and manganese nutrition of barley genotypes differing in manganese efficiency. *Annals of Botany* 79: 53-58.

Trolove, S.N., M.J. Hedley, J.R. Caradus, and A.D. Mackay (1996a). Uptake of phosphorus from different sources by *Lotus pedunculatus* and three genotypes of *Trifolium repens*. 1. Plant yield and phosphate efficiency. *Australian Journal of Soil Research* 34: 1015-1026.

Trolove, S.N., M.J. Hedley, J.R. Caradus, and A.D. Mackay (1996b). Uptake of phosphorus from different sources by *Lotus pedunculatus* and three genotypes of *Trifolium repens*. 2. Forms of phosphate utilized and acidification of the rhizosphere. *Australian Journal of Soil Research* 34: 1027-1040.

Van Overbeek, L.S., J.A. Van Veen, and J.D. Van Elsas (1997). Induced reporter gene activity, enhanced stress resistance, and competitive ability of a genetically modified *Pseudomonas fluorescens* strain released into a field plot planted with wheat. *Applied and Environmental Microbiology* 63: 1965-1973.

Von Wiren, N., H. Marschner, and V. Römheld (1995). Uptake kinetics of iron-phytosiderophores in two maize genotypes differing in iron efficiency. *Physiologia Plantarum* 93: 611-616.

Von Wiren, N., S. Mori, H. Marschner, and V. Römheld (1994). Iron inefficiency in maize mutant *ys1* (*Zea mays* L. cv yellow-stripe) is caused by a defect in uptake of iron phytosiderophores. *Plant Physiology* 106: 71-77.

Walley, K.A., M.S.I. Khan, and A.D. Bradshaw (1974). The potential for evolution of heavy metal tolerance in plants. I. Copper and zinc tolerance in *Agrostis tenuis*. *Heredity* 32: 309-319.

Walter, A., V. Römheld, H. Marschner, and S. Mori (1994). Is the release of phytosiderophores in zinc-deficient wheat plants a response to impaired iron utilization? *Physiologia Plantarum* 92: 493-500.

Warncke, D.D. and S.A. Barber (1972). Diffusion of zinc in soil. *Soil Science Society of America Proceedings* 36: 42-46.

Wei, L.C., R.H. Loeppert, and W.R. Ocumpaugh (1997). Fe-deficiency stress response in Fe-deficiency resistant and susceptible subterranean clover: Importance of induced H^+ release. *Journal of Experimental Botany* 48: 239-246.

Wei, L.C., W.R. Ocumpaugh, and R.H. Loeppert (1994). Differential effect of soil temperature on iron-deficiency chlorosis in susceptible and resistant subclovers. *Crop Science* 34: 715-721.

Welch, R.M. (1995). Micronutrient nutrition of plants. *Critical Reviews in Plant Science* 14: 49-82.

Welch, R.M., W.H. Allaway, W.A. House, and J. Kubota (1991). Geographic distribution of trace element problems. In *Micronutrients in Agriculture*, Second Edition, Eds. J.J. Mortvedt, F.R. Cox, L.M. Shuman, and R.M. Welch. Madison, WI: Soil Science Society of America, pp. 31-57.

Welch, R.M. and T.A. La Rue (1990). Physiological characteristics of Fe accumulation in the bronze mutant of *Pisum sativum* L. cultivar Sparkle E107 brz brz. *Plant Physiology* 93: 723-729.

Welch, R.M. and W.A. Norvell (1993). Growth and nutrient uptake by barley (*Hordeum vulgare* L. cv Herta): Studies using an N-(2-hydroxyethyl)ethylenedinitrilotriacetic acid-buffered nutrient solution technique. *Plant Physiology* 101: 627-631.

Welkie, G.W. (1996). Iron-deficiency stress responses of a chlorosis-susceptible and a chlorosis-resistant cultivar of muskmelon as related to root riboflavin excretion. *Journal of Plant Nutrition* 19: 1157-1169.

Wellings, N.P., A.H. Wearing, and J.P. Thompson (1991). Vesicular-arbuscular mycorrhizae (VAM) improve phosphorus and zinc nutrition and growth of pigeonpea in a vertisol. *Australian Journal of Agricultural Research* 42: 835-845.

Wenzel, A.A. and H. Mehlhorn (1995). Zinc deficiency enhances ozone toxicity in bush beans (*Phaseolus vulgaris* L. cv. *Saxa*). *Journal of Experimental Botany* 46: 867-872.

Wheal, M.S. and Z. Rengel (1997). Chlorsulfuron reduces rates of zinc uptake by wheat seedlings in solution culture. *Plant and Soil* 188: 309-317.

White, C.L., A.D. Robson, and H.M. Fisher (1981). Variation in nitrogen, sulfur, selenium, cobalt, manganese, copper and zinc contents of grain from wheat and two lupin species grown in a range of Mediterranean environments. *Australian Journal of Agricultural Research* 32: 47-59.

Wiehe, W. and G. Hoflich (1995). Survival of plant growth promoting rhizosphere bacteria in the rhizosphere of different crops and migration to non-inoculated plants under field conditions in north-east Germany. *Microbiological Research* 150: 201-206.

Xavier, L.J.C. and J.J. Germida (1997). Growth response of lentil and wheat to *Glomus clarum* NT4 over a range of P levels in a Saskatchewan soil containing indigenous AM fungi. *Mycorrhiza* 7: 3-8.

Yan, X., J.P. Lynch, and S.E. Beebe (1995). Genetic variation for phosphorus efficiency of common bean in contrasting soil types. I. Vegetative response. *Crop Science* 35: 1086-1093.

Yan, X., J.P. Lynch, and S.E. Beebe (1996). Utilization of phosphorus substrates by contrasting common bean genotypes. *Crop Science* 36: 936-941.

Yang, X., V. Römheld, and H. Marschner (1993). Effect of bicarbonate and root zone temperature on uptake of Zn, Fe, Mn and Cu by different rice cultivars (*Oryza sativa* L.) grown in calcareous soil. *Plant and Soil* 155/156: 441-444.

Yang, X., V. Römheld, and H. Marschner (1994). Effect of bicarbonate on root growth and accumulation of organic acids in Zn-inefficient and Zn-efficient rice cultivars (*Oryza sativa* L.). *Plant and Soil* 164: 1-7.

Yeo, A.R. and T.J. Flowers (1986). Salinity resistance in rice (*Oryza sativa* L.) and a pyramiding approach to breeding varieties for saline soils. *Australian Journal of Plant Physiology* 13: 161-173.

Zhang, F.S., S.J. Liu, and D.R. Mao (1996). Genotype differences in Fe accumulation and mobilization in root apoplast of Fe-deficient apple seedling. *Acta Phytophysiologica Sinica* 22: 357-362.

Zhang, F.S., V. Römheld, and H. Marschner (1989). Effect of zinc deficiency in wheat on the release of zinc and iron mobilizing root exudates. *Zeitschrift für Pflanzenernährung und Bodenkunde* 152: 205-210.

Zhang, F.S., V. Römheld, and H. Marschner (1991). Diurnal rhythm of release of phytosiderophores and uptake rate of zinc in iron-deficient wheat. *Soil Science and Plant Nutrition* 37: 671-678.

Chapter 10

Breeding Crops for Improved Nutrient Efficiency: Soybean and Wheat As Case Studies

Silvia R. Cianzio

Key Words: mineral deficiency, mineral toxicity, soybean, wheat.

INTRODUCTION

Crops continually cope with stresses that may cause yield loss. Reduction of environmental stress effects on plants involves either alteration of the environment or the plant, or both. During seed germination and early growth, seedlings depend upon minerals stored in the seed for development; upon depletion of this source, plants become dependent on soil minerals. If the required minerals are not supplied in sufficient amounts, or are in excess, the plant may exhibit visual symptoms of mineral deficiency or toxicity. Mineral stresses in soils are related to parent material and soil-forming processes and are important factors that restrict plant growth and crop production. Approximately one-fourth of the arable land in the world is affected by some kind of mineral stress (Dudal, 1976). Soil pH is also often associated with mineral stress problems (Clark, 1982). The greatest availability of most elements in organic soils is at about pH 5.5, and in mineral soils, at about 6.5.

Acid soils occur throughout the world and are frequently low in available P (phosphorus) and base exchange capacity, and high in leaching potential (Van Wambeke, 1976, cited in Clark, 1982). To maintain fertility in these soils, nearly all nutrients must be added. Acidity increases the

The author wishes to acknowledge Journal Paper No. J-17282 of the Iowa Agriculture and Home Economics Experiment Station, Ames, IA 50011, USA. Project No. 3107.

availability of Fe (iron), Mn (manganese), and Al (aluminum); these elements may be toxic to plants in acid soils.

Alkaline soils are usually adequately supplied with Ca (calcium) and Mg (magnesium), although in some locations, Mg may be high enough to limit Ca availability (Flach, 1976, cited in Clark, 1982). In others, K (potassium) may be adequate for crop production, but if vermiculite clays are present, K may be fixed and generally unavailable. Additions of P and N (nitrogen) are usually required, and S (sulfur) deficiencies also have been known. Boron (B) toxicity may occur, and plants unusually high in Mo (molybdenum) and F (fluorine) may be toxic to animals. Most alkaline soils contain enough total Fe to supply plants almost indefinitely, but available Fe in these soils is frequently insufficient to satisfy plant needs.

Changes in the environment may include use of fertilizers, pesticides, irrigation, and other practices, some of which are regarded with concern by environmentalists and others in the research community. The alternative is to improve plant adaptation to problem soils by genetic methodologies and breeding (Devine, 1982). The decision to conduct breeding in a particular crop for improving the nutrient efficiency is primarily dependent upon the economic impact the problem has in the region (Fehr, 1983). This, in turn, is a function of the size of the area and severity of the problem. Prior knowledge of the genetics of the trait (i.e., inheritance of the characteristic, extent of genetic variability, and screening techniques for assessment of genotypes) is desirable, but not required.

Concentrating on wheat and soybean, this chapter will summarize current knowledge on inheritance of nutrient efficiency traits, breeding strategies that have been used to improve these traits, and screening techniques to evaluate efficiency of micronutrient utilization by genotypes. The term nutrient efficiency has been widely used as a measure of the capacity of a plant to acquire and utilize nutrients for crop production (Gourley, Allan, and Russelle, 1994) and is used as such in this chapter. Definitions of nutrient efficiency, however, vary greatly because there are many ways in which plants can adapt or respond to nutrient stress (Blair, 1993).

NUTRIENT EFFICIENCY AND TOLERANCE TO ION TOXICITY: INHERITANCE AND BREEDING STRATEGIES

Wheat

Nutrient Efficiency

In wheat (*Triticum aestivum* L.) and other cereals of the tribe *Triticeae,* such as barley and oat, genetic diversity has been identified for tolerance to deficiencies of several micro- and macronutrients. Genetic studies of the inheritance of mineral stresses in the species have been conducted by chromosome analysis, which was possible because of the availability of aneuploid stocks. Observations of tolerance to low or excess levels of minerals indicate that in general, these traits might be simply inherited, with efficiency being generally dominant. The potential for improving nutrient efficiency in wheat was realized after examination of variation in wheat genotypes and its near relatives.

Genetic variation in species of the *Triticeae* has been identified for tolerance to deficiencies of Cu (copper), Zn (zinc), and Mn in the soil (Graham, 1988; Graham, Ascher, and Hynes, 1993) and to excess levels of B (Paull, Nable, and Rathjen, 1993). Research conducted in Adelaide, Australia, demonstrated that efficiency of rye to utilize Cu, Zn, and Mn was inherited by the amphiploid hybrid triticale and, therefore, immediately available to wheat in a genetically related background. Copper efficiency in rye and triticale is determined by a single gene carried on the 5RL chromosome arm (Graham, 1978). The 5RL translocation lines of rye were the donor parent in backcrosses to transfer the trait to locally adapted wheat genotypes. The highly Cu-efficient wheat lines developed have been top-yielding lines on Cu-deficient soils (Graham, 1988). At nondeficient sites, the lines have yielded similarly to their respective recurrent parent, indicating no suppressive effects on yield, nor in baking quality, by the presence of the 5RL chromosome segment.

Efforts to improve Zn efficiency in wheat were also attempted by transferring the trait from rye using the backcross breeding method (Graham, 1988). The wheat lines obtained from the program, however, did not attain yield levels of previously released cultivars. More recently, evaluation of the lines conducted both in field (Graham, Ascher, and Hynes, 1993) and nutrient solution tests (Rengel and Graham, 1995a, b) showed that no differences were observed in Zn efficiency between lines from the backcross program and the original wheat lines. These results could, in part, explain why the backcrossed wheat lines did not show any yield advantage over the original genotypes used in the program. The trait also appears to be inherited in a somewhat complex manner, as it was recently reported by Graham (1994). He indicated that results from a study conducted to determine inheritance of Zn utilization suggested that there might be at least three genes involved in the expression of Zn efficiency.

A breeding program to develop Mn-efficient cereals also has been implemented, following observations that widely grown wheat and barley cultivars were sensitive to Mn-deficient soils (Graham, 1988). Rye was the donor of Mn efficiency trait, which was transferred to wheat by backcrossing. A study of the rye-addition lines of wheat revealed that the rye chromosome 2R contributed significantly to Mn efficiency, with a lesser contribution from chromosome 6R. Contributions by minor genes have also been detected (Graham, 1994). Screening and characterization of existing wheat germplasm has identified genotypes with improved Mn efficiency (Graham, 1994).

Soils deficient in B cause severe yield depression of certain genotypes due to failure of grain set (Jamjod, Mann, and Rerkasem, 1993). A diallel analysis of seven wheat lines indicated the component due to general combining ability was highly significant in response to B.

No published reports on wheat indicate the presence of visual symptoms due to Fe deficiency. Symptoms of Fe deficiency, however, have been observed in a number of other graminaceous species in which cultivars and germplasm lines have been released, and inheritance studies have been conducted—oat (McDaniel and Brown, 1982), sorghum (Mikesell et al., 1973; Esty et al., 1980; Williams et al., 1987), and rice (Hoan, Rao, and Siddiq, 1992). Nevertheless, wheat has been used in physiological and biochemical studies in relation to mobilization and uptake of Fe (Awad, Römheld, and Marschner, 1995; Marschner and Römheld, 1995).

Strategies in Fe acquisition of higher plants have been studied extensively, both in dicotyledonous and monocotyledonous species (Brown and Jolley, 1989; Marschner and Römheld, 1995). Graminaceous species acquire Fe by the system usually referred to as Strategy II, which is characterized by two components: the release of phytosiderophores and a high-affinity transport system in the plasma membrane of root cells for the Fe^{3+}-phytosiderophore complex (Marschner and Römheld, 1995). Release of phytosiderophores is strongly enhanced by Fe deficiency (Takagi and Takemoto, 1984, cited in Marschner and Römheld, 1995) and is not affected by external pH (Marschner et al., 1986, cited in Marschner and Römheld, 1995).

Genotypic differences in wheat, other cereals, and species of the Graminaceae family also have been identified in nutrient efficiency of macronutrients. In wheat, grain yield of modern cultivars evaluated under both limiting and nonlimiting P supply is higher than yield of traditional cultivars, suggesting that modern cultivars are more P efficient (Palmer and Jessop, 1977; Barriga et al., 1993; Horst, Abdou, and Wiesler, 1993). The higher P efficiency of the modern cultivars may be attributed to a more efficient use of assimilates for root growth, which enhances P acquisition, and to lower P requirement for grain yield. The literature does not indicate if breeding for P efficiency is presently underway; however, results from the cited studies suggest that the trait may become another objective in some wheat breeding programs; screening techniques to evaluate P efficiency in wheat have been devised (Barriga et al., 1993; Jones, Jessop, and Blair, 1992). Chromosomes shown to carry genes directly or indirectly involved in the use of P, and also N and K, are 1A, 4A, 5A, 2B, 4B, and 1D (Gamzikova, 1994). Information in relation to P efficiency also is available for rice (Fageria, Wright, and Baligar, 1988; Majumder, Borthakur, and Rakshit, 1989; Chaubey, Senadhira, and Gregorio, 1994) and corn (Da Silva, Gabelman, and Coors, 1993).

Genetic variation was observed in the physiological efficiency index of N in triticale (Isfan, Cserni, and Tabi, 1991), oat (Isfan, 1993), and rice (Wu and Tao, 1995). In corn, seven cycles of recurrent selection conducted in field plantings under limited supply of macronutrients have demonstrated the effectiveness of selection to improve macronutrient efficiency (Malagoli, Ferrari, and Saccomani, 1993).

Tolerance to Ion Toxicity

Tolerance to heavy metals, B, and salt has also been investigated in wheat. For salt tolerance, analysis by chromosome manipulation identified a major locus in the D genome of the hexaploid bread wheat, *Kna1*, responsible for K^+/Na^+ discrimination (Dvorak, 1994). The gene was transferred to the B genome of the tetraploid macaroni (durum) wheat by backcrossing to enhance salt tolerance of genotypes.

Inheritance of tolerance to excess B in wheat appears determined by at least three major genes, *Bo1, Bo2, and Bo3* acting in an additive manner, although observations of transgressive segregants in some crosses suggest the presence of more genes (Paull, Nable, and Rathjen, 1993). A fourth gene, *Bo4*, was identified later (Campbell, Paull, and Rathjen, 1994). The four major genes act additively, causing a range in expression of B tolerance. The chromosomes involved in expression of the trait include 4A, 7B, 7D, and 7EB, confirming previous reports that chromosomes of the homologous group 7 play an important role in controlling tolerance to high concentrations of B. It also has been observed that an allele on chromosome 4D may be responsible for increased sensitivity to B (Chantachume et al., 1994). Because B-tolerant wheat lines originated from diverse regions, and transgressive segregants have been observed among progeny of tolerant parents, it is hypothesized that chromosomes other than groups 4 and 7 may also be involved in B tolerance. At the Waite Institute in South Australia, the gene *Bo1* was successfully transferred from a moderately B-tolerant wheat cultivar "Halberd," to an intolerant, but otherwise well adapted, cultivar "Schomburgk" by the backcross method of breeding (Campbell, Paull, and Rathjen, 1994). The released cultivar, "BT-Schomburgk," and another related cultivar, "Barunga," have increased average local yields up to 53 percent when planted on high-B soils.

Planting wheat in acid soils may require development of cultivars tolerant to Al, an important breeding objective in southern New South Wales, Australia (Fisher and Scott, 1993). Adequate range of tolerance to Al in wheat has been identified (Scott and Fisher, 1989), and inheritance studies of the trait have been conducted, reporting somewhat contradictory findings related to gene expression (Delhaize, Ryan, and Randall, 1993; Wheeler et al., 1993). Wheeler and colleagues (1993) indicated the trait was inherited as a single gene with incomplete dominance, while Delhaize, Ryan, and Randall (1993), using isogenic lines differing in Al tolerance, concluded the Al-tolerance gene was completely dominant.

Breeding for Al tolerance in the wheat program of New South Wales has accounted for an increase in seed yield of 3.2 percent per year, estimated over a period of ten years (Fisher and Scott, 1993). Crosses are done between high-yielding lines and Al-tolerant genotypes. After initial stages of field testing on nonacidic soils, Al tolerance of superior lines is determined using the hematoxylin stain test. Tolerant or moderately tolerant lines identified by the test are included in yield trials planted on acid soils. The hematoxylin test has contributed to the identification of major genes located on chromosomes 2D and 4D (Takagi, Namai, and Murakami, 1982, cited in Fisher and Scott, 1993).

Soybean

Nutrient Efficiency

Mineral deficiencies may cause visual symptoms in soybean (*Glycine max* L. Merr.) and reduce yield. Soybean is relatively tolerant to B deficiency, but is sensitive to deficiency of all other mineral nutrients (McGlamery and Curran, 1989). Multigenic inheritance of nutrient efficiency in soybean has been determined for most of the minerals. The most comprehensive work has been done on breeding for Fe efficiency.

The strategy for Fe acquisition in soybean and other dicotyledonous and monocotyledonous species, except the graminaceous species, relies on the mechanism referred to as Strategy I (Brown and Jolley, 1989). Strategy I is characterized by three components: a plasma membrane-bound inducible reductase, enhanced net excretion of protons, and, in many instances, enhanced release of reductants/chelators (Brown and Jolley, 1989; Marschner and Römheld, 1995). The relative importance of the three components seems to differ considerably among plant species and genotypes and also depends on the substrate used to supply Fe^{3+}. Marschner and Römheld (1995) have provided a comprehensive review of Strategies I and II used by plants for Fe acquisition.

Iron-chlorosis deficiency in soybean may occur when an Fe-inefficient genotype is grown on calcareous soil. In the United States, improvement of Fe efficiency is a breeding objective of programs of private seed companies and public institutions (Cianzio, 1991). Chlorosis in Fe-susceptible genotypes occurs in interveinal tissue of young leaves. Under severe deficiency, all leaves are affected and plants remain stunted, shedding most of their leaves at later stages (De Mooy, 1972, cited in Cianzio, 1991). Under less severe conditions, only young leaves may be discolored. Plants may recover during the growing season; however, yield is reduced whenever yellowing occurs (Niebur and Fehr, 1981). Average percentage of yield loss increases 20 percent for each unit increase in chlorosis score (Froehlich and Fehr, 1981).

The first report on the inheritance of Fe deficiency indicated that the trait was conditioned by a single gene, with efficiency being dominant over inefficiency (Weiss, 1943). Continuous variation in the expression of the trait as observed in field plantings, however, prompted additional genetic studies. Cianzio and Fehr (1980) reported that segregation from a cross of Fe-efficient and Fe-inefficient genotypes could be explained by a single major gene and several modifying genes. The presence of modifying genes would account for the range of Fe efficiencies observed among soybean cultivars evaluated in field conditions on calcareous soil. A variation to this inheritance was reported by Cianzio and Fehr (1982) in a cross between a resistant germplasm line, A2, and a highly susceptible cultivar, Pride B216. Segregation was typical of a quantitative character controlled by additive gene action. For breeding purposes, the trait is therefore considered a quantitative character (Cianzio and Fehr, 1980).

Recurrent selection has been used for the development of improved Fe-efficient soybean germplasm (Fehr and Cianzio, 1980; Prohaska and Fehr, 1981; Fehr, Voss, and Cianzio, 1984; Jessen, Fehr, and Cianzio, 1988). In the soybean breeding program at Iowa State University, Fe efficiency of high-yielding cultivars has been improved by the use of different strategies for population development, such as backcrosses. single-crosses and three- and four-way crosses (Hintz, Fehr, and Cianzio, 1987; Cianzio, 1993a, 1995; Cianzio and Voss, 1994).

Breeding to improve Fe efficiency continues to be a challenge for soybean breeders despite the progress that has been made. Selection for improved Fe efficiency must always be considered simultaneously with selection for yield to obtain genotypes that can be released for use by farmers. Additionally, there is an ever-present need for screening procedures that will discriminate genotypes more efficiently. To this end, molecular analyses have been conducted to determine markers associated with the trait and their potential use for marker-assisted selection (Lin, Cianzio, and Shoemaker, 1997). Quantitative trait loci (QTL) controlling visual scores and/or chlorophyll concentrations have been detected in one population on linkage groups B2, G, H, I, L, and N. In this population, no QTL were detected with large effects, suggesting a typical polygene mechanism. In another population, one QTL contributed an average of 73 percent of the visual score variation, and 69 percent of the chlorophyll concentration variation, and was mapped on linkage group N. In this population, the data fit a one-major-gene model. The results suggest that molecular markers detected for Fe efficiency may be population dependent and, therefore, their use in marker-assisted selection may not presently be a feasible approach. Selection for this trait in the soybean breeding program at Iowa State University continues to rely on field plantings on calcareous soil.

Tolerance to Ion Toxicity

Soybean genotypes differ markedly in their response to high Mn concentrations in nutrient solution and also in field and greenhouse plantings (Brown and Devine, 1980; Heenan, Campbell, and Carter, 1981; Reddy, 1989, 1990). Inheritance of the trait has been determined as multigenic in some crosses, with tolerance being at least partially dominant over susceptibility (Brown and Devine, 1980; Heenan, Campbell, and Carter, 1981). In other crosses, bimodal segregation was observed, indicating single-gene inheritance (Heenan, Campbell, and Carter, 1981). Presence of transgressive segregants in the populations also suggests that minor genes may be present. Reciprocal differences observed in some crosses suggest that cytoplasmic factors may be involved (Brown and Devine, 1980). However, there is no information to indicate if tolerance to Mn toxicity is an objective in current breeding programs.

Growth of soybean roots and utilization of subsoil moisture and nutrients often are limited because of Al toxicity (Foy, Fleming, and Armiger, 1969). Seed yields can be reduced appreciably, particularly if restricted root growth results in moisture stress during and after pod development (Hanson and Kamprath, 1979; Goldman, Carter, and Patterson, 1989). Soybean yield on highly Al-saturated unlimed soils can be 70 percent of yield obtained on limed soils (Dunphy and Schmitt, 1981, cited in Campbell, Carter, and Anderson, 1989). There is significant variation among cultivars of maturity groups (MG) 00 to MG VIII for tolerance to acid Bladen soil (pH 4.4) that contains a high level of KCl-extractable Al

(Armiger et al., 1968; Sapra, Mebrahtu, and Mugwira, 1982). Differential genotypic responses to drought stress and subsoil Al also have been identified (Goldman, Carter, and Patterson, 1989), as well as variation in Al tolerance of genotypes evaluated in nutrient solution (Devine et al., 1979; Foy, Duke, and Devine, 1992).

In spite of the numerous published reports on Al sensitivity in soybean, and of the general consensus that selection for Al-tolerant genotypes would be advantageous to soybean production, no information is available on whether improvement for the trait is presently an objective of breeding programs. There is, however, information on using root and seedling traits to select for Al tolerance (Hanson and Kamprath, 1979; Hanson, 1991). In these reports, inheritance of Al tolerance is assumed to be quantitative. Several studies comparing screening techniques for Al tolerance in soybean have been published and are discussed in the following section (Devine et al., 1979; Campbell, Carter, and Anderson, 1989; Campbell and Carter, 1990).

SCREENING TECHNIQUES FOR THE EVALUATION OF NUTRIENT EFFICIENCY

Techniques to evaluate and screen efficiency of use of micronutrients in a group of genotypes from collections of individuals or from segregating populations are integral parts of a breeding program. An effective screening technique simulates plant responses in the field and allows selection of the most genetically efficient types for use in breeding and, eventually, public release.

In general, screening for nutrient efficiency is done in field plantings on soils where the problem is present, as well as in controlled conditions (greenhouses or growth chambers) by use of nutrient solution media or potted soil. Genotypes are usually evaluated in replicated tests in which genotypes of known nutrient efficiency are included as controls. Ratings of visual symptoms, seed yield, plant measurements, or analysis of plant tissues are used to evaluate efficiency of genotypes.

Wheat

Nutrient Efficiency

Copper efficiency of wheat lines was evaluated in plantings on Cu-deficient soils (− Cu), to which a treatment of added Cu (+ Cu) had also been applied (Graham, 1988). Efficiency of genotypes was determined based on vegetative yield at twelve weeks and grain yield at maturity for both −Cu and +Cu treatments, and also by measuring Cu concentration in the plants. A similar technique has also been used to evaluate Zn efficiency.

Manganese efficiency was determined in growth chamber plantings using Mn-deficient calcareous soil (Graham, 1988; Marcar and Graham, 1988). Wheat genotypes grown at 15 °C during four weeks were rated for chlorosis on a scale from 0 = no yellowing to 3 = intense yellowing. Manganese concentration in the shoot, whole plant, and in the seed, and seed yield were measured. Marcar and Graham (1988) also measured fresh and dry weight of shoot and roots and determined Mn content and chlorophyll concentration. To evaluate Zn efficiency, Graham (1988) used a similar technique to that used for Mn in growth chambers, in addition to field tests. The field tests, however, are the most recommended screening method because the large spatial variability of Zn in the field is difficult to simulate in controlled conditions (Graham, Ascher, and Hynes, 1993).

Screening techniques for Fe efficiency have been reviewed previously (Williams et al., 1987; Cianzio, 1991; Hoan, Rao, and Siddiq, 1992; Ocumpaugh et al., 1992). In general, breeders rely on field testings planted on calcareous soil. Depending on the crop, nutrient solution evaluations are also used.

Tolerance to Ion Toxicity

Screening of segregating wheat populations to evaluate tolerance to B toxicity has been conducted by growing plants in plastic trays containing soil enriched with B (Paull, Rathjen, and Cartwright, 1991). Tolerance was evaluated comparing dry matter production of these plants to dry matter produced on soils low in B. Recently, testing for B tolerance has relied on field plantings (Campbell, Paull, and Rathjen, 1994). Initial evaluation of genotypes from segregating wheat populations is conducted on field sites high in B, followed by more extensive yield testing of the promising lines at later stages of the breeding program. The use of plantings at high B sites early in the breeding program has increased the proportion of tolerant lines at later generations to more than 50 percent, whereas prior to implementing this strategy, the proportion of tolerant lines was less than 10 percent.

Fisher and Scott (1993) screened wheat genotypes for Al tolerance using the hematoxylin test developed by Polle, Konzak, and Kittrick (1978). In the test, four-day-old seedlings were exposed to different total Al concentrations (0.18, 0.36, and 0.72 mM) overnight, and the intensity of staining after exposure to hematoxylin was assessed. The hematoxylin stain test is suitable for use with large populations and accounts for 48 percent of the variation in Al tolerance in nutrient solution experiments (Scott and Fisher, 1992). Nutrient solution techniques and field screenings have also been effective in identifying Al-tolerant genotypes in wheat and related species (Howeler, 1991).

Soybean

Nutrient Efficiency

Screening of individual plants for Fe efficiency in segregating populations of soybean is conducted both in field tests on calcareous soils and in controlled conditions (greenhouses or growth chambers) by use of nutrient solution media (Fehr, 1983; Chaney, Bell, and Coulombe, 1989). Field tests and nutrient solution methods are the two most reliable screening procedures, but field tests are used more extensively in breeding programs. The use of potted calcareous soil obtained from field nurseries in which chlorosis symptoms have been observed is another possible screening technique, although results have been inconclusive (Byron and Lambert, 1983; Fairbanks et al., 1987). A review of screening techniques has been published (Cianzio, 1993b). Iron efficiency in soybean is determined by visual scores of leaf symptoms, usually when the second or third trifoliate leaf is fully developed. Scores range from 1 = no yellowing to 5 = intense yellowing with some necrosis. Visual chlorosis scores are significantly correlated with total chlorophyll content and with chlorophyll *a* and *b* (Cianzio, Fehr, and Anderson, 1979). Other indicators of Fe efficiency have been proposed (e.g., root Fe-reducing capacity of genotypes), but have not been evaluated as selection criteria in practical breeding situations (Jolley et al., 1992). The nutrient solution technique for Fe efficiency used in some programs was devised to overcome two of the limitations of field screenings: (1) the inability to artificially alter, with the current technology, the severity of the test without imposing plant injury and (2) that there is only one season per year, which limits genetic progress.

The use of nutrient solution for evaluation of Fe efficiency in soybean genotypes was possible after Coulombe, Chaney, and Weibold (1984a, b) identified bicarbonate as the soil factor causing chlorosis. Although reducing the total Fe available in the solution induces chlorosis in soybean, the relative Fe efficiency of cultivars cannot be established unless bicarbonate is added (Chaney, Bell, and Coulombe, 1989).

Dragonuk, Fehr, and Jessen (1989a) evaluated the use of nutrient solution media proposed by Chaney, Bell, and Coulombe (1989). Five $NaHCO_3$ levels and three solution-change schedules were studied. The nutrient solution consisted of 7.5 liter distilled H_2O, 2 μM $Fe(NO_3)_2$, 20 μM ethylenediamine di-(*o*-hydroxyphenylacetic acid) (EDDHA), 2 μM $MgSO_4$, 3 mM $Mg(NO_3)_2$, 1 mM KNO_3, 1 mM $CaCl_2$, 4 mM $Ca(NO_3)_2$, 10 μM H_3BO_3, 50 μM KOH, 2 μM $MnCl_2$, 2 μM $CuSO_4$, 0.2 μM MoO_4, and 15 mM HCO_3^-. Daily, 10 μM KH_2PO_4 was added to the solution. Eight genotypes were evaluated for Fe efficiency in the nutrient solutions and in six replications of a randomized complete block design on calcareous soil over three years. In the nutrient solution tests, chlorosis at stage V4 increased as the level of HCO_3^- increased.

In a second study, Dragonuk, Fehr, and Jessen (1989b) used nutrient solutions to conduct experiments in which the solution was changed once at stage V2, and daily additions of nutrients in the following concentrations were made: 0.037 μM $Fe(NO_3)_2$, 0.191 μM H_3BO_3, 0.047 μM $MnCl_2$, 0.040 μM $ZnSO_4$, 0.008 μM $CuSO_4$, 21.4 μM KNO_3 and 10.6 μM $MgSO_4$. The solution was aerated continuously with 3 percent CO_2 at a flow rate of 350 ml min^{-1} $container^{-1}$. One important characteristic of these solutions was the inclusion of higher levels of Ca^{2+} and Mg^{2+} salts than used previously (Chaney, Bell, and Coulombe, 1989).

The experiments of Dragonuk, Fehr, and Jessen (1989a, b) suggested that nutrient solution and field-testing procedures may be selecting, to some extent, for different mechanisms of response to Fe stress and that the nutrient solution procedure may not re-create field environment conditions completely. Nevertheless, rates of genetic gain for improving Fe efficiency associated with field or nutrient solution tests were similar. Results from molecular mapping of genes controlling Fe efficiency evaluated in nutrient solution and field tests, however, have indicated that similar QTL were identified by both systems (Lin, Cianzio, and Shoemaker, 1997; Lin et al., 1998), thus validating findings previously reported.

In 1992, Chaney and colleagues (1992) published a description of the modified nutrient solution tests based on nutrient solution proposed initially by Chaney, Bell, and Coulombe (1989) and used since 1989. The presently recommended method uses FeDTPA (ferric diethylenetriaminepentaacetic acid) in place of FeEDDHA [ferric ethylene-di-(*o*-hydroxyphenylacetic acid)] used in earlier screenings. With this and other modifications described in the following paragraph, Chaney and colleagues (1992) indicated that the method is now "robust" enough to be insensitive to minor variations and contamination common to screening facilities. Lin and colleagues (1998) used the modified nutrient solution technique to identify QTL responsible for Fe efficiency.

The section on materials and methods of the publication by Chaney and colleagues (1992) describes in detail, and in a "step-by-step" fashion, the methodology for preparing and conducting nutrient solution test evaluations, including seed germination, seedling transplant to containers with nutrient solution, and plant stage at which the scoring of visual chlorosis symptoms can be done. A brief description of considerations on water quality and aeration is summarized. Water quality is extremely important for effective screening; solutions should be prepared using distilled-deionized or double-deionized water to minimize errors from the presence of minerals such as Fe, Ca, Mg, and salts, such as HCO_3^-, which usually are present in tap water. Another important consideration is not to use rubber in contact with the solution because it contains about 1 percent Zn, which would interfere with chelate chemistry. Polyethylene and plexiglass materials can be used for containers and plant holders. Aeration of the solution is also important and may be done by use of disposable 1 ml plastic pipettes to provide an aeration rate of 200-300 ml/min, for good aeration and recirculation of the nutrient solution in the buckets. A gas proportioner is used to obtain the desired 3 percent CO_2 concentration in air used to aerate the solution. Tables also are provided on the composition of macronutrient stock solutions for DTPA-buffered solutions, FeDTPA preparation, and weights of $NaHCO_3$ that will be added to obtain a desired level of bicarbonate in the solution, a critical component of the method for screening chlorosis susceptibility of soybean due to interference with Fe availability to plants growing in the nutrient solution.

Tissue culture techniques have also been developed for screening of Fe efficiency (Stephens, Widholm, and Nickell, 1990; Graham et al., 1992). The high correlation coefficients observed between callus growth reduction and field chlorosis ratings suggest that in vitro evaluation for Fe efficiency can be useful in breeding. Screening can be conducted in laboratory conditions. That the explant source can be grown to produce seed has been mentioned as an additional advantage for use in soybean breeding programs; however, labor and tissue sampling costs for a large number of genotypes must be considered (Cianzio, 1993b). The efficiency of the technique in identifying superior individuals in segregating populations must also be evaluated before making a final assessment for its use by soybean breeders.

Tolerance to Ion Toxicity

Screening of Mn toxicity in soybean has been conducted using a modified Steinberg solution to which additional Mn was added (Brown and Devine, 1980; Heenan, Campbell, and Carter, 1981; Leidi, Gomez, and De la Guardia, 1987; Leidi, Gomez, and Del Rio, 1989; Heenan and Campbell, 1990). The screening also has been done by plantings on potted soil in greenhouses or growth chambers (Brown and Devine, 1980; Reddy, 1989, 1990). The soil for these plantings was obtained from fields that had previously shown symptoms of Mn toxicity. Both techniques have been effective in the differentiation of genotypes according to Mn tolerance.

Several techniques have been evaluated for screening of Al tolerance in soybean (Devine et al., 1979; Campbell, Carter, and Anderson, 1989; Campbell and Carter, 1990). The techniques, however, are not reliable for the selection of Al-tolerant genotypes, which, in the authors' opinion, have deterred genetic progress in breeding for the trait. Genotypes have been evaluated in greenhouse plantings using soils with an initial Al saturation of 19 percent and a pH of 4.7 (Campbell and Carter, 1990). A portion of this soil was amended with 3.5 cmol ($^1/_3$ Al^{3+}) kg^{-1} soil applied as $Al_2(SO_4)_3$ to achieve an Al saturation of 55 percent and a pH of 4.0. Nutrient solutions with different concentrations of Al were also evaluated (Sartain and Kamprath, 1978, cited in Campbell and Carter, 1990). Differences in Al tolerance were similarly detected with both methods; however, further studies are recommended to establish the relationship between genotypic rankings determined by the nutrient solution method and field testing. Devine and colleagues (1979) also used hydroponic systems, but could not detect differences in the visual scores of genotypes. The authors concluded that seed weight appears to exert an influence on early seedling expression of Al tolerance, and that caution should be used in imputing long-term physiological tolerance to lines that express tolerance at the seedling stage. Similar conclusions were reached by Hanson (1991). Callus culture has also been used for genotypic evaluation, but the method does not seem suited for use with soybean (Campbell, Carter, and Anderson, 1989).

SUMMARY AND CONCLUSIONS

Genetic variation for nutrient efficiency has been detected in wheat and soybean in response to mineral deficiency and/or excess in soils. Similar observations have been reported in a number of other species, although not included in this chapter. The discussions on wheat and soybean were intended to serve as working examples and should not be considered as complete, but rather a summary of, available information.

In wheat and soybean, progress in breeding for improved nutrient efficiency has been accomplished using widely different research strategies. In wheat, inheritance and location of genes in the genome was studied by chromosome analysis, and the majority of the traits were transferred from donor-related species. In soybean, genetic variation was searched for and found within the species, and genetic studies were conducted in populations developed from intraspecific crosses.

Both species originated through polyploidy. Wheat is clearly an allopolyploid, whereas soybean, although probably an allopolyploid too, has undergone an evolutionary process of autopolyploidization (Shoemaker et al., 1996). Despite the differences between the two species, the backcross method was used as the primary strategy to breed for improved nutrient efficiency, at least at early stages of the breeding process. Other population development strategies are presently used, particularly in soybean. In both species, field evaluation still plays an important role in breeding for nutrient efficiency.

Progress to date indicates that breeding for improved nutrient efficiency is a feasible approach to enhance plant adaptation and production. Undoubtedly, as more is learned about the molecular nature of the traits, screening of genotypes may become more efficient, thus enhancing progress in breeding and genetics.

REFERENCES

Armiger, W.H., C.D. Foy, A.L. Fleming, and B.E. Caldwell (1968). Differential tolerance of soybean varieties to an acid soil high in exchangeable aluminum. *Agronomy Journal* 60: 67-70.

Awad, F., V. Römheld, and H. Marschner. (1995). Effect of exudates on mobilization in the rhizosphere and uptake of iron by wheat plants. In *Iron Nutrition in Soils and Plants*, Ed. J. Abadía. Dordrecht, Netherlands: Kluwer Academic Publishers, pp. 99-104.

Barriga, P.B., B.G. Carrasco, R.P. Fuentes, N.T. Manquián. and P.F. Seemann (1993). Evaluación de genotipos de trigo en su eficiencia de utilización del fósforo en solución nutritiva. *Agro Sur* 21: 19-25.

Blair, G. (1993). Nutrient efficiency—What do we really mean? In *Genetic Aspects of Plant Mineral Nutrition*, Eds. P.J. Randall, E. Delhaize, R.A. Richards, and R. Munns. Dordrecht, Netherlands: Kluwer Academic Publishers, pp. 204-213.

Brown, J.C. and T.E. Devine (1980). Inheritance of tolerance or resistance to manganese toxicity in soybeans. *Agronomy Journal* 72: 898-903.

Brown, J.C. and V.D. Jolley (1989). Plant metabolic response to iron-deficiency stress. *BioScience* 39: 546-551.

Byron, D.F. and J.W. Lambert (1983). Screening soybeans for iron efficiency in the growth chamber. *Crop Science* 23: 885-888.

Campbell, K.A.G. and T.E. Carter Jr. (1990). Aluminum tolerance in soybean: I. Genotypic correlation and repeatability of solution culture and greenhouse screening methods. *Crop Science* 30: 1049-1054.

Campbell, K.A.G., T.E. Carter Jr., and J.M. Anderson (1989). Aluminum tolerance of soybean callus cultures: Comparison with greenhouse and solution culture screening methods. *Soybean Genetics Newsletter* 16: 191-195.

Campbell, T.A., J.G. Paull, and A.J. Rathjen (1994). The breeding for boron tolerance in wheat. In *Genetics and Molecular Biology of Plant Nutrition*, Abstracts of the Fifth International Symposium, Davis, CA: July 17-24, 1994, p. 113.

Chaney, R.L., P.F. Bell, and B.A. Coulombe (1989). Screening strategies for improved nutrient uptake and use by plants. *Horticulture Science* 24: 565-572.

Chaney, R.L., B.A. Coulombe, P.F. Bell, and J.S. Angle (1992). Detailed method to screen dicot cultivars for resistance to Fe chlorosis using FeDTPA and bicarbonate nutrient solutions. *Journal of Plant Nutrition* 15: 2063-2083.

Chantachume, Y., A.J. Rathjen, J.G. Paull, and K.W. Shepherd (1994). Genetic studies on boron tolerance of wheat (*Triticum aestivum* L.). In *Genetics and Molecular Biology of Plant Nutrition*, Abstracts of the Fifth International Symposium, Davis, CA: July 17-24, 1994, p. 141.

Chaubey, C.N., D. Senadhira, and G.B. Gregorio (1994). Genetic analysis of tolerance for phosphorus deficiency in rice (*Oryza sativa* L.). *Theoretical and Applied Genetics* 89: 313-317.

Cianzio, S.R. (1991). Recent advances in breeding for improving iron utilization by plants. *Plant and Soil* 130: 63-68.

Cianzio, S.R. (1993a). Strategies in population development for the improvement of Fe efficiency in soybean. In *Genetic Aspects of Plant Mineral Nutrition*, Eds. P.J. Randall, E. Delhaize, R.A. Richards, and R. Munns. Dordrecht, Netherlands: Kluwer Academic Publishers, pp. 321-326.

Cianzio, S.R. (1993b). Case study with soybeans: Iron efficiency evaluation in field tests compared with controlled conditions. In *Iron Chelation in Plants and Soil Microorganisms*, Eds. L.L. Barton and B.C. Hemming. New York: Academic Press, pp. 387-397.

Cianzio, S.R. (1995). Strategies for the genetic improvement of Fe efficiency in plants. In *Iron Nutrition in Soils and Plants*, Ed. J. Abadía. Dordrecht, Netherlands: Kluwer Academic Publishers, pp. 119-125.

Cianzio, S.R. and W.R. Fehr (1980). Genetic control of iron deficiency chlorosis in soybeans. *Iowa State Journal of Research* 54: 367-375.

Cianzio, S.R. and W.R. Fehr (1982). Variation in the inheritance of resistance to iron deficiency chlorosis in soybeans. *Crop Science* 22: 433-434.

Cianzio, S.R., W.R. Fehr, and I.C. Anderson (1979). Genotypic evaluation for iron deficiency chlorosis by visual scores and chlorophyll concentration. *Crop Science* 19: 644-646.

Cianzio, S.R. and B.K. Voss (1994). Three strategies for population development in breeding high-yielding soybean cultivars with improved iron efficiency. *Crop Science* 34: 355-360.

Clark, R.B. (1982). Plant response to mineral element toxicity and deficiency. In *Breeding Plants for Less Favorable Environments*, Eds. M.N. Christiansen and C.F. Lewis. New York: John Wiley and Sons, pp. 71-142.

Coulombe. B.A., R.L. Chaney, and W.J. Wiebold (1984a) Use of bicarbonate in screening soybeans for resistance to iron chlorosis. *Journal of Plant Nutrition* 7: 411-425.

Coulombe, B.A., R.L. Chaney, and W.J. Wiebold (1984b). Bicarbonate directly induces Fe-chlorosis in susceptible soybean cultivars. *Soil Science Society of America Journal* 48: 1297-1301.

Da Silva, A.R., W.H. Gabelman, and J.G. Coors (1993). Inheritance studies of low-phosphorus tolerance in maize (*Zea mays* L.) grown in a sand-alumina culture medium. In *Genetic Aspects of Plant Mineral Nutrition,* Eds. P.J. Randall, E. Delhaize, R.A. Richards, and R. Munns. Dordrecht, Netherlands: Kluwer Academic Publishers, pp. 241-249.

Delhaize, E., P.R. Ryan, and P.J. Randall (1993). Aluminum tolerance in wheat (*Triticum aestivum* L.). II. Aluminum-stimulated excretion of malic acid from root apices. *Plant Physiology* 103: 695-702.

Devine, T.E. (1982). Genetic fitting of crops to problem soils. In *Breeding Plants for Less Favorable Environments,* Eds. M.N. Christiansen and C.F. Lewis. New York: John Wiley and Sons, pp. 143-174.

Devine, T.E., C.D. Foy, D.L. Mason, and A.L. Fleming (1979). Aluminum tolerance in soybean germplasm. *Soybean Genetics Newsletter* 6: 24-27.

Dragonuk, M.B., W.R. Fehr, and H.J. Jessen (1989a). Nutrient-solution techniques for evaluation of iron efficiency of soybeans. *Journal of Plant Nutrition* 12: 871-880.

Dragonuk, M.B., W.R. Fehr, and H.J. Jessen (1989b). Effectiveness of nutrient-solution evaluation for recurrent selection for Fe efficiency of soybean. *Crop Science* 29: 952-955.

Dudal, R. (1976). Inventory of the major soils of the world with special reference to mineral stress hazards. In *Plant Adaptation to Mineral Stress in Problem Soils,* Ed. M.J. Wright. Ithaca, NY: Cornell University, Agricultural Experimental Station, pp. 3-13.

Dvorak, J. (1994). The genetics and physiology of mineral stress tolerance investigated by chromosome manipulation. In *Genetics and Molecular Biology of Plant Nutrition*, Abstracts of the Fifth International Symposium, Davis, CA: July 17-24, 1994, p. 14.

Esty, J.C., A.B. Onken, L.R. Hossner, and R. Matheson (1980). Iron use efficiency in grain sorghum hybrids and parental lines. *Agronomy Journal* 72: 589-592.

Fageria, N.K., R.J. Wright, and V.C. Baligar (1988). Rice cultivar evaluation for phosphorus use efficiency. *Plant and Soil* 111: 105-109.

Fairbanks, D.J., J.H. Orf, W.P. Inskeep, and P.R. Bloom (1987). Evaluation of soybean genotypes for iron-deficiency chlorosis in potted calcareous soil. *Crop Science* 27: 953-957.

Fehr, W.R. (1983). Modification of mineral nutrition in soybeans by plant breeding. *Iowa State Journal of Research* 57: 393-407.

Fehr, W.R. and S.R. Cianzio (1980). Registration of AP9(S1)C2 soybean germplasm. *Crop Science* 20: 677.

Fehr, W.R., B.K. Voss, and S.R. Cianzio (1984). Registration of a germplasm line of soybean, A7. *Crop Science* 24: 390-391.

Fisher, J.A. and B.J. Scott (1993). Are we justified in breeding wheat for tolerance to acid soils in southern New South Wales? In *Genetic Aspects of Plant Mineral*

Nutrition, Eds. P.J. Randall, E. Delhaize, R.A. Richards, and R. Munns. Dordrecht, Netherlands: Kluwer Academic Publishers, pp. 1-8.

Foy, C.D., J.A. Duke, and T.E. Devine (1992). Tolerance of soybean germplasm to an acid Tatum subsoil. *Journal of Plant Nutrition* 15: 527-547.

Foy, C.D., A.L. Fleming, and W.H. Armiger (1969). Aluminum tolerance of soybean varieties in relation to calcium nutrition. *Agronomy Journal* 61: 505-511.

Froehlich, D.M. and W.R. Fehr (1981). Agronomic performance of soybeans with differing levels of iron deficiency chlorosis on calcareous soil. *Crop Science* 21: 438-441.

Gamzikova, O.I. (1994). Evolutionary and genetic aspects of plant mineral nutrition. In *Genetics and Molecular Biology of Plant Nutrition*, Abstracts of the Fifth International Symposium, Davis, CA: July 17-24, 1994, p. 74.

Goldman, I.L., T.E. Carter Jr., and R.P. Patterson (1989). Differential genotypic response to drought stress and subsoil aluminum in soybean. *Crop Science* 29: 330-334.

Gourley, C.P.J., D.L. Allan, and M.P. Russelle (1994). Plant nutrient efficiency: A comparison of definitions and suggested improvement. *Plant and Soil* 158: 29-37.

Graham, M.J., P.A. Stephens, J.M. Widholm, and C.D. Nickell (1992). Soybean genotype evaluation for iron deficiency chlorosis using sodium bicarbonate and tissue culture. *Journal of Plant Nutrition* 15: 1215-1225.

Graham, R.D. (1978). Nutrient efficiency objectives in cereal breeding. In *Plant Nutrition 1978*, Eds. A.R. Ferguson, R.L. Bieleski, and I.B. Ferguson. Auckland, New Zealand, pp. 165-170.

Graham, R.D. (1988). Development of wheats with enhanced nutrient efficiency: Progress and potential. In *Wheat Production Constraints in Tropical Environments: A Proceedings of the International Conference, 19-23 January 1987*, Ed. A.R. Klatt. Mexico City D.F., Mexico: International Maize and Wheat Improvement Center, pp. 305-320.

Graham, R.D. (1994). The genetics of micronutrient nutrition and efficiency. In *Genetics and Molecular Biology of Plant Nutrition*, Abstracts of the Fifth International Symposium, Davis, CA: July 17-24, 1994, p. 30.

Graham, R.D., J.S. Ascher, and S.C. Hynes (1993). Selecting zinc-efficient cereal genotypes for soils of low zinc status. In *Genetic Aspects of Plant Mineral Nutrition*, Eds. P.J. Randall, E. Delhaize, R.A. Richards, and R. Munns. Dordrecht, Netherlands: Kluwer Academic Publishers, pp. 349-358.

Hanson, W.D. (1991). Root characteristics associated with divergent selection for seedling aluminum tolerance in soybean. *Crop Science* 31: 125-129.

Hanson, W.D. and E.J. Kamprath (1979). Selection for aluminum tolerance in soybeans based on seedling-root growth. *Agronomy Journal* 71: 581-585.

Heenan, D.P. and L.C. Campbell (1990). The influence of temperature on the accumulation and distribution of manganese in two cultivars of soybean [*Glycine max* (L.) Merr.]. *Australian Journal of Agricultural Research* 41: 835-843.

Heenan, D.P., L.C. Campbell, and O.G. Carter (1981). Inheritance of tolerance to high manganese supply in soybeans. *Crop Science* 21: 625-627.

Hintz, R.W., W.R. Fehr, and S.R. Cianzio (1987). Population development for the selection of high-yielding soybean cultivars with resistance to iron-deficiency chlorosis. *Crop Science* 27: 707-710.

Hoan, N.T., U.P. Rao, and E.A. Siddiq (1992). Genetics of tolerance to iron chlorosis in rice. *Plant and Soil* 156: 327-333.

Horst, W.J., M. Abdou, and F. Wiesler (1993). Genotypic differences in phosphorus efficiency of wheat. *Plant and Soil* 155/156: 293-296.

Howeler, R.H. (1991). Identifying plants adaptable to low pH conditions. In *Plant-Soil Interactions at Low pH*, Eds. R.J. Wright, V.C. Baligar, and R.P. Murrmann. Dordrecht, Netherlands: Kluwer Academic Publishers, pp. 885-904.

Isfan, A. (1993). Genotypic variability for physiological efficiency index of nitrogen in oats. *Plant and Soil* 154: 53-59.

Isfan, D., I. Cserni, and M. Tabi (1991). Genetic variation of the physiological efficiency index of nitrogen in triticale. *Journal of Plant Nutrition* 14: 1381-1390.

Jamjod, S., C.E. Mann, and B. Rerkasem (1993). Combining ability of the response to boron deficiency in wheat. In *Genetic Aspects of Plant Mineral Nutrition*, Eds. P.J. Randall, E. Delhaize, R.A. Richards, and R. Munns. Dordrecht, Netherlands: Kluwer Academic Publishers, pp. 359-361.

Jessen, H.J., W.R. Fehr, and S.R. Cianzio (1988). Registration of germplasm lines of soybean, A11, A12, A13, A14 and A15. *Crop Science* 28: 204.

Jolley, V.D., D.J. Fairbanks, W.B. Stevens, R.E. Terry, and J.H. Orf (1992). Root iron-reduction capacity for genotypic evaluation of iron efficiency in soybean. *Journal of Plant Nutrition* 15: 1679-1690.

Jones, G.P.D., R.S. Jessop, and G.J. Blair (1992). Alternative methods for the selection of phosphorus efficiency in wheat. *Field Crops Research* 30: 29-40.

Leidi, E.O., M. Gomez, and M.D. De la Guardia (1987). Soybean genetic differences in response to Fe and Mn: Activity of metalloenzymes. *Plant and Soil* 99: 139-146.

Leidi, E.O., M.Gomez, and L.A. Del Rio (1989). Peroxidase isozyme patterns developed by soybean genotypes in response to manganese and iron stress. *Biochemie und Physiologie der Pflanzen* 185: 391-396.

Lin, S.F., J.S. Baumer, D. Ivers, S.R. Cianzio, and R.C. Shoemaker (1998). Molecular marker evaluation of the efficiency of nutrient solution test for iron deficiency chlorosis in soybeans. *Crop Science* 38: 254-259.

Lin, S., S. Cianzio, and R. Shoemaker (1997). Mapping genetic loci for iron deficiency chlorosis in soybean. *Molecular Breeding* 3: 219-229.

Majumder, N.D., D.N. Borthakur, and R.S. Rakshit (1989). Heterosis in rice under phosphorus stress. *Indian Journal of Genetics* 49: 231-235.

Malagoli, M., G. Ferrari, and M. Saccomani (1993). Assessment of a selection pressure for improved nitrate and sulfate recovery by maize. *Journal of Plant Nutrition* 16: 713-722.

Marcar, N.E. and R.D. Graham (1988). Genotypic variation for manganese efficiency in wheat. *Journal of Plant Nutrition* 10: 2049-2055.

Marschner, H. and V. Römheld (1995). Strategies of plants for acquisition of iron. In *Iron Nutrition in Soils and Plants*, Ed. J. Abadía. Dordrecht, Netherlands: Kluwer Academic Publishers, pp. 375-388.

McDaniel, M.E. and J.C. Brown (1982). Differential iron chlorosis of oat cultivars—A review. *Journal of Plant Nutrition* 5: 545-552.

McGlamery, M.D. and W.S. Curran (1989). Mineral deficiencies and toxicities. In *Compendium of Soybean Diseases*, Eds. J.B. Sinclair and P.A. Backman. St. Paul, MN: The American Phytopathological Society, pp. 78-82.

Mikesell, M.E., G.M. Paulsen, R. Ellis Jr., and A.J. Casady (1973). Iron utilization by efficient and inefficient sorghum lines. *Agronomy Journal* 65: 77-80.

Niebur, W.S. and W.R. Fehr (1981). Agronomic evaluation of soybean genotypes resistant to iron-deficiency chlorosis. *Crop Science* 21: 551-554.

Ocumpaugh, W.R., J.N. Rahmes, D.F. Bryn, and L.C. Wei (1992). Field and greenhouse screening of oat seedlings for iron-nutrition efficiency. *Journal of Plant Nutrition* 15: 1715-1725.

Palmer, B. and R.S. Jessop (1977). Some aspects of wheat cultivar response to applied phosphate. *Plant and Soil* 47: 63-73.

Paull, J.G., R.O. Nable, and A.J. Rathjen (1993). Physiological and genetic control of the tolerance of wheat to high concentrations of boron and implications for plant breeding. In *Genetic Aspects of Plant Mineral Nutrition*, Eds. P.J. Randall, E. Delhaize, R.A. Richards, and R. Munns. Dordrecht, Netherlands: Kluwer Academic Publishers, pp. 367-376.

Paull, J.G., A.J. Rathjen, and B. Cartwright (1991). Major gene control of tolerance of bread wheat to high concentrations of soil boron. *Euphytica* 55: 217-228.

Polle, E., A.F. Konzak, and J.A. Kittrick (1978). Visual detection of aluminum tolerance levels in wheat by hematoxylin staining of seedling roots. *Crop Science* 18: 823-827.

Prohaska, K.R. and W.R. Fehr (1981). Recurrent selection for resistance to iron deficiency chlorosis in soybeans. *Crop Science* 21: 524-526.

Reddy, M.R. (1989). Soil manganese stress in soybean. *Soybean Genetics Newsletter* 16: 188-190.

Reddy, M.R. (1990). Responses of various soybean genotypes to excess soil manganese. *Soybean Genetics Newsletter* 17: 150-152.

Rengel, Z. and R.D. Graham (1995a). What genotypes differ in Zn efficiency when grown in chelate-buffered nutrient solution. I. Growth. *Plant and Soil* 176: 307-316.

Rengel, Z. and R.D. Graham (1995b). What genotypes differ in Zn efficiency when grown in chelate-buffered nutrient solution. II. Nutrient uptake. *Plant and Soil* 176: 317-324.

Sapra, V.T., T. Mebrahtu, and L.M. Mugwira (1982). Soybean germplasm and cultivar aluminum tolerance in nutrient solution and Bladen Clay Loam soil. *Agronomy Journal* 74: 687-690.

Scott, B.J. and J.A. Fisher (1989). Selection of genotypes tolerant of aluminium and manganese. In *Acidity and Plant Growth*, Ed. A.D. Robson. Sydney, Australia: Academic Press, pp. 167-203.

Scott, B.J. and J.A. Fisher (1992). Tolerance of Australian wheat varieties to aluminium toxicity. *Communications in Soil Science and Plant Analysis* 23: 509-526.

Shoemaker, R.C., K. Polzin, J. Labate, J. Specht, E.C. Brummer, T. Olson, N. Young, V. Concibido, J. Wilcox, J.P. Tamulonis et al. (1996). Genome duplication in soybean (*Glycine* subgenus *soja*). *Genetics* 144: 329-338.

Stephens, P.A., J.M. Widholm, and C.D. Nickell (1990). Iron-deficiency chlorosis evaluation of soybean with tissue culture. *Theoretical and Applied Genetics* 80: 417-420.

Weiss, M.G. (1943). Inheritance and physiology of efficiency in iron utilization in soybeans. *Genetics* 28: 253-268.

Wheeler, D.M., D.C. Edmeades, R.A. Christie, and R. Gardner (1993). Comparison of techniques for determining the effect of aluminum on the growth and the inheritance of aluminum tolerance in wheat. In *Genetic Aspects of Plant Mineral Nutrition,* Eds. P.J. Randall, E. Delhaize, R.A. Richards, and R. Munns. Dordrecht, Netherlands: Kluwer Academic Publishers, pp. 9-16.

Williams, E.P., R.B. Clark, W.M. Ross, G.M. Herron, and M.D. Witt (1987). Variability and correlation of iron-deficiency symptoms in a sorghum population evaluated in the field and growth chamber. *Plant and Soil* 99: 127-137.

Wu, P. and Q.N. Tao (1995). Genotypic response and selection pressure on nitrogen-use efficiency in rice under different nitrogen regimes. *Journal of Plant Nutrition* 18: 487-500.

Chapter 11

Monitoring Water and Nutrient Fluxes Down the Profile: Closing the Nutrient Budget

Ian R. P. Fillery

Key Words: calcium, drainage, leaching, magnesium, modeling, nitrate, potassium, solute transport, suction cups, time domain reflectrometry, water transport.

INTRODUCTION

The concentration of ions in soil solution is an important determinant of nutrient uptake by plants. The amount of each ion present in solution is dependent on the balance between inputs, from processes such as mineralization of organic matter, soil weathering, exchange reactions with soil surfaces, and fertilizer additions, and outputs, including plant uptake and loss processes. Movement of ions to depth with the drainage of soil water (leaching) and reduction to gaseous forms, in the case of NO_3^- and SO_4^{2-}, are examples of loss processes that reduce ion concentrations.

Manipulation of input and output processes to ensure the optimum availability of key nutrients for crops has been a major challenge for soil scientists in the latter part of the twentieth century. Assays of nutrient availability that attempt to mimic processes determining both the quantity and intensity of nutrient supply have been widely used to advise on nutrient management. Mechanistic determinations of nutrient supply have been suggested, but have not found widespread use because of the complex solutions needed. The advent of inexpensive, powerful computers now makes the use of mechanistically based functions a realistic goal, but success is ultimately dependent on intellectual input into computer software designed to describe soil-plant processes that affect nutrient availability. Controlled environment studies with repacked soil, often involving a single treatment, have been widely used to study soil processes, chiefly to avoid the spatial and temporal variability that is inherent in field environments. Information derived from these studies has added knowledge on soil transformations, but these studies do not provide information needed to validate simulation packages that aim to describe the temporal changes in the availability of key nutrients across a range of soils and climatic conditions.

Soil water content is a critical determinant of most processes that govern the availability of nutrients in field soils, and this is a key parameter in most simulation packages. Since concurrent measurements of water flow and nutrient concentrations in soil solution are needed to calculate nutrient flow, there is merit in developing monitoring techniques that permit complementary analysis of both. Recent developments in methodologies that accurately quantify changes in soil water content have assisted in determinations of soil water fluxes. Less laborious methods have been used to quantify the concentrations of nutrients in soil solutions, although the accuracy of some methods is disputed. In an endeavor to extrapolate research findings, models have been developed to describe the processes of water and nutrient movement in soil profiles.

This chapter will examine methods that can be used to monitor the movement of nutrients in soil as prerequisite to the determination of losses by leaching, and it will evaluate the accuracy of simulation routines devised to describe both water and nutrient movement in soil.

MEASUREMENT OF SOIL PROFILE CONCENTRATIONS OF NUTRIENT IONS

Soil coring and chemical extraction methods have been widely used for analyzing soil nutrient concentrations and have been the subject of numerous reviews. Most of these methods are destructive, time-consuming, and expensive and, therefore, are not ideal for studies of leaching of nutrients that require regular, repeated measurements of concentrations of soil solution at several depths. Concentrations of ions in soil solution typically exhibit marked heterogeneity, highlighting the need for replication in estimates of nutrient movement at field sampling scales (Addiscott, 1996). As a consequence, sampling methods have been sought that are both inexpensive and rapid.

Suction Cup Samplers

Suction cup methods have found wide application in field studies of ion movement. However, the choice of cup sampler and the method of installation are critical to the accuracy of measurements of nutrient concentration (Grossmann and Udluft, 1991). The most frequently used suction cups are ceramic. This material is ideal for extracting NO_3^- and SO_4^{2-}, but the adsorptive capacity of ceramic can modify soil solution concentrations of cations, phosphates (Zimmermann, Price, and Montgomery, 1978), and trace elements (Grossmann and Udluft, 1991). Other materials used to construct suction cups include fritted glass, stainless steel, and cellulose acetate, but Teflon, with its small specific surface and a low charge density, is generally recommended where cations and trace elements are to be analyzed in extracted soil solutions (Zimmermann, Price, and Montgomery, 1978). A composite quartz-teflon soil solution sampler is produced commercially (PRENART, 1998). Pretreatment of new ceramic cups with dilute HCl is recommended to remove contaminants generated during production (Grossmann and Udluft, 1991). There is also merit in the equilibration of cups with a solution of comparable composition to the soil solution.

Numerous suction cup dimensions and designs have been produced (Grossmann and Udluft, 1991). Use of large diameter (40 mm) cups is recommended over cups with smaller (< 20 mm) diameters to increase the volume of soil sampled and to reduce the error of analysis. However, mini-ceramic discs do enable nondestructive measurements of ions in studies that aim to examine the heterogeneity of soil transformations in small soil volumes (Strong, Sale, and Helyar, 1997). Typically, ceramic cups are attached to either tubing or to a pipe that acts as a shaft between the cup and the soil surface. Solution drawn through the porous cup can either be collected in the shaft of the sampler or in a reservoir connected to the suction cup (Grossmann and Udluft, 1991). Use of a reservoir with a fixed volume is recommended where different shaft lengths, and therefore volumes, are used to accommodate different sampling depths, or where small-bore tubing is used to connect cups to the surface. Alternatively, banks of suction cups can be linked together to ensure the same suction is applied over all cups (see Anderson, Fillery, Danin, et al., 1998). Such modifications reduce the variation in suction applied between cups placed at different depths, or positions, and consequently minimize the possibility of extraction of solution from different pore volumes, which may affect the concentration of ions measured.

Use of suction to extract soil solution through cups results in short-time measures of the concentrations of ions in solution. Another approach uses a time-averaged measure of the concentration of ions in solutions in ceramic cup tensiometers after the diffusion of ions from soil solutions (Moutonnet, Paegenel, and Fardeau, 1993; Moutonnet and Fardeau, 1997). Equilibration between soil solution and solutions in tensionic samplers typically occurs within eight to ten days (Moutonnet, Paegenel, and Fardeau, 1993). Although tensionic samplers and standard ceramic suction cup samplers have given comparable measurements of NO_3^- in soil solution (Poss, Noble, et al., 1995; Moutonnet and Fardeau, 1997), the tensionic sampler can extract larger quantities of NH_4^+ and NO_2^- than standard techniques (Moutonnet and Fardeau, 1997). The sampling of NH_4^+ and NO_2^- from the entire water-filled soil pore space, in the case of the tensionic sampler, and the extraction of ions chiefly from macropores with standard suction cup methods are possible explanations for the differences in quantities of ions extracted between the two methods (Moutonnet and Fardeau, 1997). The ability to obtain more integrated measures of a suite of ions in soil solution along with measurements of soil water potential is a useful attribute of tensionic samplers. However, this strength may be of limited value in soils possessing high hydraulic conductivity because rapid pulses of ions during periods of drainage may negate equilibration between water in tensiometers and the soil solution.

Angled installation of suction cup probes is preferred where these are installed from the soil surface to ensure that the suction cup does not sample soil solution that can flow along the surface of shafts or along tubing. Alternatively, suction cups can be installed in soil at approximately 5° from a pit face (Anderson, Fillery, Dunin, et al., 1998) or vertically, provided that measures are taken to stop the flow of water along the sides of shafts or from soil above suction cups that are buried and tubing is used to evacuate air and to sample solution (Lord and Shepherd, 1993).

A range of techniques have been used to install suction cups. Anderson, Fillery, Dunin, and colleagues (1998) used steel pipe with the same outside diameter as the shaft of the ceramic cups to form the access well. A pipe with an outside diameter smaller than the diameter of the suction cup was used to form the final 0.1 m of the hole to ensure a tight fit between the suction cup and soil. Ceramic cups also can be placed into slightly oversized holes formed using an auger, but a slurry of fine silica (Lord and Shepherd, 1993) or a slurry formed from sieved augered soil needs to be placed in holes before the insertion of the suction probe (Grossmann and Udluft, 1991). The upward displacement of the slurry during probe installation ensures good hydraulic contact between the suction cup and soil and the elimination of fissures adjacent to the shaft that could channel water along the edge of sampling units. Another approach uses dry, clean, fine sand pressed around the cup to improve contact between the suction cup and soil. A slurry of sieved soil or a mix of kaolin and cement is poured about the probe shaft after a plug of bentonite or a mix of kaolin and sieved soil is rammed around the shaft just above the junction of the ceramic cup to prevent the slurry penetrating the fine sand around the suction cup.

There is general agreement that suction cups are an ideal system to extract solution from sandy soils (Lord and Shepherd, 1993; Webster et al., 1993; Djurhuus and Jacobsen, 1995). However, their use in finer-textured soils that typically exhibit heterogeneity in soil water and nutrient concentrations has been questioned (Hatch et al., 1997). Heterogeneity normally increases in soils as the percentages of clay and silt rise and soils become more structured, with distinct mobile and immobile categories of soil water. Clayey soils are also more likely to crack or to contain channels that enable water to bypass the soil matrix. Depending on the source of the nutrient under study, use of ceramic cups could either overestimate or underestimate movement in finer-textured soils. For example, Goulding and Webster (1992) found ceramic cups, placed in a field (silty clay loam with tile drains), to overestimate leaching of NO_3^- compared to measurements based on effluent from tile drains before fertilizer was applied, but underestimated leaching after N (nitrogen) fertilizer application. In this case, ceramic cups sampled soil water contained in the soil matrix where N mineralization by microorganisms is predominant. Fertilizer-derived NO_3^- would be expected to be external to the soil matrix and, therefore, to move largely by bypass flow. Such problems do not preclude the use of ceramic cups to sample soil solution in finer-textured soils; they highlight the need for supplementary measurements of soil water flow and nutrient concentrations, and caution over the interpretation of results.

Resins

Ion exchange resins have been used to measure the accumulated quantity of inorganic N (Schnabel, 1983; Schnabel, Messier, and Purnell, 1993) or base cations moved through porous media over extended time intervals (Crabtree and Kirkby, 1985). Compared to suction cup extraction of soil solution and chemical extraction of soil, exchange resins provide a time-integrated measure of anion and cation concentrations in soil, the measure that is independent of soil water flux. Consequently, use of ion exchange resins can reduce errors associated with interpolating between sampling dates and water flux estimates. However, the application of exchange resins or exchange membranes to measure the flux of ions is not without problems. Buried resin bags have been used widely to measure rates of

NO_3^- production in soil, but few studies have considered how installation affected water flow. Traps containing exchange resin or exchange membranes ideally should not modify the flow of water in soil. Resin beads are more porous than soil and therefore will increase water flow in most situations when placed in direct contact with soil (Torbert and Elkins, 1992). However, the need to confine exchange resin within structures that physically separate resin from soil, to ensure complete recovery of resin, can result in a break in the flow of soil water. This disruption to water flow can result in the perching of water in soil above the trap and redirection of water flow away from the trap toward soil that is in contact with underlying soil with lower matric potentials. Underestimation of nutrient fluxes is likely, should this occur.

Lysimeters

Lysimeters are the preferred method to determine ion movement in fine-textured soils in which heterogeneity in concentration of nutrients in soil solution reduces the reliability of suction cup extracts (Goulding and Webster, 1992), but they have been used to study solute flow in a range of soil types. Lysimetry is appealing because the inputs and outputs of water and solutes can be measured precisely. Nevertheless, the application of lysimeters to measurements of water and nutrient flows is not without difficulty or criticism (Wagner, 1962). Soil disturbance during installation, preferential flow along lysimeter walls (Cameron et al., 1990), and changes in the water potential at the base of lysimeters are problems encountered with this technique (Simmonds and Nortcliff, 1998). Soil solution at the bottom of lysimeters can be collected under gravity or under tension (suction lysimeters), using an array of porous ceramic discs or plates, sometimes supplemented with porous ceramic cups of several sizes or with porous Teflon cylinders. Tension may be applied by means of a hanging water column, by manual or automatic vacuum pumping systems. The choice of tension used to extract soil water is critical to the volume removed and to determinations of water flow (Haines, Waide, and Todd, 1982). Use of gravity or zero-tension lysimeters is not recommended, but for practical reasons, these systems have been widely applied.

Simmonds and Nortcliff (1998) concluded that in the case of structureless sandy soils with small macropore water flow, lysimeters of the order of 1 m deep and 1 m in diameter were sufficiently large to be representative of field environments. The existence of macropore flow in structured finer-textured soils should require lysimeters with larger diameters than those just specified. Use of drainage and solute samplers with an area less than 0.1 m^2 is likely to result in misleading estimates of solute fluxes at a field scale, if analysis of the variation in the magnitude of the solute flux density is not conducted through a large number of replicates (Simmonds and Nortcliff, 1998).

Mole-Pipe Drained Fields

Mole or pipe drains remove most of the excess water during periods of drainage and therefore are convenient methods to measure solute movement to the depth of the tile or mole in fine-textured soils. This approach has one main advantage over the measuring systems previously described in that it overcomes the effects of field-scale spatial variability in both water and solute concentrations. Devices to measure water flow are needed, and automatic samplers to collect solution are recommended (Magesan, White, and Scotter, 1996).

MEASUREMENT OF SOIL WATER CONTENT

Reliance in the past on manual methods to measure soil water content has limited the scope of studies of nutrient flows in undisturbed soil under field conditions. Techniques are needed that can continuously monitor soil water content, are precise yet durable, have low power requirements to assist remote application, facilitate automatic data acquisition, and are cheap enough to aid replication. Gardner (1986) and Smith and Mullins (1991) have described the range of techniques available and their applications.

Application of Time Domain Reflectrometry to Soil Water Measurements

The time domain reflectrometry (TDR) technique (Zegelin, White, and Jenkins, 1989; Dalton, 1992; Whalley, 1993; Topp, Watt, and Hayhoe. 1996) is the method that meets most of the criteria identified previously for remote automated analysis of soil water. TDR methods, and the less-developed and -tested capacitance probe (Dean, Bell, and Baty, 1987), use changes in the dielectric constant of soil to determine volumetric soil water content. In the case of the TDR technique, measurements are made of the velocity of a pulse that travels along an electromagnetic transmission line as a guided wave. Pulse velocity is used to calculate the soil dielectric constant that is dominated by the contribution from soil water. Two systems have been used—a balanced two-wire transmission line

(Topp and Davis, 1985) and an unbalanced three-wire line design (Zegelin, White, and Jenkins, 1989). In conventional equipment, an expensive Tektronix 1502 cable tester, or comparable oscilloscope (TRASE, 1997), is used to measure the time of travel of each pulse. The need to reduce the cost of TDR equipment has encouraged experimentation with alternate techniques to measure pulse velocity. Methodologies based on either frequency domain reflectrometry (FDR) or the use of algorithms to interrogate specific points of TDR pulse signals at distinct voltage levels, the patented TRIME method (Time Domain Reflectrometry with Intelligent Micromodule Elements; Fundinger, Köhler, and Stacheder, 1997), have been introduced to replace cable testers. The FDR-based equipment uses measures of the number of pulses sent over a period of seconds to determine wave travel time, or pulse velocity, which is calculated as the inverse of the frequency (Campbell Scientific, 1996).

Early TDR studies suggested that a single empirically derived relationship between soil dielectric constant and volumetric water content may be used across a range of mineral soils (Topp, Davis, and Annan, 1980; Topp and Davis, 1985). Later work has emphasized the need for a separate calibration for organic soils (Roth, Malicki, and Plagge, 1992); it is now recognized that the relationship can be influenced by bulk density, clay content (Zegelin, White, and Russell, 1992; Dirksen and Drasberg, 1993; Jacobesen and Schjonning, 1993), the type of Fe (iron) minerals in the soil (Robinson, Bell, and Batchelor, 1994), and electrical conductivity. The universally derived relationship between soil dielectric constant and volumetric water content, initially described by Topp, Davis, and Annan (1980), appears to work best in coarse-textured soils; individual calibration curves are recommended where the TDR is used to measure soil water content in fine-textured soils or in soils with high electrical conductivity (White and colleagues (1994). The magnitude of errors encountered with the use of the TDR method, including probe geometry, are discussed by Walley (1993; 1994) and White and colleagues (1994). Errors associated with the incorrect installation of probes are discussed by Gregory and colleagues (1995).

In their initial configuration, capacitance probes had two stainless steel electrodes that acted as the two sides of a capacitor across which an alternating electric field at a known frequency was applied. A resonant frequency, which is dependent on the dielectric constant of soil, results when the electrode is inserted into soil (Robinson and Dean, 1993). In comparison to TDR equipment, capacitance probes are largely untested. Evett and Steiner (1995) concluded that capacitance probes had poor precision when tested against the neutron probe method. Electrical conductivity in soil appears to have a larger effect on the determination of soil dielectric constant for calibration of capacitance probes than is the case with TDR (Robinson, Bell, and Batchelor, 1994).

TDR Applications

Gregory and colleagues (1995) have described TDR applications in which transmission lines are kept buried in soil and a moveable head is fitted for measurements. This approach overcame the need for replication of expensive TDR heads, cabling, and a multiplexer. However, this method does not permit frequent measurement of soil water content and automatic data acquisition, which are the strengths of TDR systems. Several examples of application of TDR equipment at remote locations have been reported (Herkelrath, Hamburg, and Murphy, 1991; Zegelin, White, and Russell, 1992).

The TDR probes can be installed vertically, obliquely, or horizontally. Vertical installation is easiest to accomplish, but heat conduction into soil and preferential flow of water along the probe wires could result in local modification of the soil environment. Horizontally placed probes require excavation of a pit. Anderson, Fillery, Dunin, and colleagues (1998) used lined access wells dug into sandy soil to install three-pronged TDR probes (PyeLabs, CSIRO, Canberra, Australia) at eight soil depths to 1.5 m. Each probe was connected to a cable (maximum length of 12 m) that was in turn connected to a multiplexer that facilitated sequential analysis of traces from sixteen probes. Computer software controlled the scanning, scaling, analysis, and storage of traces and the switching between probes. Smith and colleagues (1998) installed similar equipment in a Red Kandosol in eastern Australia. The robustness of 6-mm-diameter stainless probes used as part of the PyeLab equipment is an important asset that facilitates installation into hard-setting or gravelly soils. Metal guide wires used on Campbell Reflectometers are less robust. Nevertheless, these have been installed horizontally into a hard-setting gradational duplex soil after the excavation of 1-m-deep pits (I. R. P. Fillery and F. X. Dunin, unpublished data). Use of a proforma made of stainless steel rods with identical dimensions and a spacing to that used on the Reflectrometer 2-prong probes (30 cm long) was needed to make channels for probe insertion. Experimentation showed that it was preferable to use three lengths of rod (0.1, 0.2, and 0.3 m) sequentially, with a guide to ensure that channels were parallel along the full length of probe installation. Recompaction of soil to bulk densities close to that in adjacent soil should be undertaken after TDR or FDR probe installation from pit faces, to avoid changing the flow of water in soil adjacent to probes or the artificial ponding of water in texture-contrast soils. Manual removal of soil in prescribed layers and separate storage are practical procedures that help the repacking of pits to a soil condition close to that in adjacent undisturbed soil.

CALCULATION OF SOIL WATER FLUXES AND DRAINAGE

Drainage from individual soil layers can be calculated from:

$$D = P - ET - \Delta S \quad (1)$$

where D is drainage of water below a specified soil depth (mm); P is precipitation (mm); ET is evapotranspiration (mm); ΔS is the change in stored water in a specified depth of soil (mm)

ΔS can be calculated from TDR measurements of soil water content (v/v) and precipitation measured using a tipping rain gauge. Estimates of ET should preferably be obtained using micrometeorological techniques (Dunin et al., 1989) or weighing lysimeters, but Penman estimates of evapotranspiration are suitable where these have been calibrated against actual measurements (Anderson, Fillery, Dunin, et al., 1998).

In the absence of measurements of changes in stored soil water and ET, estimates of drainage volume can be obtained from weather-based models. The MORECS calculation system and the computer software IRRI-GUIDE, as well as the WATBAL model, have been used for determining drainage from bare soil in the United Kingdom (Lord and Shepherd, 1993; Vinten et al., 1994).

CALCULATION OF NUTRIENT FLUXES

Ideally, nutrient movement within soil and leaching losses should be determined from temporal changes in nutrient concentration measured at more than one depth. Such measurements should also be concurrent with studies of other soil processes that either add nutrients to, or remove them from, soil solution. Procedures adopted by Anderson, Fillery, Dunin, and colleagues (1998) in studies of NO_3^- movement in sandy soil are an excellent example of approaches needed to quantify nutrient movement. In this case measurements of N mineralization, plant uptake of N, and root length distribution were concurrent with measurements of NO_3^- in soil, and in soil solution, and this enabled the content of NO_3^- in each soil layer to be adjusted for inputs and losses not associated with NO_3^- movement. The concentration of NO_3^- in each layer was interpolated from measured NO_3^- concentrations in soil or in soil solution. Inorganic N was added to the surface layer in accordance with measured rates of net N mineralization, and daily N uptake was calculated from the measured aboveground plant N derived from soil N (Anderson, Fillery, Dunin, et al., 1998). Uptake of NO_3^- from the soil was in proportion to root length density and distribution. Leaching of NO_3^- from the surface soil layer was set at zero when evapotranspiration was greater than rainfall. The amount of NO_3^- leached (kg N ha^{-1}) from each soil layer (N_L) was computed using Equation 2:

$$N_L = ((N_i + N_a - N_u) / (W_i + W_a)) \times D_L \qquad (2)$$

where N_i is the initial NO_3^- content (kg N ha^{-1}) within a given soil layer, L; N_a is the amount of N mineralization (kg N ha^{-1}) for soil layer L_1 or the amount of NO_3^- leached from the above soil layer to the layer in question during 24 h (hours); N_u is N uptake by the plant (kg N ha^{-1}) from the specified soil layer over 24 h; W_i is the initial amount of water (mm) within the specified soil layer; W_a is daily rainfall (mm) for L_1 or daily drainage (mm) from the preceding soil layer to the layer in question during the current day; D_L is the current day drainage (mm) through the specified soil layer.

In the study undertaken by Anderson, Fillery, Dunin, and colleagues (1998), drainage through each of eight soil layers was determined from Equation 1 using (1) Bowen ratio measurements to calculate evapotranspiration and (2) TDR measurements to determine changes in soil water content by depth. The quantities of NO_3^- moved each day were calculated using a spreadsheet and the interpolated soil profile NO_3^- concentrations compared against measured values. The advantage of this approach is that NO_3^- movement could be assessed for multiple depths within the soil profile, enabling assessments of NO_3^- losses from the root zone and the contribution of NO_3^- leaching in surface soil layers to the process of soil acidification. Although the example shown is for NO_3^-, the same procedures could be used to determine the rate of movement of other nutrients, provided that processes contributing ions to soil solution, and causing the depletion of ions from the same solution, are measured and represented in Equation 2. For cations, this would include exchange reactions.

The trapezoidal rule has been used to calculate the quantity of nutrient leached in studies in which the concentration of a specified nutrient is determined at only one depth. For example, Lord and Shepherd (1993) used the areas under a plot of the NO_3^- concentration against cumulative drainage to calculate loss. The area was calculated as the sum of the areas of the trapezia, resulting from successive pairs of sampling occasions (c_1, c_2; mg N dm^{-1}), and the drainage volume (v, mm). The total N leached in each sampling interval was:

$$\text{N leached (kg N ha}^{-1}) = 0.5 \times (c_1 + c_2) \times v/100 \quad (3)$$

The total for the drainage period was the sum of these trapezia. Additions of mineral N through N mineralization can be ignored where soil solution is collected from subsoil that has negligible organic matter content. One significant drawback of this approach is that the size of nutrient fluxes could be underestimated in situations in which the peak concentration of nutrient is not determined because of infrequent extractions of soil solutions.

FIELD STUDIES OF NUTRIENT FLUXES

Movement of Inorganic Nitrogen

Concern about the inefficient use of N fertilizers in agriculture and the subsequent pollution of ground water, waterways, and marine ecosystems has resulted in numerous studies of soil NO_3^- movement and leaching losses. A wide array of techniques have been used to determine quantities of NO_3^- leached, and these are summarized in Table 11.1, along with pertinent soil and crop information. Documented rates of NO_3^- leaching range from 152 to 215 kg N ha^{-1} for irrigated corn fertilized with between 178 to 268 kg N ha^{-1} when grown on sandy soil in the Midwest of the United States (Martin et al., 1994) to $<$ 10 kg N ha^{-1} for grazed pasture on finer-textured soils fertilized with large amounts of N (Barraclough et al., 1992). Most estimates of NO_3^- movement below 0.6 to 1 m depth are between 30 to 70 kg N ha^{-1}, with fertilizer management, time of cultivation, and the presence of cover crops as major factors determining the quantity leached during periods of drainage (see Table 11.1).

Leached NO_3^- in soil is often linked to N fertilizer use because of the large increase in inorganic N in soil solution post application. However, growing evidence suggests that NO_3^- derived from the mineralization of soil organic matter, and crop and legume residues, is also subject to movement, especially when this NO_3^- accumulates in soil before periods of drainage (MacDonald et al., 1989; Anderson, Fillery, Dunin, et al., 1998).

TABLE 11.1. Summaries of Nitrate Leaching (kg N ha^{-1}) for Different Production Systems, N Additions, Soil and Crop Management

Production system	Soil type	Measurement system/drainage	Additions of N kg N ha^{-1}	Country	Nitrate leached	Reference
• Barley • Spring wheat • Oilseed rape • Sugar beet • Grass ley	Coarse sand	Tile drains at ~1 m, outflow recorded automatically	• 78-141 • 110-120 • 190 • 112-140 • 92	Sweden	• 20-60 • 1-22 • 20	Gustafson (1987)
• Winter wheat • Bare soil during intercrop period • Ryegrass catch drop during intercrop • Corn	Calcareous brown soil	1 m × 1 m wide × 1 m deep lysimeters filled with soil; data collected for 11 years after setup	200 in NO_3 form	France	• 11 • 110 • 40 • 0.2-14	Martinez and Guiraud (1990)
• Spring barley • Spring barley	• Sand over clay • Clay loam	Tile drains in both soils; outflow measured automatically	• 150 • 150	Scotland	• 30-38 • 24-27	Vinten et al. (1994)
• Wheat • Ploughed-fallow • Ryegrass • Volunteer weeds	Sandy clay loam	Ceramic suction cups at 0.9 m	• 200, post drainage • 0 • 0 • 0	United Kingdom	• 50 • 77 • 52 • 19	Webster and Goulding (1995)
• Barley • Barley • Grass ley • Lucerne ley	Clay over sand and clay	Disturbed soil in lysimeters	• 0 • 120 as $Ca(NO_3)_2$ • 120 + 80 split as $Ca(NO_3)_2$ • 0	Sweden	• 2-23 • 3-36 • 0.4-45 • 0.1-14	Bergstrom (1987)

TABLE 11.1 (*continued*)

Production system	Soil type	Measurement system/drainage	Additions of N kg N ha^{-1}	Country	Nitrate leached	Reference
• Corn, 153-347 mm irrigation	Sandy loam	Disturbed soil in lysimeters (1.2 m × 1.5 m × 1.8 m deep)	• 178-268, anhydrous NH_3 preplant	Michigan, United States	• 152-215	Martin et al. (1994)
• Corn, 49-267 mm irrigation			• 123-130, anhydrous NH_3 preplant		• 76-79	
• Wheat after peas, fall cultivation	Shallow limestone soil	Suction cups at 0.6 m, drainage calculated from IRRI guide program	• N as NH_4NO_3, late February	United Kingdom	• 34	Johnson, Shepherd, and Smith (1997)
• Peas after wheat, December cultivation			• 0		• 77	
• Oilseed rape after barley, August cultivation			• 30 N applied at seeding and in an equal split in late February and late March		• 40	
• Barley after oilseed rape, September cultivation			• N applied late February		• 53	
• Winter fallow after summer pea crop	Silt loam	Soil sampling; leaching determined from knowledge of plant N uptake and changes in soil water	0 in all treatments	Canterbury, New Zealand	• 90	Adams and Pattinson (1985)
• Spring wheat after winter fallow and summer pea crop					• 60	
• Lightly grazed white clover pasture					• 10	
Pasture-ley systems:	Silt loam	Suction cups at 0.6 m; drainage calculated from water balance model	0 in all treatments	Canterbury, New Zealand		Francis, Haynes, and Williams (1995)
• Pastures plowed early fall, fields then left fallow through winter					• 72-106	
• Pastures plowed late fall, fields then left fallow through winter					• 8-52	
• Winter wheat after May plowing (wheat seeded in May)					• 14-102	

Production system	Soil type	Measurement system/drainage	Additions of N kg N ha^{-1}	Country	Nitrate leached	Reference
Cultivation of a 2-year-old pasture in late summer, then different cover crops:		Undisturbed monolith lysimeters, 200 mm dia. and 250 mm in depth, zero suction applied at base	0 in all treatments	Canterbury, New Zealand		McLenaghen et al. (1996)
• Rye-corn during winter					• 4	
• Mustard during winter					• 6	
• Winter field beans during winter					• 30	
• Fallow during winter					• 33	
Spring barley after different cultivation, with/without catch crops:	Coarse sand	Suction cups at 0.8 m, drainage calculated using models based on either soil water content and root depth or meteorological data	120 as calcium ammonium nitrate, applied to spring barley	Denmark (Nitrate leaching values are averages for 5 years)		Hansen and Djurhuus (1996, 1997)
• Plowed late fall; no catch crop					• 67	
• Plowed late fall; crop—ryegrass					• 42	
• Plowed spring; crop—ryegrass					• 29	
• Plowed spring; no catch crop					• 68	
• Direct drill; no catch crop					• 56	
Crop rotations on different soils:		Ceramic or teflon suction cups at 1 m		Denmark		Simmelsgaard (1998)
• sequence of corn, spring barley, clover, and spring barley	• Coarse sand	• 695 mm drainage	• 22 as fertilizer and 121 as manure each year		• 104	
• Spring barley in each of 6 years	• Fine sand	• 361 mm drainage	• 120		• 92	
• Spring barley in each of 6 years	• Sandy loam	• 305 mm drainage	• 106		• 74	
• Spring barley in each of 6 years	• Sandy loam	• 226 mm drainage	• 80		• 49	
• Sequence of 3 years spring barley, rape, winter wheat, and spring barley	• Sandy loam	• 325 mm drainage	• 65 fertilizer and 94 of manure N each year		• 49	

TABLE 11.1 (*continued*)

Production system	Soil type	Measurement system/drainage	Additions of N kg N ha^{-1}	Country	Nitrate leached	Reference
		Ceramic suction cups at 0.9 m		NSW, Australia		Smith et al. (1998)
• Wheat	Red earth with texture contrast with depth	• 97 mm drainage	• 17 as DAP		• 4	
• Alfalfa		• 112 mm drainage, from $D=P-ET-\Delta S$	• 0		• 12	
Sequences of pasture and crops:	Sandy loam	Ceramic cups at 7 depths, leaching calculated at 1.5 m, 114-214 mm of drainage calculated using $D=P-ET-\Delta S$	0 applied to all treatments	Western Australia		Anderson, Fillery, Dunin, et al. (1998)
• Grain lupin after wheat					• 25-35	
• Wheat after lupin					• 43-59	
• Wheat after 2 years of annual pasture					• 44	
• Second year of annual pasture					• 17-28	
Grazed fertilized pastures studied for:		Suction cups at 0.6m	Total of 250 applied at regular intervals each season	United Kingdom		Barraclough et al. (1992)
• 3 winter seasons	• Silt loam	• 167-230 mm drainage			• 0.1-5.9	
• 2 winter seaons	• Loam	• 167-177 mm drainage			• 6.5-12	
• 2 winter seasons	• Clay	• 223-244 mm drainage			• 0.5-4.3	
Grazed white clover-ryegrass pastures:	Silt loam	Tile drains at 0.45 m, volume of effluent from tiles measures automatically		Palmerston North, New Zealand		Magesan, White, and Scotter (1996)
• Pasture, no fertilizer N, in 1988			• 0		• 9-13	
• Pasture, with N applied in spring 1989			• 50		• 15-19	
• Pasture, no N applied in 1990			• 0		• 50	
• Pasture, with N applied in fall 1990 before drainage			• 120		• 44	

An example of temporal changes in soil NO_3^- that are possible in sandy soils under different legume-based cropping rotations is outlined in Figure 11.1. At least 70 percent of the soil profile NO_3^- was detected in the top 0.5 m when sampled three days after rainfall that established the onset of a growing season in a region with a Mediterranean-type climate (predominantly dry summer/fall and a wet winter/spring). The larger quantities of NO_3^- in soil profiles (97 to 106 kg ha^{-1}) at the end of legume phases (lupin plus one year of subterranean-clover-based pasture or two years of subterranean-clover-based pasture), compared to wheat (62 to 68 kg N ha^{-1}), are consistent with the increased net N mineralization in soil after legumes. The quantities and distribution of this NO_3^- in soil solution after a period of rainfall are shown in Figure 11.1c and d (measured on July 26) and Figure 11.1e and f (measured on August 15). Drainage at 1.5 m depth during these periods amounted to 114 mm, resulting in losses of NO_3^- below 1.5 m of 28 kg N ha^{-1} under subterranean-clover-based pasture (second year of pasture), 23 kg N ha^{-1} under lupin, 40 kg N ha^{-1} under wheat following lupin and 44 kg N ha^{-1} for wheat after pasture. The quantities of NO_3^- estimated to have been moved below 0.08, 0.2, 0.5, and 0.9 m for the same treatments are shown in Table 11.2. Rotation treatment did not affect the quantity of NO_3^- leached from the top 0.08 m of soil, but differences in uptake of NO_3^- by pasture and crop species resulted in marked differences in amounts leached below 0.2, 0.5, and 0.9 m, as well as below 1.5 m.

The efficiency of utilization of mineralized N by wheat, lupin, and pasture species and the proportion of this N leached or retained in soil in the studies of Anderson and colleagues (Anderson, Fillery, Dolling, et al., 1998; Anderson, Fillery, Dunin, et al., 1998) are summarized in Table 11.3. Wheat was particularly inefficient in the use of NO_3^-, while N uptake by a volunteer pasture species, capeweed (*Arctotheca calendula*), was largely responsible for the low contents of NO_3^- in soil under annual pasture treatments (Anderson, Fillery, Dolling, et al., 1998). The faster growth of roots in the top 0.5 m of soil in pasture, compared to wheat treatments, explains some of the differences in NO_3^- utilization evident between species in the studies described by Anderson and colleagues (Anderson, Fillery, Dolling, et al., 1998; Anderson, Fillery, Dunin, et al. et al., 1998). However, by the end of the growing season, wheat had produced a larger measurable root length density below 0.5 m than pasture species. Despite the presence of roots below 0.5 m, 19 to 35 kg N ha^{-1} (43 to 52 percent of total soil NO_3^-) remained in soil below 0.5 m at wheat grain harvest, highlighting the potential for leaching losses of NO_3^- over the next growing season (Anderson, Fillery, Dolling, et al., 1998; Anderson, Fillery, Dunin, et al., 1998).

FIGURE 11.1. Measured Soil Profile NO_3^- (kg N ha^{-1}) Under Phases of the AP/W (Annual Pasture/Wheat) Rotation: $W_{96}/2AP_{95}$, $1AP_{96}/W_{95}$ and $2AP_{96}/1AP_{95}$ (a, c, e, g), and Phases of the L/W (Lupin/Wheat) Rotation: W_{96}/L_{95} and L_{96}/W_{95} (b, d, f, h), at June 18, 1996 (a, b), July 26, 1996 (c, d), August 15, 1996 (e, f) and December 5, 1996 (g, h) Sampling Dates

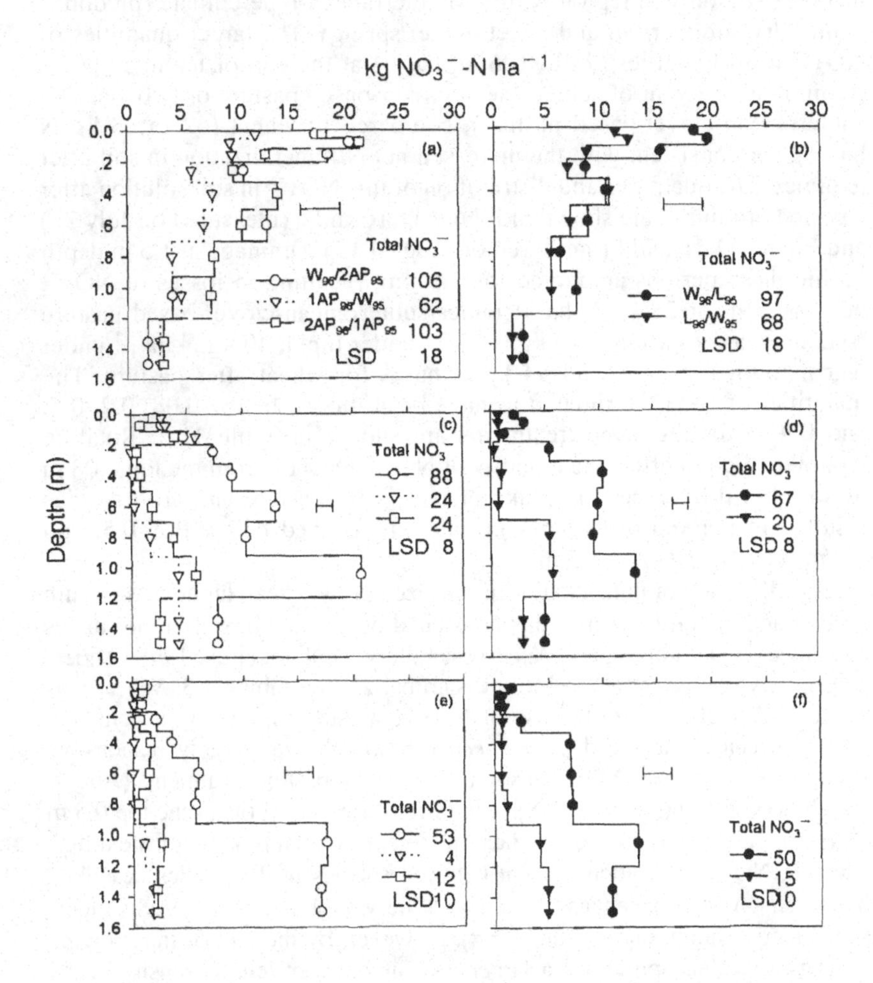

Note: The horizontal bars represent the LSD for the interaction treatment × depth at each sampling date. Total NO_3^- is the quantity (kg N ha^{-1}) to 1.5 m depth for each treatment and the corresponding LSD value.

FIGURE 11.1 (*continued*)

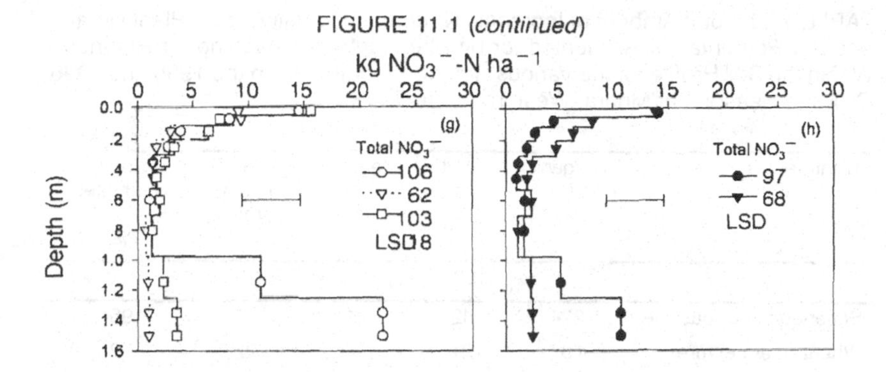

Source: Adapted from Anderson, Fillery, Dunin, et al. (1998).

Note: The horizontal bars represent the LSD for the interaction treatment × depth at each sampling date. Total NO_3^- is the quantity (kg N ha^{-1}) to 1.5 m depth for each treatment and the corresponding LSD value.

TABLE 11.2. Quantity (kg N ha^{-1}) of NO_3^- Leached to Different Depths in a Sandy Soil in Relation to Phase of Legume-Cereal Rotations Grown on a Deep Sand at Moora, Western Australia

Treatment	Soil depth, m				
	0.08	0.2	0.5	0.9	1.5
Wheat after lupin	38	56	69	70	40
Wheat after pasture	36	54	73	80	44
Lupin after wheat	36	42	41	35	23
Second year of pasture	40	29	22	35	28

Source: Calculated from data summarized by Anderson, Fillery, Dunin, et al. (1998).

Overall, the picture emerging from Australian studies is not encouraging. Crop species intentionally introduced into sandy soils of southern Australia appear to be poor users of mineralized NO_3^-, especially when this N is moved below a 1-m depth. On the other hand, the introduction of effective N_2-fixing legumes to Australian soils has resulted in large inputs of organic N (Unkovich, Pate, and Sanford, 1997; Anderson, Fillery, Dolling, et al., 1998).

TABLE 11.3. Total Amount of Inorganic N (kg ha^{-1}) Available for Plant Uptake and the Percentages Accounted for by Plant Uptake, Leaching, or Retention Within the Soil Profile for the Various Rotation Treatments in the 1995 and 1996 Growing Seasons at Moora, Western Australia

Treatment	Inorganic N kg ha^{-1}	Plant uptake %	Leached %	Retained in soil profile %	Total accounted for %
		1995			
Second year of pasture	224	42	12	41	95
Wheat after pasture	137	21	29	22	72
Wheat after lupin	139	28	42	27	97
Lupin after wheat	121	28	24	55	97
		1996			
First year pasture after wheat	177	55	10	25	90
Second year of pasture	272	57	8	24	89
Wheat after pasture	213	34	22	39	95
Wheat after lupin	168	36	17	37	90
Lupin after wheat	165	42	26	36	104

Source: Adapted from Anderson, Fillery, Dolling, et al. (1998).

Subsequent mineralization of this N can result in large amounts of NO_3^- in soil immediately prior to periods of drainage. Nitrate leaching in southern Australian soils is exacerbated by the low demand for NO_3^- during early winter when plant demand for N, especially in cereal crops, is low. This asynchrony between supply and demand for N is the underlying cause of leaching losses of NO_3^- in this region. Asynchrony in N supply and N use by crops can also occur in intensive agriculture when N in fertilizers, or animal manures, is applied in the fall and subsequent mineralization and nitrification results in NO_3^- production during periods of drainage when soils are often fallowed (see references listed in Table 11.1).

Although most interest in NO_3^- movement has been generated from concerns of ground water and other off-site pollution issues, the gain in H^+ commensurate with movement of NO_3^- from the zone of nitrification can cause soil acidification (Poss, Smith, et al., 1995). This acidification is largely unnoticed in well-pH-buffered soils or in agricultural systems where lime is regularly applied. However, soil acidification is viewed as a serious land degradation problem in many Australian soils (Helyar and Porter, 1989) and could also pose a major risk to sustainable crop production in other regions with weakly pH-buffered soils and a history of low lime use.

Movement of Calcium, Potassium, and Magnesium

Displacement of NO_3^- and SO_4^{2-} within soil must be coupled with the movement of a cation. Surprisingly, few studies of anion movement in soil have determined which cation species are coupled to anion movement and what effect this movement has on the long-term capacity of soil to supply cations for crop growth. The lysimeter studies of Wong and Rowell (1994) on the effect of crop and fertilizer application on cation and anion movement in a tropical soil at Onme, Nigeria, provide an excellent example of the relationships between NO_3^- and cation movement (see Table 11.4). The amounts of Ca^{2+} leached accounted for 27 to 34 percent of Ca^{2+} in soil in unlimed treatments, while 29 to 37 percent of initial Mg^{2+} and 8 to 10 percent of initial K^+ in soil was found in leachate collected at 1.35 m. Since NO_3^- was produced in the surface soil, this loss of cations must also occur from surface soil. When product removal of cations was included as an outflow with cation leaching, it was evident that inputs of Ca^{2+} and Mg^{2+} were needed to sustain crop yields in the corn-rice systems examined by Wong and Rowell (1994). The relevance of these findings for other crop-soil systems in high rainfall zones requires evaluation. Use of ceramic cups to sample soil solution in sandy soil in Senegal showed that losses of Ca^{2+} (11 kg Ca^{2+} ha^{-1}) and K^+ (1 tp 2 kg K^+ ha^{-1}) under millet were small (Pieri, 1979) in comparison to those described by Wong and Rowell (1994). Use of K^+ fertilizer increased Ca^{2+} losses, while incorporation of straw raised the concentrations of K^+ and NO_3^- in soil solutions (Pieri, 1979). Lysimeter studies of K^+ leaching in soils with 500 mm drainage under corn monocultures in Brittany, France, showed this to be variable (25 to 50 kg K ha^{-1}), but much less than annual loss of Ca^{2+} (250 to 430 kg ha^{-1}). The quantities of Ca^{2+} and Mg^{2+} leached were closely related to NO_3^-, Cl^- and SO_4^{2-} leached (Simon and Le Corrie, 1989). Leaching losses of cations and anions in effluent from a mole-drained soil under pasture that had been recently fertilized with KCl and superphosphate measured in (kg ha^{-1}) 40 of Ca^{2+}, 65 of Na^+, 11 of K^+, and 10 of Mg^{2+} in outflow at 0.45 m. In this study, Cl^- (100 kg ha^{-1} per year) was the major anion that was associated with cation movement, while NO_3^- (20 kg ha^{-1} per year) and SO_4^{2-} (13 kg ha^{-1} per year) were leached in relatively small amounts (Heng et al., 1991).

TABLE 11.4. Leaching of Cations and Anions (kg ha^{-1}) from a Typic Paleudult at Onne, Nigeria, for Contrasting Soil and Crop Management

Treatment (fertilizer additions as kg ha^{-1})	Amounts leached					Amount leached as % in soil after fertilizer addition		
	Ca	Mg	K	NO_3^-	Cl	Ca	Mg	K
Corn fertilized with 138 N as urea, 35 P, 67 K, and 30 Mg, followed by upland rice with 92 N as urea, 35 P, 42 K, and 20 Mg	312	31	33	144	110	27	29	10
Upcropped, but same fertilizer as for corn in the first season, and no fertilizer in the second season	399	40	29	251	121	34	37	9
Uncropped and given no fertilizer in the first season; received same fertilizer as upland rice in the second season	316	27	20	154	99	27	35	8

Source: Adapted from Wong and Rowell (1994).

Small quantities of K^+ (4.5 kg K ha^{-1} $year^{-1}$ without K [potassium] fertilizer; 7.5 kg ha^{-1} $year^{-1}$ following the application of 137 kg K ha^{-1} $year^{-1}$) were leached in a Ferralsol cropped to corn in Togo, where drainage at 1.8 m ranged from 216 to 436 mm in the two seasons studied (Poss, Fardeau, and Saragoni, 1997). In this case, low soil solution concentrations of K^+ limited the potential for K^+ leaching. Concurrent measurements of K^+ uptake by crops, recycling of K^+ in residue, K^+ adsorption, and K^+ fixation enabled Poss, Fardeau, and Saragoni (1997) to conclude that K^+ removal in grain and crop residues, and K^+ fixation after fertilization, were the major causes of "loss" of K^+ in their studies. Replacement of Ca^{2+} by K^+ after lime application can increase retention of K^+ (Sparks, 1980).

Management to Minimize Solute Leaching

Overcoming the buildup of anions in soil solution, particularly NO_3^-, ahead of periods of drainage would appear necessary to minimize solute leaching and, in weakly pH-buffered soils, topsoil acidification. Rescheduling of fall applications of fertilizer N to spring, changes to grazing strategies in pasture land, and the storage of excreta from housed animals for dispersal on land when drainage is minimal are practical management options to reduce NO_3^- (and associated cations) movement in northern hemisphere crop-livestock agriculture systems (Shepherd, Davies, and Johnson, 1993; Cuttle and Scholefield, 1995; Goss, Beauchamp, and Miller, 1995). From the summaries of NO_3^- leaching shown in Table 11.1, it is also evident that the seeding of catch crops, including ryegrass, barley with undersown grass, rye, and forage rape, in fall to otherwise fallow crop land reduces NO_3^- loss in temperate climates and is a practical method to minimize movement of NO_3^- mineralized from organic sources. Use of legume monocultures as catch crops is not recommended, since these appear to use less NO_3^- than nonleguminous species (Anderson, Fillery, Dolling, et al., 1998) or may increase N mineralization after cultivation because of recent N inputs through N_2 fixation (Campbell et al., 1994). Delaying cultivation from fall to spring, and the tailoring of fertilizer N inputs to meet the shortfall between N mineralization and crop demand for N, will also reduce NO_3^- leaching (see Table 11.1). Fewer management options are apparent for controlling the buildup of NO_3^- in regions where winter cropping follows dry summer/fall seasons (Anderson, Fillery, Dunin, et al., 1998). In this case, the deficit of soil water during the summer/fall period probably precludes seeding of short-season catch crops. Replacement of annual pasture with either perennial legumes or perennial grasses is being advocated for southern regions of Australia to increase use of deep soil water. These species could also act as sinks for soil NO_3^- should growth occur in the fall. Use of minimum tillage to reduce N mineralization, and the seeding of cereals and oilseed crops on the first substantial late fall-early winter rain to increase the time for root growth before drainage, are other practical management tools that could be used in Mediterranean-type climates to reduce NO_3^- leaching in winter.

MODELING OF NUTRIENT MOVEMENT

Numerous soil-plant models have been developed that include routines for simulating water and nutrient movement. Many of these models have been compared using common data sets; published reports on the performance of different models are excellent sources of information on the range of models available and the types of routines used in each to describe water and nutrient movement (see De Willigen, 1991; Vereecken, Jansen, et al., 1991; Vereecken, Swerts, et al., 1991; Diekkrüger et al., 1995; Bond et al., 1996; Bristow et al., 1996). In the case of solute movement, most progress has been achieved with modeling the movement of NO_3^-, as well as nonreactive ions such as Br^- and Cl^-. In all cases, a description of water flow is essential. Procedures adopted to model water movement include (1) simple cascading or "tipping bucket" routines that move water through soil layer by layer (functional models) and (2) more complex mechanistically based solutions that use Richard's equation (see De Willigen, 1991; Vereecken, Swerts, et al., 1991) to describe water movement and the convection-dispersion equation to explain the behavior of dissolved ions (mechanistic or deterministic models). A third category of solute transfer models, although not strictly water movement models, is based on stochastic or transfer function approaches (see Jury, Sposito, and White, 1986; White, 1989). Stochastic models do not use the physics of fluid flow to describe solute flow; rather, they use a previously determined probability density function of travel times of solutes through soil to predict solute movement. Stochastic models take into account the spatial variability in solute movement in soil, while this property is generally ignored in functional and mechanistic models. However, probability density functions need to be characterized for each situation in which the stochastic model is applied, reducing the practicality of this approach.

Functional models are largely based on the principle of mass conservation of water within designated layers of soil. The amount of water moved layer by layer depends on the pore volume available for mobile water, and this can be defined by the upper and lower drainage limits or by moisture release characteristics. Water is typically moved on a daily time step.

Although functional models are less rigorous than mechanistic models, the input requirements are simpler, and needed initialization information is more readily available for a range of soils. Capacity-based parameters are also less spatially variable than rate parameters required in mechanistic models. Experience has shown that functional models can accurately describe ET (evapotranspiration) and drainage in a deep sandy soil (see Figure 11.2). In this case, the soil water simulation subroutine, SOILWAT (Probert et al., 1997), was part of a more comprehensive soil-plant model described as the Agricultural Production SIMulator or APSIM (McCown et al., 1996). Subroutines that describe the perching of water in texture-contrast soils have been added to the APSIM SOILWAT submodel to enable the APSIM model to describe soil water flows in these soils (Asseng, Keating, et al., 1998). The Solute Leaching Intermediate Model (SLIM), developed by Addiscott and Whitmore (1991), categorizes soil water as either mobile or immobile and can be described as a hybrid functional-mechanistic model. Hall (1993) further partitioned mobile water flow in SLIM into

FIGURE 11.2. Rainfall (a) and Observed and Predicted Rates of Evapotranspiration (ET) and Drainage Obtained from the Functional SOILWAT Simulation Routine (b) when Applied to Deep Sand

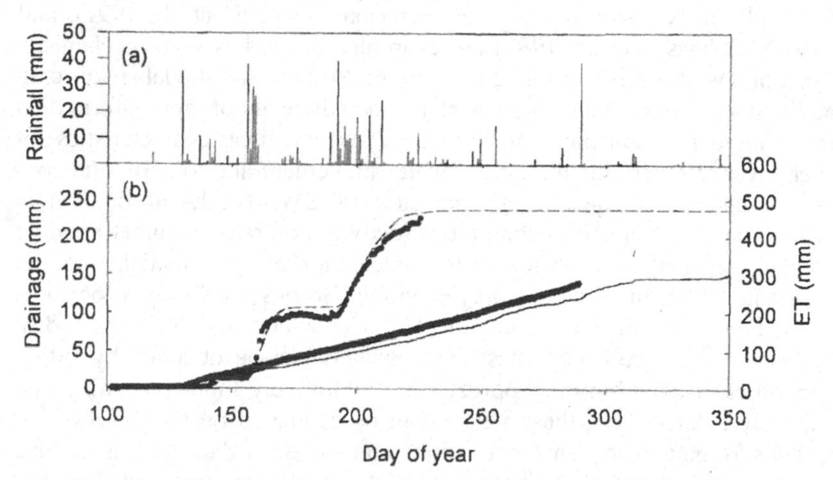

Source: Adapted from Asseng, Fillery, et al. (1998).

Note: Solid lines are evapotranspiration (ET), with the bold solid line representing measured values and dashed lines being drainage below 1.5 m depth (symbols are actual measured values).

"slow" and "fast" flow domains to allow for flow through smaller water-filled pores and rapid preferential flow through larger macropores and fissures. These modifications to SLIM resulted in good matches between predicted and measured discharge of both water and solute in a sand and a clay soil (Hall and Webster, 1993). Both SLIM and the Hall-modified SLIM do not contain routines that describe N mineralization, precluding evaluation of these models against field-generated NO_3^- data, without the addition of relationships to describe net N mineralization, nitrification, and plant uptake of N (Addiscott et al., 1991; Whitmore, 1995). The APSIM wheat model has mineralization, nitrification, and N uptake subroutines as part of its SOILN subroutine, enabling evaluation of soil NO_3^- movement from soil- and fertilizer-derived N sources during periods of crop growth. Although Asseng, Fillery, and colleagues (1998) found the accuracy of prediction of NO_3^- concentrations in sandy soil to be poorer than observed for either drainage or ET, discussed earlier, some of the inaccuracy in prediction of the temporal patterns of NO_3^- was attributed to problems with simulation of net N mineralization.

SWIM (Soil Water Infiltration Model; Ross, 1990), LEACHM (Leaching Estimation and Chemical Model; Hutson and Wagenet, 1992), DAISY model (Danish Simulation Model for the Transformation and Transport of Matter and Energy in the Soil Plant System; Hansen et al., 1990), ANIMO (Agricultural NItrogen MOdel; see Vereecken, Swerts, et al., 1991), and SOILN (Johnsson et al., 1987) are examples of models with mechanistic descriptions of water and solute movement. Mechanistic models require, in addition to weather and environmental data, a diversity of input information for a number of soil physical, hydraulic, and chemical characteristics for each layer to determine the rate of water movement and to apply the convection-dispersion equation. In the case of SWIMv2.1, information is needed on the saturated hydraulic conductivity, saturated volumetric water content, residual volumetric water content, air entry potential, and the constants b from the Brooks-Corey equation (Brooks and Corey, 1966) or α from the Van Genuchten equation (Van Genuchten and Nielsen, 1985). Values for these soil properties and constants can be obtained by fitting experimental data, but this approach is prohibitively time-consuming and expensive. Alternately, these values can be estimated on the basis of so-called surrogate pedo-transfer functions that use other data such as particle size and bulk density (see Smettem et al., 1994; Smettem and Gregory, 1996; Verburg, Ross, and Bristow, 1996). Less success has been achieved using surrogate pedo-transfer functions to estimate functions needed to apply the convection-dispersion equation. Values for ion diffusion in free water, tortuosity, and dispersivity of the medium can be obtained from the literature. For reactive solutes, terms that describe adsorption must be included. As is the case with functional models, additional information is needed to initialize subroutines that describe surface runoff, ET, and plant uptake of soil water to complete the description of water flow.

The better description of water flow, inherent in mechanistic models, might be expected to improve fits of observed against predicted NO_3^- leaching, but this is not necessarily the experience of studies in which both types of models have been evaluated against common data sets (De Willigen, 1991). In a comparison of the SOILWAT and SWIMv2.1 simulation routines, Bond and colleagues (1996) and Bristow and colleagues (1996) found both to give good descriptions of soil water content and solute movement. Mechanistic models did outperform functional models in another comparison, but only one model (DAISY) consistently reproduced experimental data (Diekkrüger et al., 1995). Poor descriptions of NO_3^- leached below 1.2 m for soil under corn have been reported for LEACHM (Jabro et al., 1993). In this case, the poor characterization of water flow through macropores and/or inaccurate characterization of N rate constants for net N mineralization and nitrification were identified as possible weaknesses in LEACHM, as applied. A subsequent comparison of standard LEACHM (incorporating solutions for Richard's and the convection-dispersion equations), with another version using the capacity approach of SLIM to model water movement, failed to demonstrate an advantage of the new version over the original (Jabro et al., 1995). LEACHM has given good representations of soil profile volumetric water content, together with drainage and NO_3^- leached at 0.25 and 0.9 m, for a Red Earth sown to either wheat or alfalfa (Smith et al., 1998). However, knowledge of evaporation and transpiration and the distribution of roots in soil was critical to the successful application of this model (Smith, 1998). The lack of routines that describe plant growth and N uptake hinder the application of LEACHM to agricultural environments involving crops.

Except for SLIM and SLIM derivatives, movement of water and solutes in macropores, or the preferential flow of water and solutes, is largely ignored in most models. SWIMv2.1 includes routines that enable bypass flow when runoff occurs; otherwise, water flow is assumed to be uniform through the soil continuum, and the effect of macropores on flow is essentially handled through the description of the water retention and hydraulic properties (Verburg, Ross, and Bristow, 1996). The existence of mobile and less mobile water could have a marked effect on the rate of solute transport and leaching. Advection and dispersion only occur in the mobile phase, while adsorption occurs in both. Exchange of ions between the two phases of soil water will affect the concentration in the mobile phase and the potential for solute movement. Some progress has been made with the measurement of immobile water in soil (Clothier, Kirkham, and McLean, 1992) and the exchange of solute between the mobile and immobile water (Jaynes, Logsdon, and Horton 1995). However, insufficient information has been collected to enable a detailed assessment of the significance of omission of macropore flow on the performance of mechanistic models.

Diekkrüger and colleagues (1995) concluded that the experience of operators using models was often as important as the differences between model approaches in the final analysis of model performance. One explanation for this finding is that models have been tested on data sets that are deficient in essential information needed to initialize different models. In this case, prior knowledge of the operator in the selection of appropriate input parameters could affect the accuracy of prediction and the perceived success of one model compared to another. Interpretation of model performance is also complicated by the variability in water and solute movement in soil (Addiscott, 1996). The level of precision used to define model input parameters should be viewed in the context of the variability of the processes being simulated.

All of the models described here can be considered as pedon or plot scale models (Wagenet, 1998). In this case, a pedon is defined as a three-dimensional natural body, large enough to represent the nature and arrangement of its horizons and variability. Methods are being sought to enable these pedon-scale models to be used to predict water and solute movement at a polypedon (field), catena (farm), and catchment or watershed scales (see Wagenet, 1998). Radcliffe, Gupte, and Box (1998) concluded that good agreement existed between average convection-dispersion equation parameters for a sandy loam measured at a pedon scale and those estimated at a polypedon scale. It is noteworthy that these authors also concluded that a mechanistic model based on the convection-dispersion equation more accurately predicted the estimated polypedon breakthrough curve than a stochastic approach based on the convective lognormal transfer function. Use of mechanistic models to predict water and solute flows at catchment scales, by evaluating these processes across a sequence of grids or cells, is less certain (Beven, 1989; Smettem et al., 1994). Uncertainty about the accuracy of inferred initialization parameters, a widely adopted approach when mechanistic models are applied across grids within a catchment, appears to be the major point of debate. The availability of more rapid and user-friendly methods to measure hydraulic properties, and the use of pedotransfer functions to determine hydraulic properties from widely measured soil data, such as texture, should enable more relevant initialization information to be collected over catchment scales. In this context, the mapping of catchment soil clay content on the basis of geomorphological knowledge, using conditional probability equations (Bayesian methods), could produce surrogate hydraulic properties needed for catchment scale modeling using mechanistic models (Smettem et al., 1994). Strong arguments can also be advanced for the use of functional models in place of mechanistic models when water and solute movement are being predicted over large areas. These models have characteristics that are conducive to use over large scales, including linearity in respect to input parameters, input parameters that display low variability (soil water content), and parameters that are nonscalar and therefore can be treated as additive in the event of spatial dependence (Addiscott, 1998).

CONCLUSIONS

Substantial advances have been made in the automatic measurement of soil water content and the determination of ET, and these now enable routine measurements of drainage in field soils. The availability of simple and low-cost methods for extracting soil solution have made field studies of solute movement more feasible. Nevertheless, few detailed studies of cation and anion movement have been undertaken. Most studies of solute movement have involved a single ion and one depth of measurement. The temporal changes in ion concentration in soil or soil solution have typically been made without measurements of ET, drainage, root distribution, and mineralization in the case of NO_3^- and SO_4^{2-}. Considerable progress has been made with the development of computer models that describe water and solute movement in soil. Testing of these models has been frustrated by the lack of data sets containing measurements of key processes that determine the availability and movement of soil water and solutes, including ET, drainage, and root distribution. Concurrent measurements are needed of anion and cation concentrations in soil solution for a range of soil types for which ET, changes in soil water content, drainage, crop or pasture growth, root length densities, and key soil physical and chemical properties are determined. The problem of scale is an important challenge for soil and water modelers, and it is evident that methods for applying water and solute movement models to scales larger than soil pedons are in their infancy.

Different strategies for managing soils and crops to minimize NO_3^- movement and leaching have been evaluated for temperate climates. However, NO_3^- leaching is also important in some soils in Mediterranean-type climates and in tropical environments. Practical management strategies to reduce NO_3^- leaching in these environments are needed.

Differences among plant species in the extraction of NO_3^- from soil profiles indicate that there is scope for improving the rooting characteristics of major crop species. However, the root architecture and root activity needed to capture deep-placed nutrients remain to be characterized.

REFERENCES

Adams, J.A. and J.M. Pattinson (1985). Nitrate leaching losses under a legume-based crop rotation in Central Canterbury, New Zealand. *New Zealand Journal of Agricultural Research* 28: 101-107.

Addiscott, T.M. (1996). Measuring and modelling nitrogen leaching: Parallel problems. *Plant and Soil* 181: 1-6.

Addiscott, T.M. (1998). Modelling concepts and their relation to the scale of the problem. *Nutrient Cycling in Agroecosystems* 50: 239-245.

Addiscott, T.M., N.J. Bailey, G.J. Bland, and A.P. Whitmore (1991). Simulation of nitrogen in soil and winter wheat crops: A management model that makes the best use of limited information. In *Nitrogen Turnover in the Soil-Crop System*, Eds. J.J.R. Groot, P. De Willigen, and E.L.J. Verberne. Dordrecht, Netherlands: Kluwer Academic Publishers, pp. 141-149.

Addiscott, T.M. and A.P. Whitmore (1991). Simulation of solute leaching in soils of differing permeabilities. *Soil Use and Management* 7: 94-102.

Anderson, G.C., I.R.P. Fillery, P.J. Dolling, and S. Asseng (1998). Nitrogen and water flows under pasture-wheat and lupin-wheat rotations in deep sands in Western Australia. 1. Nitrogen fixation in legumes, net N mineralization, and utilization of soil-derived nitrogen. *Australian Journal of Agricultural Research* 49: 329-343.

Anderson, G.C., I.R.P. Fillery, F.X. Dunin, P.J. Dolling, and S. Asseng (1998). Nitrogen and water flows under pasture-wheat and lupin-wheat rotations in deep sands in Western Australia. 2. Drainage and nitrate leaching. *Australian Journal of Agricultural Research* 49: 345-361.

Asseng, S., I.R.P. Fillery, G.C. Anderson, P.J. Dolling, F.X. Dunin, and B.A. Keating (1998). Use of the APSIM wheat model to predict yield, drainage and NO_3^- leaching for a deep sand. *Australian Journal of Agricultural Research* 49: 363-377.

Asseng, S., B.A. Keating, I.R.P. Fillery, P.J. Gregory, J.W. Bowden, N.C. Turner, J.A. Palta, and D.G. Abrecht (1998). Performance of the APSIM wheat model in Western Australia. *Field Crops Research* 57: 163-179.

Barraclough, D., S. Jarvis, G.P. Davies, and J. Williams (1992). The relation between fertilizer nitrogen applications and nitrate leaching from grazed grassland. *Soil Use and Management* 8: 51-56.

Bergstrom, L. (1987). Nitrate leaching and drainage from annual and perennial crops in tile-drained plots and lysimeters. *Journal of Environmental Quality* 16: 11-18.

Beven, K.J. (1989). Changing ideas in hydrology—The case of physically-based models. *Journal of Hydrology* 105: 157-172.

Bond, W.J., M.E. Probert, H.P. Cresswell, N.I. Huth, B.A. Keating, and K. Verburg (1996). Simulation of the Wagga data set. In *Methodology in Soil Water and Solute Balance Modelling: An Evaluation of the APSIM-SoilWat and SWIMv2 Models*, Ed. K. Verburg. Glen Osmond, South Australia: CSIRO Land and Water, Divisional Report No. 131, pp. 29-41.

Bristow, K.L., J.P. Dimes, J.E. Turpin, and P.S. Carberry (1996). Simulation of the van Bavel data set. In *Methodology in Soil Water and Solute Balance Modelling: An Evaluation of the APSIM-SoilWat and SWIMv2 Models*, Ed. K. Verburg. Glen Osmond, South Australia: CSIRO Land and Water, Divisional Report No. 131, pp. 55-82.

Brooks, R.H. and A.T. Corey (1966). Properties of porous media affecting fluid flow. *Proceedings of the American Society of Civil Engineers. Irrigation and Drainage Division* 92: 555-560.

Cameron, K.C., D.F. Harrison, N.P. Smith, and C.D.A. McLay (1990). A method to prevent edge-flow in undisturbed soil cores and lysimeters. *Australian Journal of Soil Research* 28: 879-886.

Campbell, C.A., G.P. Lafond, R.P. Zentner, and Y.W. Jame (1994). Nitrate leaching in a udic haploboroll as influenced by fertilization and legumes. *Journal of Environmental Quality* 23: 195-201.

Campbell Scientific (1996). *Instruction Manual for CS615 Water Content Reflectometer*. Logan, UT: Campbell Scientific, Inc.

Clothier, B.E., M.B. Kirkham, and J.E. McLean (1992). *In situ* measurement of the effective transport volume for solute moving through soil. *Soil Science Society of America Journal* 56: 733-736.

Crabtree, R.W. and M.J. Kirkby (1985). Ion-exchange resin samplers for the *in situ* measurement of major cations in soil water solute flux. *Journal of Hydrology* 80: 325-335.

Cuttle, S.P. and D. Scholefield (1995). Management options to limit nitrate leaching from grassland. *Journal of Contaminant Hydrology* 20: 299-312.

Dalton, F.N. (1992). Development of time-domain reflectometry for measuring soil water content and bulk soil electrical conductivity. In *Advances in Measurement of Soil Physical Properties: Bringing Theory into Practice*, Special Publication 30, Eds. G.C. Topp, W.D. Reynolds, and R.E. Green. Madison, WI: Soil Science Society of America. pp. 143-167.

Dean, T.J., J.P. Bell. and A.J.B. Baty (1987). Soil moisture measurement by an improved capacitance technique. Part I. Sensor design and performance. *Journal of Hydrology* 93: 67-78.

De Willigen, P. (1991). Nitrogen turnover in the soil-crop system: Comparison of fourteen simulation models. In *Nitrogen Turnover in the Soil-Crop System*, Eds. J.J.R. Groot. P. De Willigen, and E.L.J. Verberne. Dordrecht, Netherlands: Kluwer Academic Publishers, pp. 141-149.

Diekkrüger, B., D. Söndgerath, K.C. Kersebaum, and C.W. McVoy (1995). Validity of agroecosystems models. A comparison of results of different models applied to the same data set. *Ecological Modelling* 81: 3-29.

Dirksen, C.E. and S. Drasberg (1993). Four component mixing model for improved calibration of TDR soil water content measurements. *Soil Science Society of America Journal* 57: 660-667.

Djurhuus, J. and O.H. Jacobsen (1995). Comparison of ceramic suction cups and KCl extraction for the determination of nitrate in soil. *European Journal of Soil Science* 46: 387-395.

Dunin, F.X., W.S. Meyer, S.C. Wong, and W. Reyenga (1989). Seasonal change in water use and carbon assimilation of irrigated wheat. *Agricultural and Forest Meteorology* 45: 231-250.

Evett, S.R. and J.L. Steiner (1995). Precision of neutron scattering and capacitance type soil water content gauges from field calibration. *Soil Science Society of America Journal* 59: 961-968.

Francis, G.S., R.J. Haynes, and P.H. Williams (1995). Effects of the timing of ploughing-in temporary leguminous pastures and two winter cover crops on nitrogen mineralization, nitrate leaching and spring wheat growth. *Journal of Agricultural Science (Cambridge)* 124: 1-9.

Fundinger, R., K. Köhler, and M. Stacheder (1997). *Measurement of Material and Soil Moisture with the TRIME-Method.* Ettlingen, Germany: INKO.

Gardner, W.H. (1986). Water content. In *Methods of Soil Analysis.* Part 1. *Physical and Mineralogical Methods,* Ed. A. Klute. Madison, WI: American Society of Agronomy, pp. 493-544.

Goss, M.J., E.G. Beauchamp, and M.M. Miller (1995). Can farming systems approach minimize nitrogen losses to the environment. *Journal of Contaminant Hydrology* 20: 285-297.

Goulding, K.W.T. and C.P. Webster (1992). Methods for measuring nitrate leaching. *Aspects of Applied Biology* 30: 63-70.

Gregory, P.J., R. Poss, J. Eastham, and S. Micin (1995). Use of time domain reflectometry (TDR) to measure the water content of sandy soils. *Australian Journal of Soil Research* 33: 265-276.

Grossmann, J. and P. Udluft (1991). The extraction of soil water by the suction-cup method: A review. *Journal of Soil Science* 42: 83-93.

Gustafson, A. (1987). Nitrate leaching from arable land in Sweden under four cropping systems. *Swedish Journal of Agricultural Research* 17: 169-177.

Haines, B.L., J.B. Waide, and R.L. Todd (1982). Soil solution nutrient concentrations sampled with tension and zero-tension lysimeters: Report of discrepancies. *Soil Science Society of America Journal* 46: 658-661.

Hall, D.G.M. (1993). An amended functional leaching model applicable to structured soils. I. Model description. *Journal of Soil Science* 44: 579-588.

Hall, D.G.M. and C.P. Webster (1993). An amended functional leaching model applicable to structured soils. II. Model application. *Journal of Soil Science* 44: 589-599.

Hansen, E.M. and J. Djurhuus (1996). Nitrate leaching as affected by long-term N fertilization on a coarse sand. *Soil Use and Management* 12: 199-204.

Hansen, E.M. and J. Djurhuus (1997). Yield and N uptake as affected by soil tillage and catch crop. *Soil and Tillage Research* 42: 241-252.

Hansen, S., H.E. Jensen, N.E. Neilsen, and H. Svendsen (1990). DAJSY: A soil plant system model. In *Danish Simulation Model for Transformation and Transport of Energy and Matter in the Soil Plant Atmosphere System.* Copenhagen, Denmark: The National Agency for Environmental Protection, NPO Report No. A10, 272 pp.

Hatch, D.J., S.C. Jarvis, A.J. Rook, and A.W. Bristow (1997). Ionic contents of leachate from grassland soils: A comparison between ceramic suction cup samples and drainage. *Soil Use and Management* 13: 68-74.

Helyar, K.R. and W.M. Porter (1989). Soil acidification, its measurement and the processes involved. *In Soil Acidity and Plant Growth*, Ed. A.D. Robson. Sydney, Australia: Academic Press, pp. 61-101.

Heng, L.K., R.E. White, N.S. Bolan, and D.R. Scotter (1991). Leaching losses of major nutrients from a mole-drained soil under pasture. *New Zealand Journal of Agricultural Research* 34: 325-334.

Herkelrath, W.N., S.P. Hamburg, and F. Murphy (1991). Automatic, real-time monitoring of soil moisture in a remote field area with time domain reflectometry. *Water Resources Research* 27: 857-864.

Hutson, J.L. and R.J. Wagenet (1992). *LEACHM—Leaching Estimation and Chemistry Model: A Process Based Model of Water and Solute Movement Transformations, Plant Uptake and Chemical Reactions in Unsaturated Zone.* Version 3. Ithaca, NY: Department of Agronomy, Cornell University.

Jabro, J.D., J.M. Jemison Jr., L.L. Lengnick, R.H. Fox, and D.D. Fritton (1993). Field validation and comparison of LEACHM and NCSWAP models for predicting nitrate leaching. *Transactions of the American Society of Agricultural Engineers* 36: 1651-1657.

Jabro, J.D., J.D. Toth, A. Dou, R.H. Fox, and D.D. Fritton (1995). Evaluation of nitrogen version of LEACHM for predicting nitrate leaching. *Soil Science* 160: 209-217.

Jacobesen, O.H. and P. Schjonning (1993). A laboratory calibration of time-domain reflectometry for soil water measurement including effects of bulk density and texture. *Journal of Hydrology* 151: 147-157.

Jaynes, D.B., S.D. Logsdon, and R. Horton (1995). Field method for measuring mobile/immobile water content and solute transfer rate coefficient. *Soil Science Society of America Journal* 59: 352-356.

Johnson, P.A., M.A. Shepherd, and P.N. Smith (1997). The effects of crop husbandry and nitrogen fertilizer on nitrate leaching from a shallow limestone soil growing a five course combinable crop rotation. *Soil Use and Management* 13: 17-23.

Johnsson. H., L. Bergström, P.-E. Jansson, and K. Paustian (1987). Simulated nitrogen dynamics and losses in a layered agricultural soil. *Agricultural Ecosystems and Environment* 18: 333-356.

Jury, W.A., G. Sposito, and R.E. White (1986). A transfer function model of solute transport through soil. 1. Fundamental concepts. *Water Resources Research* 22: 243-247.

Lord, E.I. and M.A. Shepherd (1993). Developments in the use of porous ceramic cups for measuring nitrate leaching. *Journal of Soil Science* 44: 435-449.

MacDonald, A.J., D.S. Powlson, P.R. Poulton, and D.S. Jenkinson (1989). Unused fertilizer nitrogen in arable soils—Its contribution to nitrate leaching. *Journal of the Science of Food and Agriculture* 46: 407-419.

Magesan, G.N., R.E. White, and D.R. Scotter (1996). Nitrate leaching from a drained, sheep-grazed pasture. I. Experimental results and environmental implications. *Australian Journal of Soil Research* 34: 55-67.

Martin, E.C., T.L. Loudon, J.T. Ritchie, and A. Werner (1994). Use of drainage lysimeters to evaluate nitrogen and irrigation management strategies to minimize nitrate leaching in maize production. *Transactions of the American Society of Agricultural Engineers* 37: 79-83.

Martinez, J. and G. Guiraud (1990). A lysimeter study of the effects of a ryegrass catch crop, during a winter wheat/maize rotation, on nitrate leaching and on the following crop. *Journal of Soil Science* 41: 5-16.

McCown, R.L., G.L. Hammer, J.N.G. Hargreaves, D.P. Holzworth, and D.M. Freebairn (1996). APSIM: A novel software system for model development, model testing and simulation in agricultural system research. *Agricultural Systems* 50: 255-271.

McLenaghen, R.D., K.C. Cameron, N.H. Lampkin, M.L. Daly, and B. Deo (1996). Nitrate leaching from ploughed pasture and the effectiveness of winter catch crops in reducing leaching losses. *New Zealand Journal of Agricultural Research* 39: 413-420.

Moutonnet, P. and J.C. Fardeau (1997). Inorganic nitrogen in soil solution collected with tensionic samplers. *Soil Science Society of America Journal* 61: 822-825.

Moutonnet, P., J.F. Paegenel, and J.C. Fardeau (1993). Simultaneous field measurement of nitrate-nitrogen and matric pressure head. *Soil Science Society of America Journal* 57: 1458-1462.

Pieri, C. (1979). A study of the soil solution in a cultivated sandy soil in Senegal, using porous ceramic cup samplers. *Agronomie Tropicale* 34: 9-22.

Poss, R., J.C. Fardeau, and H. Saragoni. (1997). Sustainable agriculture in the tropics: The case of potassium under maize cropping in Togo. *Nutrient Cycling in Agroecosystems* 46: 205-213.

Poss, R., A.D. Noble, F.X. Dunin, and W. Reyenga (1995). Evaluation of ceramic cup samplers to measure nitrate leaching in the field. *European Journal of Soil Science* 46: 667-674.

Poss, R., C.J. Smith, F.X. Dunin, and J.F. Angus (1995). Rate of acidification under wheat in a semi-arid environment. *Plant and Soil* 177: 85-100.

PRENART (1998). *Technical Information on PRENART Equipment*. Frederiksberg, Denmark: PRENART.

Probert, M.E., J.P. Dimes, B.A. Keating, R.C. Dalal, and W.M. Strong (1997). APSIM's water and nitrogen modules and simulation of the dynamics of water and nitrogen in fallow systems. *Agricultural Systems* 56: 1-28.

Radcliffe, D.E., S.M. Gupte, and J.E. Box (1998). Solute transport at the pedon and polypedon scales. *Nutrient Cycling in Agroecosystems* 50: 77-84.

Robinson, D.A., J.P. Bell, and C.H. Batchelor (1994). Influence of iron minerals on the determination of soil water content using dielectric techniques. *Journal of Hydrology* 161: 169-180.

Robinson, M. and T.J. Dean (1993). Measurement of near surface soil water content using a capacitance probe. *Hydrological Processes* 7: 77-86.

Roth, C.H., M.A. Malicki, and R. Plagge (1992). Empirical evaluation of the relationship between soil dielectric constant and volumetric water content and the basis for calibrating soil moisture measurements by TDR. *Journal of Soil Science* 43: 1-13.

Ross, P.J. (1990). *SWIM—A Simulation Model for Soil Water Infiltration and Movement, Reference Manual to SWIMv1*. Glen Osmond, Australia: CSIRO Division of Soils.

Schnabel, R.R. (1983). Measuring nitrogen leaching with ion exchange resin: A laboratory assessment. *Soil Science Society of America Journal* 47: 1041-1042.

Schnabel, R.R., S.R. Messier, and R.F. Purnell (1993). An evaluation of anion exchange resin used to measure nitrate movement through soil. *Communications in Soil Science and Plant Analysis* 24: 863-879.

Shepherd, M.A., D.B. Davies, and P.A. Johnson (1993). Minimizing nitrate losses from arable soils. *Soil Use and Management* 9: 94-99.

Simmelsgaard, S.E. (1998). The effect of crop, N-level, soil type and drainage on nitrate leaching from Danish soil. *Soil Use and Management* 14: 30-36.

Simmonds L.P. and S. Nortcliff (1998). Small scale variability in the flow of water and solutes, and implications for lysimeter studies of solute leaching. *Nutrient Cycling in Agroecosystems* 50: 65-75.

Simon, J.C. and L. Le Corrie (1989). Leaching of mineral elements (except nitrogen) in maize monoculture on granitic soil in south-west Brittany, France. *Fourrages* 118: 127-148.

Smettem, K.R.J., K.L. Bristow, P.J. Ross, R. Haverkamp, S.E. Cook, and A.K.L. Johnson (1994). Trends in water balance modelling at field scale using Richard's equation. *Trends in Hydrology* 1: 383-402.

Smettem, K.R.J. and P.J. Gregory (1996). The relationship between soil water retention and particle size distribution parameters for some predominantly sandy Western Australian soils. *Australian Journal of Soil Research* 34: 695-708.

Smith, C.F., F.X. Dunin, S.J. Zegelin, and R. Poss (1998). Nitrate leaching from a Riverine clay soil under cereal rotation. *Australian Journal of Agricultural Research* 49: 379-389.

Smith, C.J. (1998). Personal communication. Dr. Smith is a Senior Principal Research Scientist in Land and Water, Australia.

Smith, K.A and C.E. Mullins (1991). *Soil Analysis: Physical Methods*. New York: Marcel Dekker, Inc.

Sparks, D.L. (1980). Chemistry of soil potassium in Atlantic coastal plain soils: A review. *Communications in Soil Science and Plant Analysis* 11: 435-449.

Strong, D.T., P.W.G. Sale, and K.R. Helyar (1997). A technique for the non-destructive measurement of nitrate in small soil volumes. *Australian Journal of Soil Research* 35: 571-578.

Topp, G.C. and J.L. Davis (1985). Measurement of soil water content using TDR: A field evaluation. *Soil Science Society of America Journal* 49: 19-24.

Topp, G.C., J.L. Davis, and A.P. Annan (1980). Electromagnetic determination of soil-water content: Measurement in coaxial transmission lines. *Water Resources Research* 16: 574-582.

Topp, G.C., M. Watt, and H.N. Hayhoe (1996). Point specific measurement and monitoring of soil water content with an emphasis on TDR. *Canadian Journal of Soil Science* 76: 307-316.

Torbert, H.A. and C.B. Elkins (1992). Determining differential water movement through ion exchange resin for nitrate leaching measurements. *Communications in Soil Science and Plant Analysis* 23: 1043-1052.

TRASE (1997). *Instruments for the Measurement of Moisture and Dielectric Properties Using Time Domain Reflectrometry*. Santa Barbara, CA: Soil Moisture Equipment Corporation.

Unkovich, M.J., J.S. Pate, and P. Sanford (1997). Nitrogen fixation by annual legumes in Australian Mediterranean agriculture. *Australian Journal of Agricultural Research* 48: 267-293.

Van Genuchten, M.T. and D.R. Nielsen (1985). On describing and predicting the hydraulic properties of unsaturated soils. *Annals de Geophysique* 3: 615-628.

Verburg, K., P.J. Ross, and K.L. Bristow (1996). *SWIMv2.1 User Manual*. Glen Osmond, Australia: CSIRO Division of Soils, Divisional Report No. 130.

Vereecken, H., E.J. Jansen, M.J.D. Hack-ten Broeke, M. Swerts, R. Engelke, S. Fabrewitz, and S. Hansen (1991). Comparison of simulation results of five nitrogen models using different datasets. In *Nitrate in Soils, Soil and Groundwater*, Research Report II. EUR 13501 EN, pp. 321-338.

Vereecken, H., M. Swerts, M.J.D. Hack-ten Broeke, E.J. Jansen, J.F. Kragt, R. Engelke, S. Fabrewitz, and S. Hansen (1991). Systematic comparison of model requirements and transformation processes. In *Nitrate in Soils, Soil and Groundwater*, Research Report II, EUR 13501 EN, pp. 237-248.

Vinten, A.J.A., B.J. Vivian, F. Wright, and R.S. Howard (1994). A comparative study of nitrate leaching from soils of differing textures under similar climatic and cropping conditions. *Journal of Hydrology* 159: 197-213.

Wagenet, R.J. (1998). Scale issues in agroecological research chains. *Nutrient Cycling in Agroecosystems* 50: 23-34.

Wagner, G.H. (1962). Use of porous ceramic cups to sample soil water within the profile. *Soil Science* 94: 379-386.

Webster, C.P. and K.W.T. Goulding (1995). Effect of one year rotational set-aside on immediate and ensuing nitrogen leaching loss. *Plant and Soil* 177: 203-209.

Webster, C.P., M.A. Shepherd, K.W.Y. Goulding, and E. Lord (1993). Comparisons of methods for measuring the leaching of mineral nitrogen from arable land. *Journal of Soil Science* 44: 49-62.

Whalley, W.R. (1993). Considerations on the use of time-domain reflectometry (TDR) for measuring soil water content. *Journal of Soil Science* 44: 1-9.

Whalley, W.R. (1994). Response to comments on "Considerations on the use of time-domain reflectometry (TDR) for measuring soil water content" by I.

White, J.H. Knight, S.J. Zegelin, and G.C. Topp. *European Journal of Soil Science* 45: 509-510.

White, I., J.H. Knight, S.J. Zegelin, and G.C. Topp (1994). Comments on "Considerations on the use of time-domain reflectometry (TDR) for measuring soil water content" by W.R. Whalley. *European Journal of Soil Science* 45: 503-508.

White, R.E. (1989). Prediction of nitrate leaching from a structured clay soil using transfer functions derived from externally applied or indigenous solute fluxes. *Journal of Hydrology* 107: 31-42.

Whitmore, A.P. (1995). Modelling the mineralization and leaching of nitrogen from crop residues during three successive growing seasons. *Ecological Modelling* 81: 233-241.

Wong, M.T.F. and D.L. Rowell (1994). Leaching of nutrients from undisturbed lysimeters of a cleared Ultisol, an Oxisol collected under rubber plantation and an Inceptisol. *Intenciencia* 19: 352-355.

Zegelin, S.J., I. White, and D.R. Jenkins (1989). Improved field probes for soil-water content and electrical conductivity measurement using time domain reflectrometry. *Water Resources Research* 25: 2367-2376.

Zegelin, S.J., I. White, and G.F. Russell (1992). A critique of the time domain reflectometry technique for determining soil-water content. In *Advances in Measurement of Soil Physical Properties: Bringing Theory into Practice*, Special Publication 30, Eds. G.C. Topp, W.D. Reynolds, and R.E. Green. Madison, WI: Soil Science Society of America, pp. 187-208.

Zimmerman, C.F., M.T. Price, and J.R. Montgomery (1978). A comparison of ceramic and teflon *in situ* samplers for nutrient pore water determinations. *Estuarine Coastal Marine Science* 7: 93-97.

Chapter 12

Mechanistic Simulation Models for a Better Understanding of Nutrient Uptake from Soil

Norbert Claassen
Bernd Steingrobe

Key Words: depletion zone, diffusion, mass flow, mechanistic models, nutrient uptake, root hairs, sensitivity analysis, uptake kinetics, validation.

INTRODUCTION

Nutrient uptake includes desorption, nutrient transport in soil toward the root, transport across root membranes and transport to the shoot: it is a complex process influenced by many parameters. The use of models can improve understanding of this process. Two general categories of models are available. The first category is empirical models that describe facts by statistical means and regressions (Chanter, 1981; Ross, 1981). These models are often called black-box models, since they only describe the relation between input and output, without taking the underlying mechanisms into account. Models of this kind are often used successfully for practical purposes because they usually work with a low number of easily obtainable parameters, and results are sufficient for practical use. The other category includes mechanistic models aimed at explaining a phenomenon by means of basic biophysical, biochemical, and physiological mechanisms (Nye and Tinker, 1977; Claassen, Syring, and Jungk, 1986; Barber, 1995; review by Rengel, 1993). These models are mainly built to understand processes and, thus, are more suitable for scientific use. Most models are in between the two types mentioned, using partly empirical and partly mechanistic approaches. Obviously, there are different ways to describe a given phenomenon; models are therefore seldom right or wrong, but, depending on the purpose, only more or less suitable.

What can be the scientific benefit of using mechanistic models in plant nutrition? First of all, mechanistic models provide the opportunity to test the correctness of the underlying concept. In this way, it is possible to confirm ideas or to show the necessity to reject or improve them. A lack of agreement between model results and observations can reveal where our knowledge is insufficient and further research is needed. Second, a validated model can be used to calculate data that are not measurable or are difficult to get, such as nutrient concentration around a root. Third, in contrast to empirical models, it is possible to extrapolate the data by means of mechanistic models because these are based on basic mechanisms. Finally, by changing a single input parameter of a model (sensitivity analysis), it is possible to evaluate these parameters even in complex processes and systems, such as nutrient uptake and transport, and assess the importance of various parameters in uptake of a specific nutrient.

In this chapter, we will concentrate on mechanistic models of different complexity used by our group. As such, it will not be an overall review of the topic of modeling nutrient uptake (for this, see Rengel, 1993), but a recount of the experience and knowledge acquired by applying specific models in a theoretical, but practical, way. We start by describing the soil-root system, its properties and processes related to nutrient uptake. Based on these principles, several models of different complexity are developed, and their ability to describe the reality is tested. By changing soil and plant parameters, we are able to assess their significance. Finally, we present an overall evaluation of modeling and its significance for the advancement of knowledge.

THE SOIL-ROOT SYSTEM

As the root takes up water, nutrients are moved to the root by mass flow. However, nutrient uptake is usually either higher or lower than the quantity transported by mass flow and, consequently, the concentration at the root surface is either decreased or increased. Figure 12.1 shows a concentration decrease at the root surface that will enhance the diffusion of nutrients toward the root surface and the release or desorption of ions from the solid phase. This results in an asymptotic increase in the concentration of the ion(s) with increasing distance from the root (Figure 12.1b).

FIGURE 12.1. (a) The Soil-Root System Consisting of the Root Surrounded by the Soil Solid Phase with Sorbed Ions and the Pore Space Filled with Liquid (and Air, Not Shown Here); (b) Concentration Profile of an Ion in the Rhizosphere As Shown in (a).

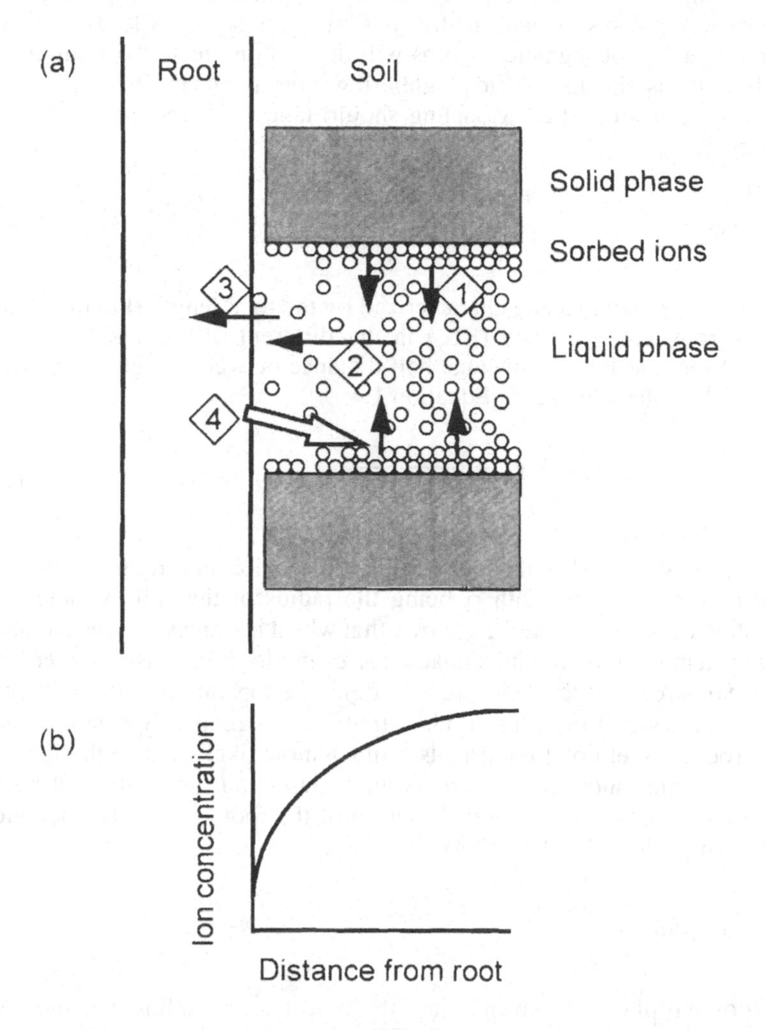

Note: Circles symbolize ions. The processes related to nutrient uptake are (1) release of sorbed ions into the liquid phase (desorption), (2) transport to the root in the liquid phase, and (3) uptake into the root. Process (4) symbolizes the release of root exudates that may change the solubility of ions bound to the solid phase.

Root Properties

The amount of nutrient a plant takes up from soil will depend on the size of the root system and its distribution in the soil profile. How much nutrient each root segment absorbs will depend on the soil volume it can exploit, that is, the distance to neighboring roots and its morphological and physiological properties. Modeling should take into account all of these factors.

Root Systems

The root system can be characterized by the total length (km m^{-2}) and by the root length density (RL_v) in the different soil layers (see Table 12.1). From the RL_v, an average half distance between neighboring roots (r_1) can be calculated using Equation 1:

$$r_1 = \frac{1}{\sqrt{\pi \, RL_v}} \qquad (1)$$

This calculation makes the simplifying assumption that roots are distributed regularly in soil, with r_1 being the radius of the soil cylinder surrounding each root. Table 12.1 shows that wheat has an extensive and deep root system compared with spinach, for example. It can also be seen that 70 to 80 percent of the roots are located in the topsoil layer (0 to 30 cm), where roots are close to each other, that is, r_1 is relatively small, so that inter-root competition for nutrients is much more likely than in the subsoil. Soil properties such as moisture, compactness, and distribution of nutrients may affect the size and distribution of the root system (Kuchenbuch and Barber, 1988; Müller, 1988).

Root Morphology

Root morphological properties of interest in modeling are the root radius (r_0) and root hair growth. The radius is used to calculate root surface for nutrient absorption. The radius of absorbing roots may vary widely, but average r_0 is about 0.1 to 0.2 mm in most cases, while that of root hairs is around 0.005 mm, and that of mycorrhiza hyphae, 0.003 mm. The volume of soil, V_S, at a given distance, Δr, from a surface is related to

TABLE 12.1. Vertical Changes in the Root Length Density (RL_v) of Four Crops and the Average Half Distance Between the Roots (r_1)

Soil layer	Root length density, RL_v				Average half distance, r_1			
	Wheat[a]	Corn[b]	Kohlrabi[c]	Spinach[c]	Wheat	Corn	Kohlrabi	Spinach
cm	cm cm^{-3}				cm			
0-30	8.2	3.8	1.4	2.3	0.2	0.3	0.5	0.4
30-60	1.7	1.5	0.03	0.06	0.4	0.5	3.2	2.3
60-90	1.0	0.4	—[e]	—[e]	0.6	0.9	—[e]	—[e]
90-120	0.7	0.1	—	—	0.7	1.8	—	—
120-150	0.27	0.01	—	—	1.1	5.6	—	—
150-180	0.03	—[e]	—	—	3.2	—[e]	—	—
Total root length[d] km m^{-2}	36	17	4	7				

[a] Adapted from Kuhlmann (1988); or [b] Wiesler (1991); or [c] Heins (1989).
[d] In the soil layers 0-180 cm, 0-150 cm, 0-60 cm, 0-60 cm for wheat, corn, kohlrabi, and spinach, respectively.
[e] No roots were determined in deeper soil layers.

its curvature, that is, its radius, r_0 (Equation 2).

$$V_s = \Delta r + \frac{\Delta r^2}{2r_0} \qquad (2)$$

For example, for a root hair with $r_0 = 0.005$ mm, the amount of soil per unit surface area within a distance of 0.2 mm is about 100 times that of a root cylinder of $r_0 = 0.1$ mm. This means that root hair surfaces can draw nutrients from a larger soil volume than the root cylinder. Measurements of root morphological properties show that the root hair surface area is similar to that of the root cylinder in some plant species (see Table 12.2), but, as model calculations will show, root hairs are more efficient at absorbing nutrients of low mobility, such as phosphorus (P), from the soil. Furthermore, since root hairs grow perpendicularly to the root axis into "undepleted" soil, they increase the volume of soil from which a root can draw nutrients, particularly for nutrients with low mobility in the soil.

Kinetics of Ion Uptake

The uptake kinetics describe the relationship between the influx (I_n) of an ion (uptake per unit root and unit time, e.g., mol cm^{-1} s^{-1}) and its concentration at the root surface (C_{L0}). For the low concentration range, as found in the soil solution, that is, mostly below 1 μmol cm^{-3}, this relationship is a saturation curve that can be described by a modified Michaelis-Menten function (Nielsen, 1972; see Figure 12.2):

$$I_n = \frac{I_{max}(C - C_{min})}{K_m + C - C_{min}} \qquad (3)$$

where I_{max} is the maximum influx theoretically achieved at an infinite concentration and C_{min} is the minimum concentration at which no net influx occurs

The Michaelis-Menten constant, K_m, is often stated as the concentration that allows uptake at half the maximal rate, but as K_m is used in Equation 3, it is actually the concentration equal to $K_m + C_{min}$ that allows uptake at half I_{max}.

Root Exudates and Microorganisms

Root exudates may affect the solubility of nutrients in the rhizosphere or increase microbial growth, which in turn may affect the solubility of

TABLE 12.2. Root Morphological Properties (Root Radius, r_0, and Root Hairs) of Seven Plant Species

Species	r_0	Root hairs		
		Number	Average length	SA[a]
	mm	per mm	mm	$mm^2\ mm^{-2}$
Onion	0.23	1	0.05	0.006
Ryegrass	0.07	45	0.34	1.2
Wheat	0.08	46	0.33	1.2
Rape	0.07	44	0.31	1.3
Tomato	0.10	58	0.17	0.6
Spinach	0.11	71	0.62	1.9
Bean	0.15	49	0.20	0.4

Source: Adapted from Föhse, Claassen, and Jungk (1991).

Note: Assumed root hair radius = 0.005 mm.

[a] SA = surface area of root hairs in mm^2 per mm^2 of the root cylinder surface area.

nutrients. Even though root exudates may be of large significance for nutrient uptake from soil, they have seldom been included in uptake models because little is known about quantities released and the exact mode(s) of action, that is, desorption/adsorption, chelation, reduction/oxidation or altering microbial activity. Knowledge of these quantities and mechanisms are essential for modeling.

Soil Properties

The soil is a heterogeneous medium consisting of solid, liquid, and gaseous phases in various proportions. Plant nutrients essentially move in the liquid phase only, at least during the life span of a root segment. A proportion of soil nutrients are bound to the soil matrix and, as such, are immobile. However, these may be released when the equilibrium between the solid and liquid phase is disturbed, for example, due to nutrient uptake by roots (see Figure 12.2).

For modeling nutrient uptake from soil, the soil properties of interest are the concentration of ions in the soil solution; its buffering by the solid phase, that is, desorption and sorption of ions; and the geometry of the liquid phase, that is, the amount of water in soil and the tortuosity of the pores filled with water through which ions will move.

FIGURE 12.2. Relation Between Soil Solution Concentration at Root Surface, C_L, and net Uptake Rate, I_n, Described by the Michaelis-Menten Kinetics

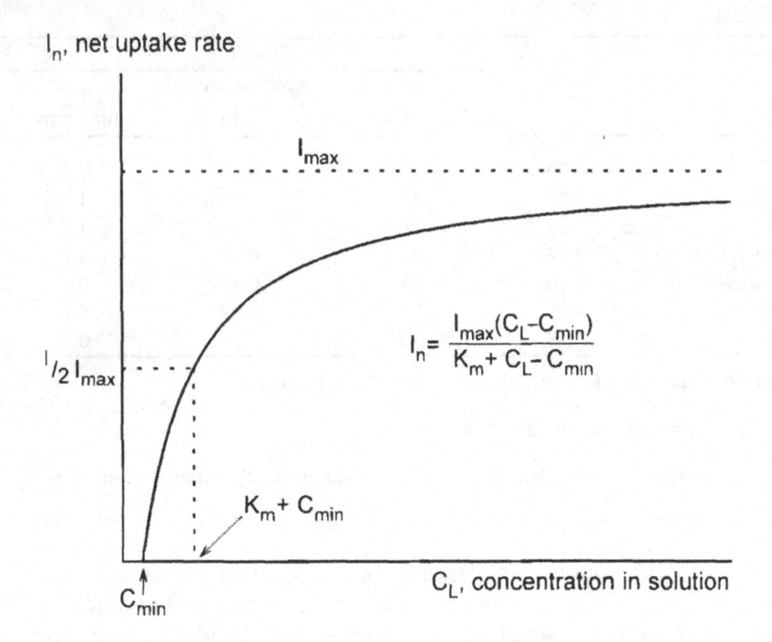

Note: Parameters are maximum uptake rate, I_{max}, Michaelis-Menten constant, K_m, and minimum concentration, C_{min}.

Transport Processes in Soil

Nutrient transport in soil to the root is by mass flow, F_M, and diffusion, F_D (Barber, 1962). Mass flow is the convective transport of nutrients dissolved in the soil solution and moving to the plant roots as a result of shoot transpiration. Diffusion is the movement of a substance from one region to adjacent regions where that substance has a lower concentration. It results from the spontaneous oscillation of ions and molecules driven by thermal agitation.

The total flux of nutrients from soil to root, F_T, is the sum of mass flow and diffusion (Equation 4):

$$F_T = F_M + F_D \tag{4}$$

Mass flow is given by the product of water flux, v, and the concentration of the nutrient in soil solution, C_L (Equation 5):

$$F_M = C_L v \tag{5}$$

Diffusive flux is governed by the principles of diffusion established by Fick (1855). Fick's first law (Equation 6) states that the flux by diffusion, F_D, is proportional to the concentration gradient, dC/dx, with the proportionality constant being the diffusion coefficient D:

$$F_D = -D \frac{dC}{dx} \tag{6}$$

The minus sign is a convention which indicates that the flux proceeds down the gradient. Applying Equation 6 to soils, C is the concentration of all ions per volume of soil that participate in diffusion. These include the ions in the soil solution plus those ions on the solid phase that may be released into the soil solution. The diffusion coefficient is then designated as the effective diffusion coefficient, D_e.

Most studies show that diffusion in soil occurs only in the liquid phase. Therefore, D_e is given by the diffusion coefficient in water, D_L, multiplied by the volumetric soil water content, Θ, the fraction of the soil volume in which diffusion occurs (Equation 7). Furthermore, ions in soil have to move through the tortuous soil pore system, which is longer than the straight line distance between two points. This is accounted for by the impedance factor, f. The reciprocal of the buffer power, b, allows for the fact that not all ions, but only the fraction of the ion in solution, are actually diffusing.

$$D_e = D_L Q f \frac{1}{b} \tag{7}$$

The buffer power, b, is generally defined by:

$$b = \frac{dC}{dC_L} \quad \text{or simplified} \quad b = \frac{\Delta C}{\Delta C_L} \tag{8}$$

The buffering behavior of a soil is usually described by buffer curves, as shown for P in Figure 12.3. It can be seen that at high P saturation of the soil, b is only 50, but increases to almost 5,000 at low P. Since many simulation models work with a single b value, an average value has to be estimated.

MODELING APPROACHES

The functioning of the soil-root system is rather complex if all possible processes and interactions are considered. Therefore, most models only describe part of the system because of either seeking an answer to one aspect only (e.g., the extension of the depletion zone around roots), or assuming some factors or processes are not important, or realizing that some processes are insufficiently defined to be included in the model. The degree of discrepancy between calculated and measured values may indicate the significance of those processes not included in the model.

Transport by Mass Flow

Water flux to the root is given by water consumption or transpiration in liter m^{-2}. Furthermore, knowing the concentration of the nutrient in soil solution, the mass flow can be estimated (Equation 5) and is often expressed as a fraction or percentage of total uptake (see Table 12.3).

Table 12.3 shows that only a small fraction of the total P and K (potassium) taken up by crop plants (1 to 4 percent of P and 1 to 14 percent of K) reaches the root by mass flow. The major proportion of these nutrients, therefore, reaches the root by diffusion. For Ca (calcium) and Mg (magnesium), on the other hand, mass flow is the predominant process of transport to the root, since their concentration in solution is high and uptake is relatively small. For Ca, there is even a strong accumulation at the root surface, which is in agreement with measurements of Barber and Ozanne (1970) and Hendriks and Jungk (1981).

Nitrate is not sorbed to the soil solid phase and remains entirely in the soil solution. It is therefore often assumed to move to the root by mass flow. However, this view is inconsistent with measurements. At adequate, but not excessive, NO_3^- supply, mass flow only accounted for 15 to 33 percent of total transport to roots (Strebel and Duynisveld, 1989). Similar results were obtained by Gregory, Crawford, and McGowan (1979) and Retzer (1995).

These model calculations, which show that at low P and K concentrations are still adequate for crop growth, mass flow does not play a major

FIGURE 12.3. The P Buffer Curve of a Sandy Soil (Psammudept: 85% Sand, 8.5% Silt, and 6.5% Clay)

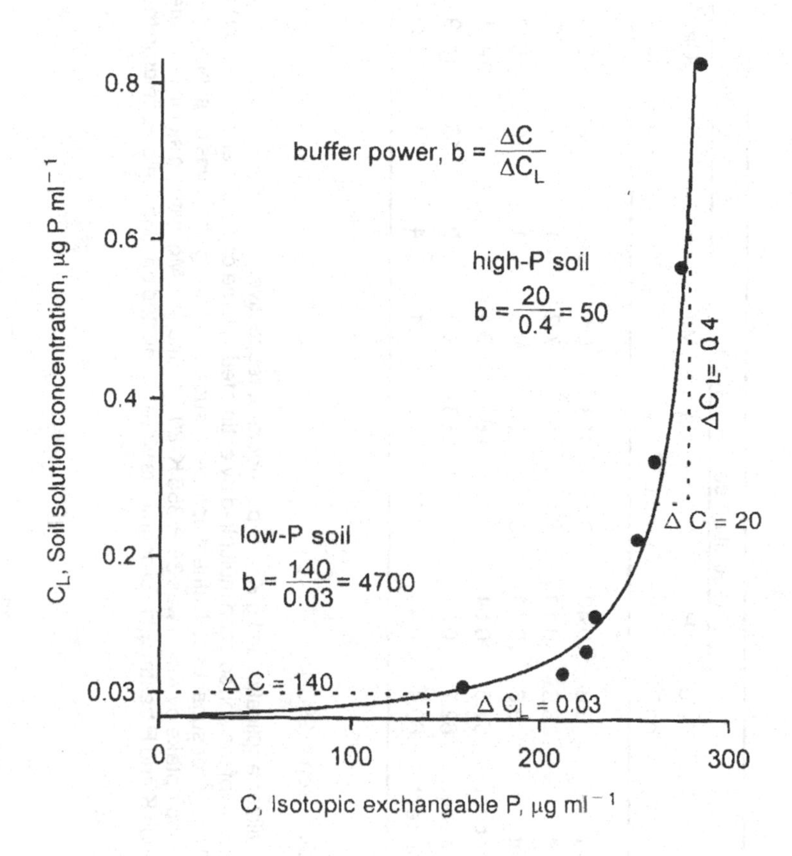

Source: Adapted from Hendriks, Claassen, and Jungk (1981).

Note: The relation between P concentration in solution, C_L, and isotopically exchangeable P denoted as C, was determined by desorption. The soil was equilibrated 48 hours with a 5 mmol dm^{-3} $CaCl_2$ solution at soil-to-solution ratios varying from 1:2 to 1:2,000.

TABLE 12.3. Soil Solution Concentration and Resulting Mass Flow of Nutrients from Soil to the Roots of Winter Wheat and Sugar Beet

Site	Crop	Yield	Soil solution concentration				Mass flow[b]			
			P	K	Ca	Mg	P	K	Ca	Mg
		t ha^{-1}	mg L^{-1}				% of uptake			
Oesselse K_0[a]	Wheat	8.8	0.40	3.5	113	6.5	3	2	848	115
Klein Ilde K_0	Wheat	8.6	0.10	1.1	97	7.2	1	1	728	127
Klein Ilde K_{240}	Wheat	8.6	0.17	15.3	92	9.3	1	10	690	164
Dinklar P_0	Sugar beet	51.5	0.04	12.1	263	9.4	1	14	1481	94
Dinklar P_{200}	Sugar beet	62.7	0.21	12.1	263	9.4	2	13	1169	78
Dinklar P_{400}	Sugar beet	60.4	0.46	12.1	263	9.4	4	14	1422	87

Source: Adapted from Claassen (1990a).

[a] Subscripts refer to fertilizer application in kg K or P per hectare, respectively.

[b] Mass flow = concentration of the soil solution multiplied by estimated volume of water transpired, 300 l m^{-2} for winter wheat and 400 l m^{-2} for sugar beet. Nutrient uptake by sugar beet, in kg ha^{-1}, was on average 40 P, 400 K, 40 Ca, and 17 Mg, and uptake by wheat was 35 P, 350 K, 80 Ca, and 456 Mg, with no significant differences recorded among various K and P treatments. Experiments were conducted on high-yielding fields in the region of Göttingen, Germany.

role in transport of these nutrients to roots. It is important to stress that the underlying principles are the same for all nutrients. Thus, for a given concentration in soil solution, the potential for nutrient transport by diffusion is higher than that for transport by mass flow. Therefore, diffusion is a more efficient mechanism of nutrient transport to roots than mass flow. This will be treated in a more quantitative manner later (see "Diffusion versus Mass Flow").

Extension of the Depletion Zone

Roots extract nutrients from their immediate surroundings. If mass flow does not supply enough of a given nutrient to satisfy the plant demand, the concentration of the nutrient will be reduced at the root surface. The resulting concentration gradient will induce ions from further away to move toward the root by diffusion, increasing the zone of ion depletion around the root. As time proceeds, this zone will extend further, depending on the mobility of the nutrient in the soil. Ions that are strongly bound to the soil solid phase, such as P, are largely immobile, and roots only extract them from very close proximity. In contrast, poorly bound or nonsorbed ions, such as NO_3^-, are highly mobile and move relatively large distances. Soil water content also affects the mobility of nutrients in the soil. All these parameters are included in the effective diffusion coefficient, D_e (Equation 7), which is a measure of the mobility of an ion in soil. The extension of the depletion zone, Δx, can be calculated using Equation 9 (Syring and Claassen, 1995):

$$\Delta x = \sqrt{\pi \quad D_e \, t} \tag{9}$$

where Δx is the distance at which the decrease of concentration is 20 percent of the maximum decrease at the root surface (see Figure 12.4)

Theoretically, the depletion will continue infinitely, and in order to obtain a measurable distance, a certain degree of depletion has to be defined; 20 percent was used for Equation 9. The concentration profile of NO_3^- in soil next to a mat of roots (which simulates a planar root surface, cf. Kuchenbuch and Jungk [1982]) shows apparent NO_3^- depletion to about 8 cm (Figure 12.4). The distance at which 20 percent depletion occurs is about 3.5 cm, which is close to the value of 3.6 cm calculated using Equation 9. Measurements with K gave similar agreements to calculated values (Kuchenbuch, 1983).

FIGURE 12.4. Nitrate Depletion in the Rhizosphere of Ryegrass Grown in a Sandy Soil After Ten Days

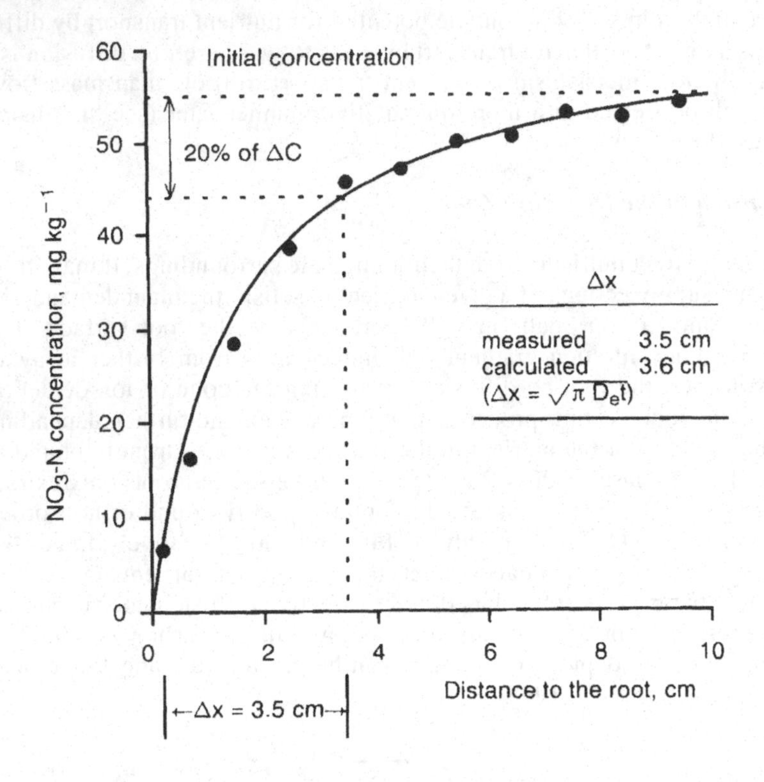

Note: Calculation of Δx according to Equation 8 with $D_e = D_L f$; $D_L = 1.9 \times 10^{-5}$ $cm^2\ s^{-1}$; $f = 1.58\ \Theta - 0.17$; and $\Theta = 0.18$.

Even though Equation 9 is applicable to planar conditions, the values for cylindrical geometry are not very different (Syring and Claassen, 1995). Therefore, the results show that with a very simple model, the extension of the zone of nutrient depletion around a root can be estimated, and thereby the occurrence of inter-root competition can be predicted.

The extension of depletion of NO_3^-, P, and K under common conditions was calculated using Equation 9 (see Table 12.4). Comparing these data with the half distance among roots in soil, r_1, shown in Table 12.1, it can be seen that for NO_3^- the extension of the depletion zone is much larger than r_1. Thus, the depletion zones will overlap extensively and the whole soil

TABLE 12.4. Calculated Extension of the Depletion Zone (Δx) for NO_3^-, P, and K for a Range of D_e Values after Twenty Days of Uptake

b	D_e $\times 10^{-9}$ cm^2 s^{-1}	Δx cm
	NO_3^-	
0.25	4275	4.8
	K	
3	371	1.4
10	111	0.8
30	37	0.4
100	11	0.2
	P	
50	10	0.233
500	1	0.073
1000	0.50	0.052
2000	0.25	0.037

Note: $D_e = D_L \Theta f/b$; D_L for $NO_3 = 1.9 \times 10^{-5}$, for K = 1.98×10^{-5}, and for P = 0.89×10^{-5} cm^2 s^{-1}; $\Theta = 0.25$ and $f = 1.58\ \Theta - 0.17 = 0.225$; $\Delta x = (\pi\ D_e\ 20 \times 24 \times 60 \times 60)^{0.5}$

volume may be emptied of NO_3^-, even in the deep soil layers where only a few roots are present. This is the theoretical basis for the N_{min} procedure to determine N fertilizer demand (Wehrmann and Scharpf, 1986) where the mineral N in the whole root zone, for most crops down to 1 m, is determined and considered as completely available to crops.

On the other extreme is P as immobile nutrient. The extension of the depletion zone will usually be below 0.1 cm, and so even at the highest root length densities of wheat in the topsoil layer, there will be no inter-root competition. It can furthermore be recognized that only a small portion of the chemically available P is also spatially available and can be used by one crop. It also explains the general observation that the use of fertilizer P in the year of application only amounts to 10 percent, or at most 20 percent, of the amount applied. Potassium has an intermediate position between NO_3^- and P.

Steady State Models

The model presented here allows calculation of the concentration difference in soil solution, ΔC_L, needed to drive a given flux by diffusion. In many cases, as shown earlier, diffusion is the main mechanism of ion transport to roots. In the following approach, diffusion is assumed to be the only transport mechanism of nutrients to roots. The flux by diffusion is driven by the concentration gradient (Equation 6). The higher the concentration difference in soil solution, ΔC_L, between the soil and the root surface, the higher the concentration gradient and therefore the flux from soil toward the root. The flux to the root is equal to the influx because of mass conservation. If the influx, I_n, is known, then the concentration difference between the soil and the root surface needed to drive the diffusive flux can be calculated using Equation 10 (Barraclough, 1986), based on the steady state model of Baldwin, Nye, and Tinker (1973):

$$nC_L = \overline{C}_L - C_{L0} = -\frac{I_n}{4\pi D_L \Theta f}\left(1 - \frac{1}{1-\pi r_0^2 RL_v} \ln \frac{1}{\pi r_0^2 RL_v}\right) \tag{10}$$

where $\overline{C}_L$ is the average soil solution concentration of the bulk soil and C_{L0} is the solution concentration at the root surface

Using Equation 10, Barraclough (1986) found that winter wheat needed a soil solution concentration of 165 μM NO_3^-, 14 μM P, and 56 μM K to sustain the influx observed.

If the influx, I_n, and the concentration in soil solution, $\overline{C}_L$, are known, Equation 10 can be solved for the concentration at the root surface, C_{L0}. Having the same soil with several levels of a nutrient, one will obtain data pairs of I_n with its corresponding C_{L0} values. Using this approach, Seeling and Claassen (1990) estimated the K uptake kinetics for corn in soil. The results obtained, $I_{max} = 52 \times 10^{-14}$ mol cm^{-1} s^{-1} and $K_m = 67$ μM, were comparable to those from experiments in solution culture.

Diffusion versus Mass Flow

In the section "Transport by Mass Flow," we stated that the transport potential by diffusion is higher than that by mass flow. Now this can be treated in a more quantitative manner.

Equation 10 can be solved for I_n, which is equal to the transport by diffusion, since this was the only transport mechanism assumed. Since plants can reduce the concentration at root surface, C_{L0}, close to zero $\Delta C_L \approx \overline{C}_L$, that is, the maximum concentration difference is about equal to soil solution concentration $\overline{C}_L$, and therefore, the transport potential by diffusion is proportional to $\overline{C}_L$. Using the following parameters: $D_L = 2 \times 10^{-5}$ $cm^2\, s^{-1}$, $\Theta = 0.25$, $f = 0.225$ (from Table 12.4), $r_0 = 0.01$ cm, and $RL_v = 3$ cm cm^{-3}, the diffusive flux is $I_n = F_D = \overline{C}_L \times 2.4 \times 10^{-6}$.

Mass flow, F_M, is given by the product of water influx, v, and soil solution concentration, $\overline{C}_L$ (Equation 5). Assuming a transpiration rate of 5 mm per day (5 l m^{-2} day^{-1}) and a root length of about 20 km m^{-2} (see Table 12.1), water flux equals 2.9×10^{-8} $cm^3\, cm^{-1}\, s^{-1}$. In this case, mass flow is:

$$F_M = \overline{C}_L \times 2.9 \times 10^{-8} \text{ and } \frac{F_D}{F_M} = \frac{\overline{C}_L \times 2.4 \times 10^{-6}}{\overline{C}_L \times 2.9 \times 10^{-8}} = 82.8$$

For the conditions shown, transport by diffusion could be more than 80 times higher than by mass flow, which is in agreement with the data in Table 12.3. This ratio may vary because F_D may be smaller due to a smaller D_L (about one-half for P, Ca, and Mg, in comparison to the one assumed in previous calculations), or somewhat smaller Θ and f. However, the latter will not be much smaller for the present comparison because in a dry soil (small Θ and f) there is no mass flow either. Diffusion to the root (Equation 10) is not very sensitive to RL_v.

In contrast, F_M may be somewhat higher due to a higher transpiration rate (2 to 3 times) and if fewer roots absorb the water. So, in extreme cases, the ratio F_D/F_M may decrease, but will still be much larger than 1, that is, in the soil-root system, diffusion is a much more efficient transport mechanism of nutrients than mass flow.

The potential of diffusion, however, is often not used by the plant, mainly if $\overline{C}_L$ is high, because the plant does not need so many nutrients, or mass flow already supplies a large portion of, or even enough to cover completely, the plant needs.

Transient Models

As nutrients are taken up by the roots, their concentration in soil solution changes continuously. Transient models try to describe this dynamic process of nutrient uptake and concomitant changes in ion concentration and distribution in the rhizosphere. The models described in this section are often called mechanistic models because they are based on the mechanisms of ion transport to the root surface by mass flow and diffusion and the subsequent uptake into the root according to the Michaelis-Menten kinetics. They are also called deterministic models because, once the soil and plant parameters are set, the calculated nutrient uptake is "determined," that is, the calculated value is independent of the actual uptake.

Based on the above approach—transport by mass flow and diffusion and uptake following the Michaelis-Menten kinetics—Nye and Marriott (1969) developed a model that was later applied and modified in various ways by Claassen and Barber (1976), Cushman (1979, 1980), Barber and Cushman (1981), Itoh and Barber (1983), Claassen, Syring, and Jungk (1986), and Claassen (1990a, b). These models are based on the transport equation (Nye and Marriott, 1969), which was extended by a sink term, A, (Claassen, 1990a) to take into account uptake by root hairs (Equation 11):

$$b \frac{\partial C_L}{\partial t} = \frac{1}{r} \frac{\partial}{\partial r}\left(rD_e b \frac{\partial C_L}{\partial r} + v_0 r_0 C_L \right) - A \qquad (11)$$

where C_L is the concentration of the nutrient in soil solution that, multiplied by the buffer power, b, gives the total concentration of diffusible nutrient in soil; t is the time; r is the radial distance from the root axis; D_e is the effective diffusion coefficient; v_0 is the water flux across the soil-root interface; r_0 is the root radius

The sink term A calculates uptake by root hairs using a quasi-steady state approach that allows for the Michaelis-Menten uptake kinetics for root hairs as well (Claassen, 1990a).

To integrate Equation 11 and run the model, the following parameters are needed. For a better illustration, an example was taken from Föhse, Claassen, and Jungk (1991) who used a low-P soil planted to spinach.

Input Parameters

Soil parameters:	C_{Li}	soil solution concentration:	1,400 pmol cm^{-3}
	D_e	effective diffusion coefficient:	3.72×10^{-10} cm^2 s^{-1}
	b	buffer power:	1,500
Root parameters:	I_{max}	maximum influx, without root hairs:	1.7 pmol cm^{-2} s^{-1}
		with root hairs:	0.6 pmol cm^{-2} s^{-1}

(I_{max} with root hairs is lower than without them because the total root surface is greater.)

K_m	Michaelis-Menten constant	400 pmol cm^{-3}
C_{min}	minimum concentration:	100 pmol cm^{-3}
v_0	water influx:	$2.7 \times 10^{-7} cm^3 cm^{-2} s^{-1}$
r_0	root radius:	0.011 cm
r_1	half distance among roots:	0.21 cm
L_0	initial root length:	607 cm
k	relative root growth rate:	0.14 d^{-1}

Root hairs: I_{max}, K_m, C_{min} as shown above for the root cylinders

r_{h0} root hair radius 5×10^{-4} cm

Root hair distribution:

half distance among hairs of ($\times 10^{-3}$ cm) 8.7, 15, 23, 33, 51, 86, and 147 in the compartments with 0-0.025, 0.025-0.05, 0.05-0.075, 0.075-0.1, 0.1-0.125, 0.125-0.150, 0.150-0.175 cm distance from the root, respectively.

Results

As the root absorbs P, its concentration in the rhizosphere changes in space and with time (see Figure 12.5a). For the case simulated here, soil solution concentration decreased to a very low value at the root surface, while the zone affected, that is, the extension of P depletion, was only 0.5 mm without, and 1.0 mm with, root hairs. The quantitative effect of root hairs is seen more clearly in Figure 12.5b and the inserted table, showing that more than 80 percent of P uptake was by root hairs. The significance of root hairs was especially high in this case because spinach is a species with many and long root hairs (see Table 12.2). The model also calculates the depletion profiles around root hairs in the different compartments (data not shown).

The results demonstrate that modeling helps to understand the dynamics of nutrients in the rhizosphere that are difficult or impossible to measure. This is the case for the extension of the depletion zone of single roots growing in soil. It helps to assess whether inter-root competition will occur. For a depletion zone of 1 mm, no inter-root competition for P will occur, even at the highest root length density of 8 cm cm^{-3} for wheat (see Table 12.1) in the topsoil layer.

FIGURE 12.5. Calculated Data for P Depletion in Soil and Uptake by Spinach

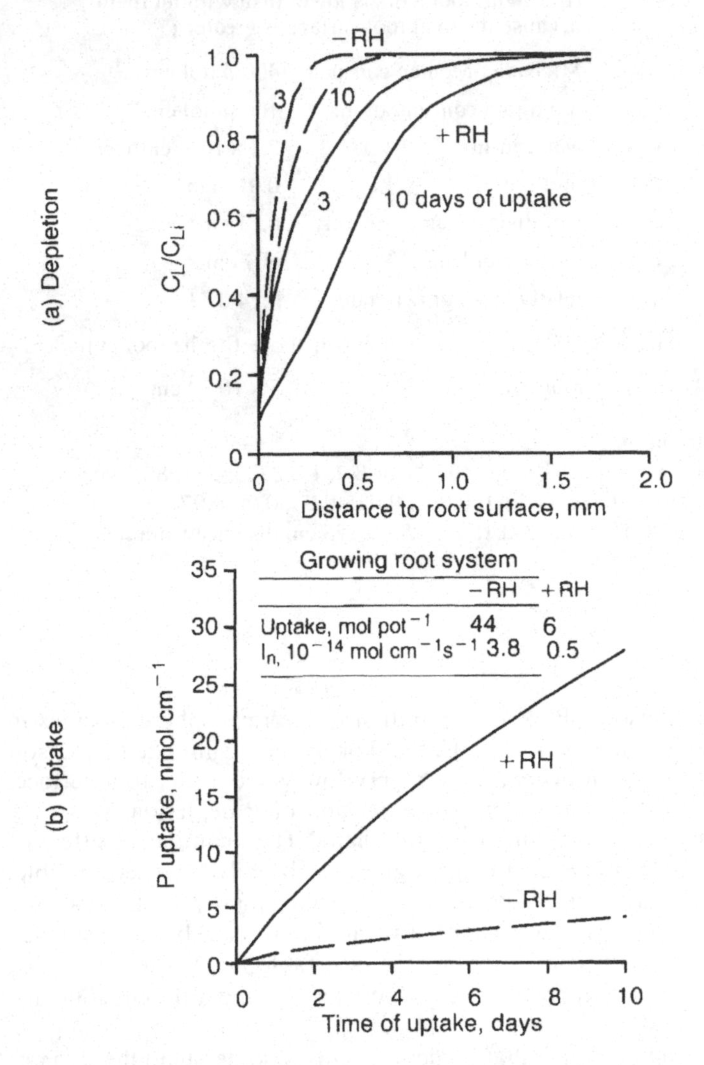

Source: Adapted from Föhse, Claassen, and Jungk (1991).

Notes: (a) Phosphorus concentration profiles around a root without root hairs (−RH) or with root hairs (+RH) after three and ten days of uptake. (b) Cumulative P uptake by 1-cm-long root without root hairs (−RH) or with root hairs (+RH). The table shows total P uptake by the growing root system during ten days and its average influx (I_n).

The time-course of P uptake by a root segment (see Figure 12.5b) is difficult to measure. It is shown here that uptake increases almost linearly with time, that is, the rate of uptake is almost constant over time. This finding is against the expectation of a decreasing transport to the root, and thereby uptake, as the depletion zone increases with time (for more details, see also Figure 12.10). Total uptake by, and average influx into, a growing root system (see Figure 12.5, inset table) are quantities that can be measured easily in an experiment and are therefore useful measures to test whether the model properly describes reality or not.

Simplifications, Assumptions, and Comments

The model assumes that a root segment can exploit only a limited volume of soil given by a cylinder of radius r_1, that is, the average half distance of neighboring roots. The root is at the center of this cylinder. At the outer border of the cylinder, nutrient flux is zero, and at the root surface, transport by mass flow and diffusion is equal to the influx given by Michaelis-Menten kinetics (for the equations and further comments, see Nye and Marriott, 1969; Barber and Cushman, 1981; Claassen, 1990a, b).

The model assumes that roots are distributed evenly in the whole soil volume, and no allowance is given for a changing distance among roots (i.e., r_1) as roots grow. This is addressed elegantly by Hoffland and colleagues (1990). No changes in soil water content are assumed during the period of calculation; in addition, in most models, buffer power, b, and therefore D_e are also constant. In such cases, average values of those parameters are taken. However, in the model presented here, b and D_e are allowed to change with nutrient content of the soil. The buffer curve is being described by a Freundlich equation. It is further assumed that root uptake kinetic parameters (I_{max}, K_m, C_{min}) are constant over time, and the root radius, r_0, is constant.

The soil is assumed homogeneous and isotropic. For distances of several mm, this may be a good approximation, but not necessarily for depletion zones below 1 mm, as shown in the previous example, and more so around root hairs. In general, the heterogeneity of the soil-plant system is simplified by taking average values, but this will not solve the problem in all cases, as will be seen in the next section.

A further assumption is that the root is a sink for nutrients only, that is, no root exudates or their action on nutrient availability and microorganism activity is considered. However, some model developments do address this (Nye, 1984; Jones, Darrah, and Kochian, 1996).

VALIDATION OF THE MODELS

The models used in this chapter are a system of postulates presented as a mathematical description of the soil-root system as related to nutrient uptake. To test whether the models describe reality properly, a validation is needed, via a comparison of calculated and observed or measured values. This has been done by several authors. Among others, Claassen and Barber (1976); Claassen, Syring, and Jungk (1986); or Barber (1995). Some results from other sources are shown in Figures 12.4 and 12.6.

The results of Figure 12.4 show a close agreement between the calculated and measured extension of the depletion zone of nitrate. This indicates that the very simple model (Equation 9) properly describes the process of diffusion in the rhizosphere and that the parameters needed were estimated correctly. The agreement for the more complex situation described by the transient model (Equation 11) is not always so close.

When calculated and measured average K influx values of a growing root system of corn are compared in a K-limiting situation (at the ample K supply, the influx was twice the highest value shown), the model described K flux in soil and uptake by corn roots in a realistic way at low soil bulk density (Figure 12.6a). However, at high bulk density, and mainly at high soil moisture content, that is, the highest values, the model overestimated K influx. Seiffert and colleagues (1995) speculated that this could be due to an irregular root distribution in the compacted soil. With this hypothesis, that is, assuming roots would grow in one-quarter of the soil volume only, model calculations were run again. The result was a close agreement of calculated and measured values, because only the highest value decreased sharply. This was because at high soil water content (highest influx), the diffusion coefficient is high and the extension of the depletion zone is large, causing inter-root competition. However, calculated influx was not affected at low soil water content because no inter-root competition occurred, even when roots were restricted to one-quarter of the soil volume. This, of course, is not hard proof of the stated hypothesis, but shows how mechanistic models can be used to explain and understand observed phenomena.

The model has been used to calculate the P influx of spinach grown in a subsoil from loess fertilized to different P concentrations in soil solution (Figure 12.6b, the same experiment as described in Figure 12.5). It can be seen that the ratio of calculated to measured P influx was close to one at all P levels, but only if root hairs were included in the model calculations.

In contrast to the two examples discussed previously, the model calculates only 25 percent of the actual P uptake of sugar beet, even at 220 kg P ha^{-1}, which was enough for maximum P uptake as well as maximum

FIGURE 12.6. Validation of the Model

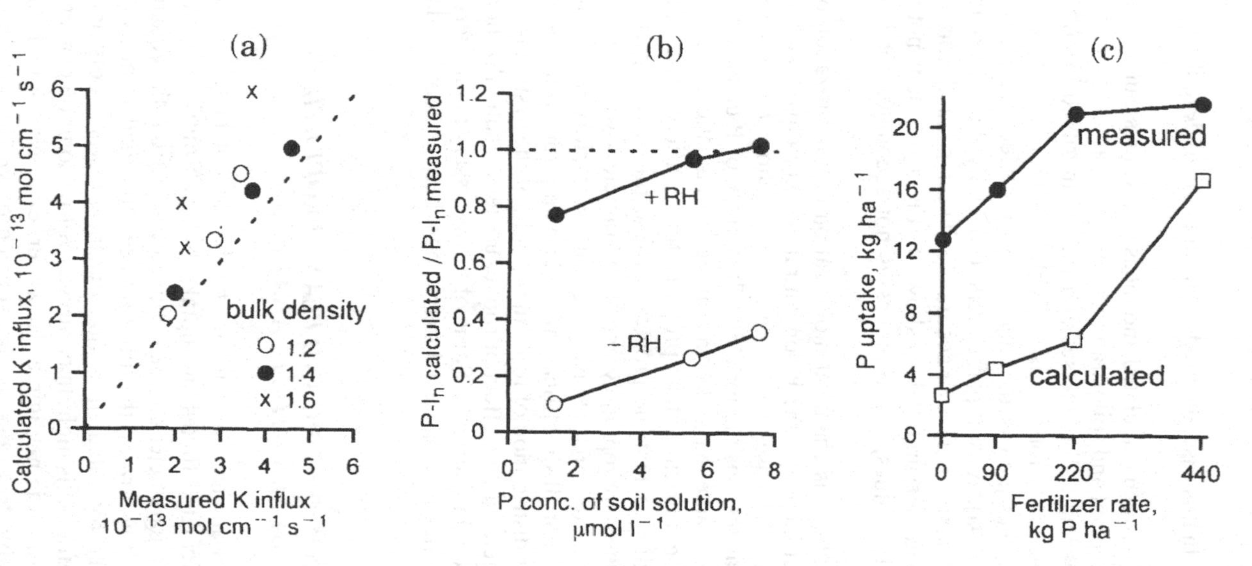

Sources: (a) Adapted from Seiffert et al. (1995); (b) adapted from Föhse, Claassen, and Jungk (1991); (c) adapted from Claassen (1990b).

Notes: (a) Comparison of measured and calculated K influx into corn grown on a Luvisol with different bulk density and three water contents. Dashed line is the 1:1 line. (b) Effect of the P concentration in soil solution on the ratio between measured and calculated P influx for spinach. Calculation was done with (+RH) or without (−RH) root hairs. The same experiment as in Figure 12.5. (c) Effect of P fertilization on measured and calculated P uptake of field-grown sugar beet.

yield (Figure 12.6c). In this case, the model was not able to describe reality properly.

The discrepancy between calculated and measured values can be due to incorrectly estimated plant and soil parameters or to the fact that the model does not take into account all processes and factors involved in nutrient uptake from soil. The value for I_{max} is difficult to obtain for field-grown crops, but when I_{max} was increased, it had almost no effect on calculated uptake (Claassen, 1990a). As soil parameters, especially the soil solution concentration, can be estimated rather accurately, it seems that the discrepancy of calculated and measured values shown in Figure 12.6c is not due to incorrect parameter values, but to processes not accounted for in the model.

The process postulated is chemical mobilization or solubilization of P by root exudates that increased the P concentration very close to the root surface. If this were the case, model calculations show that chemical mobilization of P was responsible for about 75 percent of P uptake.

The results of this chapter have shown that the mechanistic models used were able to describe nutrient uptake properly for a variety of soil-plant conditions. This indicates that the processes (desorption, transport, and uptake) were described appropriately and the parameters were estimated correctly. However, model predictions were not satisfactory in all cases. This may be because not all simplifications and assumptions were met, for example, the regular distribution of roots in the soil, or the system is more complex than assumed, and the effect of root exudates on solubilization of nutrients was not taken into account. Some of the necessary parameters are difficult to obtain, which may cause a disagreement between observed and calculated values.

EVALUATION OF SOIL AND PLANT PROPERTIES BY SENSITIVITY ANALYSIS

Soil and plant parameters used in the model do not have the same consequence on calculated fluxes and uptake for each nutrient. In some cases, relevance of a parameter can be deduced easily from the equation. For example, water flow toward the root (v_0) used to calculate mass flow can be of importance for calculating total transport toward roots only if soil solution concentration is high. In contrast, for low solution concentration, the influence of v_0 is negligible. However, even with a high mass flow, it is not possible to assess the importance of v_0 for total uptake, without taking into account uptake physiology and diffusion conditions. Thus, although the basic equations (Equations 3 to 7) to calculate diffusive and mass flow, as well as uptake rate, are relatively simple, it is not possible to evaluate the importance of each plant or soil property directly because of the nonlinearity of the overall model. However, such an evaluation of soil and plant properties can be done by using the model in a sensitivity analysis. In such an analysis, several calculations are performed and only one input parameter or a set of combined parameters is changed, while keeping all other input data constant. This enables evaluation of the influence of an input parameter on the calculated uptake and depletion profile.

Soil Parameters

As described earlier, diffusion is the main transport process for most nutrients. The contribution of mass flow to total transport of K and P is negligible (see Table 12.3). Even for NO_3^- and Mg, usually stated as moving by mass flow, Strebel and Duynisveld (1989) showed that the portion of total transport by diffusion can reach 67 to 85 percent and 77 percent, respectively. Therefore, this evaluation of soil parameters will concentrate on diffusion processes. Diffusive flux is dependent on D_e and the concentration gradient, the necessary parameters being C_{Li}, b, Θ, f, and D_L.

Initial Soil Solution Concentration, C_{Li}, and Buffer Power, b

A sensitivity analysis of one of these parameters could be misleading, as they are related to each other and also to the total amount of diffusible nutrients, ΔC (Equation 8). Increasing C_{Li} by keeping the buffer power constant results in an increased ΔC (see Table 12.5), as does a higher buffer power at unchanged C_{Li}. Also, a higher buffer power lowers D_e, according to Equation 7 (see Table 12.5). Therefore, a sensitivity analysis of b and C_{Li} is better done together, with changes in both ΔC and D_e taken into account for interpretation of the results.

The K data set used for this calculation was derived from a pot experiment with corn grown in a silty loam (Claassen, Syring, and Jungk, 1986). Values above and below the measured values of C_{Li} (0.27 μmol cm^{-3}) and b (17) were used in the analysis (Table 12.5).

Calculated K uptake is strongly influenced by initial solution concentration, C_{Li}, whereas buffer power has only a small effect, if root length density is low enough that no inter-root competition for K uptake occurs (see Figure 12.7). With increasing root competition, the influence of buffer power on K uptake increases, while the effect of C_{Li} remains high.

TABLE 12.5. Input Parameter for a K Sensitivity Analysis Derived from a Pot Experiment with a Silty Loam Soil, and Influence of Varied Initial Solution Concentration, C_{Li}, and Buffer Power, b, on the Effective Diffusion Coefficient, D_e, and Total Amount of Diffusible K, ΔC

C_{Li}	b			
μmol K cm^{-3}	3	10	17	30
		D_e, 10^{-6} cm^2 s^{-1}		
0.10	54.7	16.4	9.7	5.5
0.27	54.7	16.4	9.7	5.5
0.50	54.7	16.4	9.7	5.5
1.00	54.7	16.4	9.7	5.5
		ΔC, μmol cm^{-3}		
0.10	0.3	1.0	1.7	3.0
0.27	0.8	2.7	4.6	8.1
0.50	1.5	5.0	8.5	15.0
1.00	3.0	10.0	17.0	30.0

Source: Data from Claassen (1990a).

Note: Other parameters:

Θ	0.29	I_{max}	20×10^{-12} mol cm^{-2} s^{-1}
f	0.29	K_m	0.04 μmol cm^{-3}
r_0	0.015 cm	C_{min}	0.0002 μmol cm^{-3}
r_1	0.56 cm at root length density of 1 cm cm^{-3}		
r_1	0.18 cm at root length density of 10 cm cm^{-3}		

These results can be explained by means of the depletion curves around the roots (see Figure 12.8). In the case of highest uptake (C_{Li} = 1 μmol cm^{-3}, b = 30, no root competition), solution concentration at root surface dropped to about 0.5 μmol cm^{-3} after ten days of uptake. Comparing this value with the Michaelis-Menten constant, K_m (0.04 μmol cm^{-3}) indicates that the concentration at the root surface was sufficient for plants to achieve a near maximum rate of uptake. To ensure this high uptake rate, diffusive flux of K toward the root must be in the same order as uptake. For the given diffusive conditions, flux depends on the concentration gradient in soil solution (dC_L/dx or simplified $\Delta C_L/\Delta x$). If Δx is taken as the extension of the depletion zone around the root, ΔC_L is the difference between initial soil solution concentration and concentration at root surface (C_{Li} − C_{L0}). In this example, a ΔC_L of about 0.5 µmol cm^{-3} is needed to ensure that the diffusive flux toward the root is high enough for a near maximal rate of uptake (see Figure 12.8). As plants are able to decrease the concentration at root surface to nearly zero, ΔC_L will be about 0.5 µmol cm^{-3}, providing C_{Li} is greater than 0.5 µmol cm^{-3}. If C_{Li} is lower than 0.5 µmol cm^{-3}, then the required ΔC_L cannot be established and uptake depends mainly on the rate of transport of K in the soil. In this case, ΔC_L will be in the same order as C_{Li} because C_{L0} is nearly zero. Therefore, transport in the soil and uptake rate are strongly influenced by C_{Li}.

In contrast to C_{Li}, buffer power influences calculated K uptake only in the case of root competition (Figure 12.7). This seems surprising, as b is part of the effective diffusion coefficient D_e, which describes the diffusive conditions in the soil. According to Equation 6, a change in D_e at a constant gradient dC/dx will change the diffusive flux in the same manner. However, diffusion happens mainly in the liquid phase and, thus, the "driving force" is not the gradient of available nutrients, dC/dx, but the gradient in solution, dC_L/dx. The relation between both gradients is described by the buffer power. Thus, to calculate F_D in terms of the gradient in solution, Equations 6 and 7 can be reduced to $F_D = - D_L \Theta f \, dC_L/dx$. Consequently, buffer power has no influence on F_D, and calculated K uptake (see Figure 12.7) is not affected by buffer power if no inter-root competition occurs.

In the case of root competition, another characteristic of D_e becomes important, that is, the influence on the extension of the depletion zone around the roots (Equation 9). Depletion curves at low buffer power are more extended than those calculated with a greater b, resulting in an earlier overlapping of the depletion zones of two roots. This effect is illustrated by the lower concentration between the roots in cases of lower buffer power, even at a low root density (see Figure 12.8). As discussed previously, this decrease in concentration between two roots has no great influence on uptake as long as roots are able to decrease the concentration at root surface to maintain the concentration gradient necessary for a high rate of transport. However, if concentration at root surface is already low, the rate of transport in soil, and hence uptake, will decrease. This inability to maintain the necessary gradient caused by overlapped depletion zones will occur faster with higher root densities and lower buffer power. This is illustrated by the calculated uptake (see Figure 12.7b) and depletion curves (see Figure 12.8b).

Buffer power describes the amount of available nutrients at a given concentration. Therefore, buffer power will gain in importance when the soil volume supplying a unit root is restricted by root competition. In this case, a high buffer power results in a large amount of available nutrients;

FIGURE 12.7. Influence of Buffer Power, b, and Initial Solution Concentration, C_{Li} (μmol cm^{-3}), on K Uptake after Ten Days Calculated (a) Without and (b) with Root Competition (Root Length Density 1 and 10 cm cm^{-3}, Respectively)

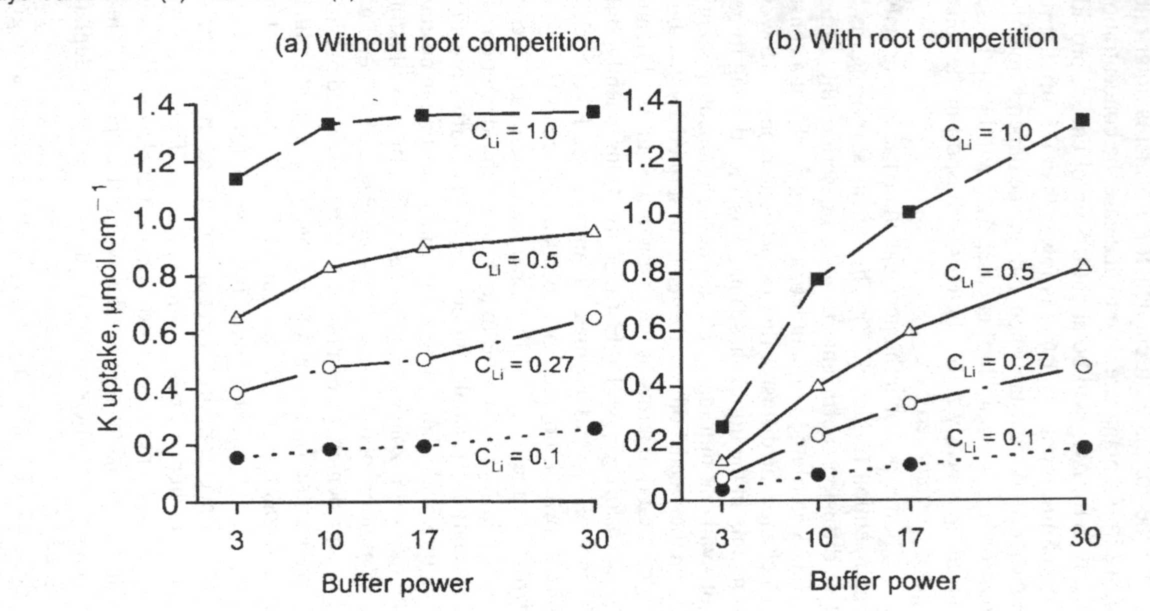

Note: The input data were based on Table 12.5.

FIGURE 12.8. Potassium Concentration in Soil Solution Around Roots (Depletion Curves) After Ten Days of Uptake Calculated with Varied Buffer Power, b, and C_{Li} (a) Without and (b) with Root Competition (Root Length density 1 and 10 cm cm^{-3}, Respectively)

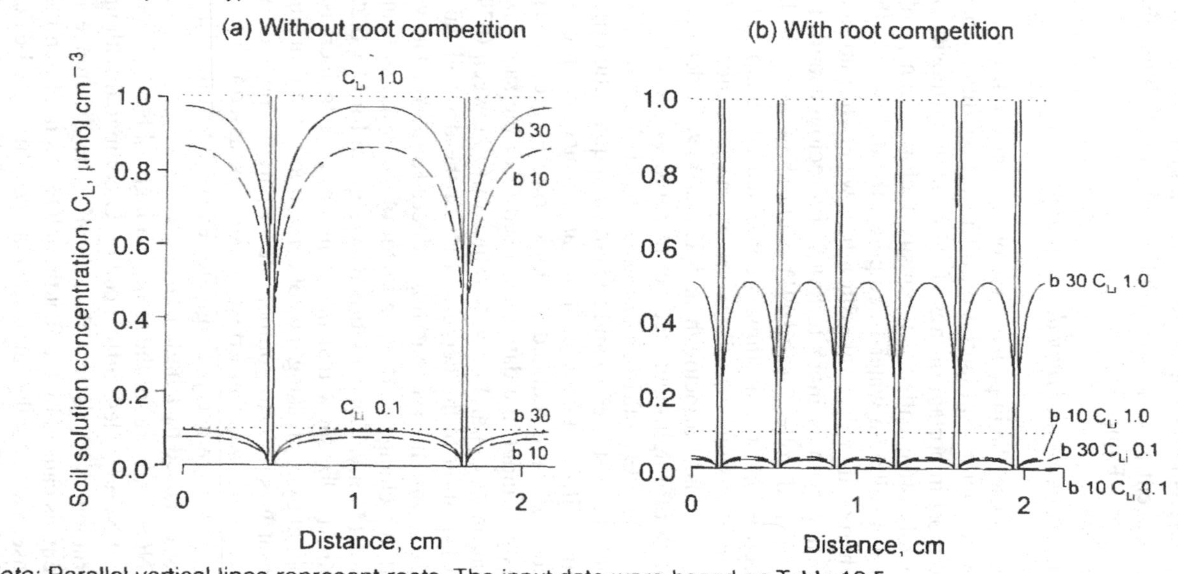

Note: Parallel vertical lines represent roots. The input data were based on Table 12.5.

hence, nutrient shortage for the plant will occur later than at a low buffer power. Although plants had taken up more K at the high buffer power (see Figure 12.7b), the concentration in soil solution remained higher than at the low buffer power (see Figure 12.8b).

Volumetric Water Content, Θ, and Impedance Factor, f

In a mathematical sense, the impedance factor, f, is only a fitting parameter that describes all factors influential on diffusion conditions besides D_L, Θ, and b. The most important of these factors is the tortuosity of the diffusion path, so f is often called a tortuosity factor. It can easily be imagined that in a wet soil with water-filled pores, the diffusion path will be more direct, whereas in a dry soil with a thin water film around the particles, the diffusion path is more tortuous. Barraclough and Tinker (1981) and Bhadoria and colleagues (1991) could show a nearly linear relationship between Θ and f, and, thus, a combined evaluation of both parameters is sensible. Besides Θ, soil texture has an influence on f too. For a wide range of soils, Barraclough and Tinker (1981) derived two equations to calculate f for light- and medium-textured soils and for heavy and organic soils ($f = 1.58\,\Theta - 0.17$ and $f = 0.99\,\Theta - 0.17$, respectively).

The close relationship between Θ and f results in a quadratic influence of Θ on the effective diffusion coefficient D_e and, hence, to an overproportional influence of Θ on calculated K uptake (see Figure 12.9a). For this sensitivity analysis input, the data were used from the same experiment as for evaluation of C_{Li} and b, with values of Θ between 0.29 and 0.1. According to Equation 6, the decreased D_e of the dry soil causes a lower diffusive flux toward the root and this reduces K uptake. The values of Θ used resulted in a change of D_e by a factor of 42, whereas K uptake varied by a factor of 15 only. The depletion curves given for $\Theta = 0.29$ and $\Theta = 0.1$ (see Figure 12.9b) show that the plant was able to increase the concentration gradient by decreasing the concentration at the root surface; in this way, the root was able to compensate in part for poor diffusion conditions in the dry soil. Another effect already discussed can be seen from the depletion curves, namely the greater extension of the depletion zone in the moist soil caused by a higher D_e.

Calculation of NO_3^- uptake for the same soil highlights two important points: (1) uptake of NO_3^- is less influenced by Θ compared with uptake of K, and (2) variation of D_e is less than for K. The difference in D_e between the nutrients is caused by the fact that NO_3^- is not bound to soil particles. In this case, b will get the same value as Θ, which can be derived from Equation 8 and the different reference values of dC and dC_L. Thus, the calculation of D_e for NO_3^- reduces to $D_e = D_L\,f$. This results in a

FIGURE 12.9. Influence of Volumetric Water Content, Θ, on (a) Uptake and (b) Relative Depletion Curves of K and NO_3^- After Ten Days

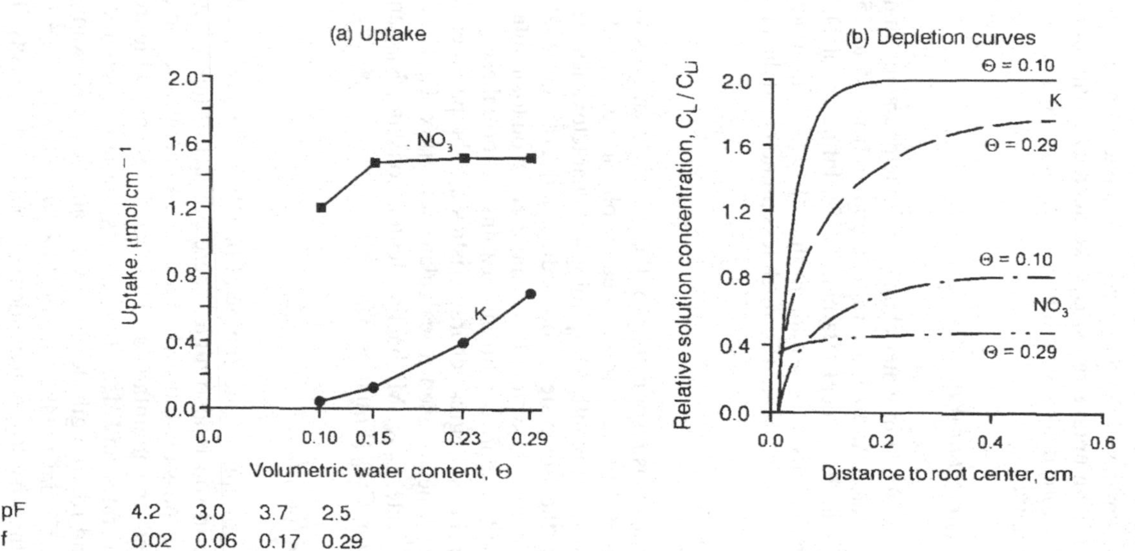

pF	4.2	3.0	3.7	2.5	
f	0.02	0.06	0.17	0.29	
D_e (K)	0.23	1.05	4.55	9.80	$cm^2 s^{-1}$
D_e (NO_3)	39	117	332	566	$cm^2 s^{-1}$

Note: The input data were according to Table 12.5, with root length density 1 cm cm^{-3}. ΔC (NO_3) = 2.8 µmol cm^{-3}; b (NO_3) = Θ (see text) and C_{Li} (NO_3) = ΔC (NO_3)/b.

smaller range of values of D_e for NO_3^-. In addition, the values of D_e for NO_3^- are higher than for K, and, therefore, concentration gradients necessary to secure high fluxes toward roots can be small. The concentration at the root surface remains high compared to the concentration in bulk soil and this allows for high uptake rates. Uptake decreases only in a dry soil where the plant is forced to lower the concentration at root surface to nearly zero (see Figure 12.9).

The Diffusion Coefficient in Water, D_L

The diffusion coefficient in water changes with temperature, mainly due to the viscosity of water. An increase in temperature from 15 to 25°C will increase D_L by about 30 percent (Barber, 1995). Influence of D_L on D_e, uptake, and depletion curves is similar to that of Θ or f in principle, but because influence of D_L is linear, it is much less pronounced than that of Θ.

Plant Parameters

Ion uptake by plants is responsible for the gradient that drives diffusive transport to the root. Uptake is influenced by physiological and morphological characteristics of plants. Morphological characteristics are the size and growth rate of the root system, root radius, and root hair development. Size and growth of the root system has a major effect on total uptake, but if we concentrate on transport conditions to a unit of root, these parameters will play no role. Physiological characteristics are the parameters of the Michaelis-Menten kinetics used to calculate net influx, I_n. They are the maximum uptake rate, I_{max}, Michaelis-Menten constant, K_m, and minimum concentration, C_{min} (Equation 3).

Maximum Uptake Rate, I_{max}

A high maximum uptake rate can be of benefit to the plant, but only if the transport rate in soil is also high. Therefore, achieving a high maximum uptake rate is possible only when a sufficient concentration gradient can be established and maintained. This can be seen in Figure 12.10, where net uptake rates and depletion curves are compared for various values of I_{max} and root length density by means of the K example used previously. A maximum uptake rate of 20×10^{-12} mol cm^{-2} s^{-1} was achieved during the first day, independent of root length density. This was possible by developing a steep gradient. The depletion zone extends with time, but in cases of low root density, almost no inter-root competition

FIGURE 12.10. Influence of I_{max} and Root Length Density on Development of (a) Depletion Curves and (b) K Influx over Thirty Days

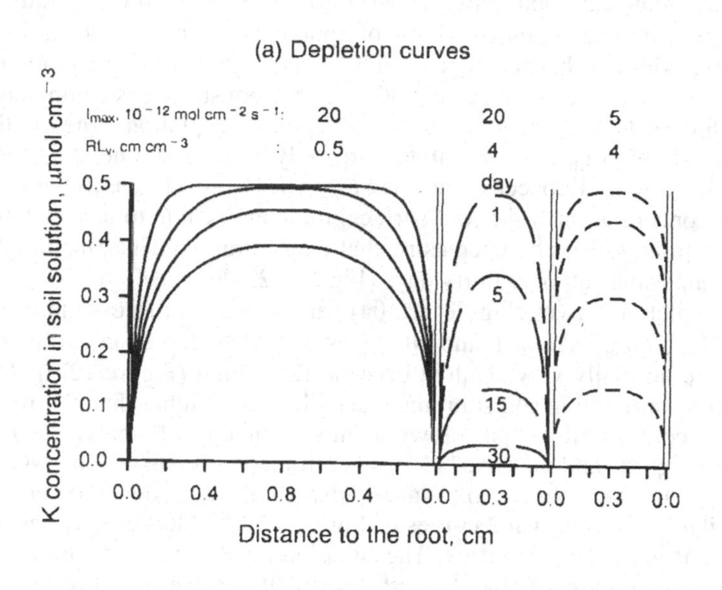

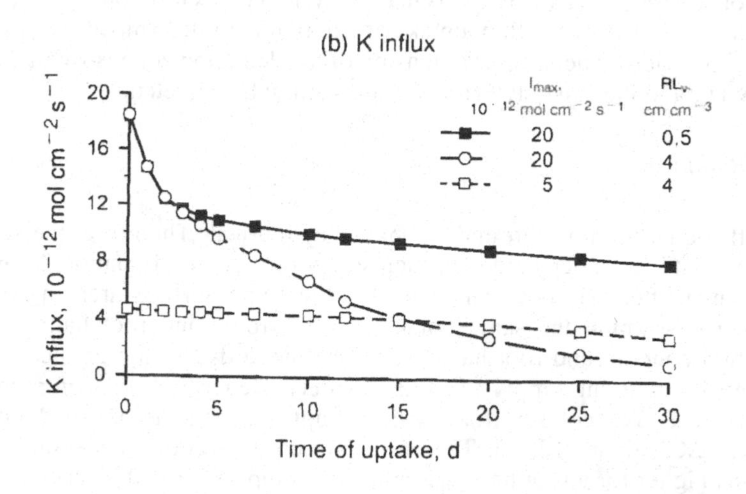

Note: The input data not specified were from Table 12.5.

occurs, so the concentration gradient is maintained and the net influx remains high. A high root density leads to an early overlapping of depletion zones, concentration between two roots decreases, and the gradient necessary to maintain a maximal rate of uptake cannot be sustained.

Uptake with low I_{max} can occur with a smaller concentration gradient, and, hence, the decrease in concentration at the root surface is comparably small (Figure 12.10a, right curves). Extension of depletion zones is the same as with high I_{max} because it depends only on D_e and time; therefore, root competition will occur at day five as well. However, due to the high starting concentration at the root surface, the plant is able to maintain the concentration gradient by decreasing that concentration. Thus, the uptake rate remains constant over thirty days (Figure 12.10b).

The depletion curves (Figure 12.10a) can be used to discuss the influence of I_{max} on uptake of P and NO_3^- as well. The depletion curves for NO_3^- are normally very shallow because D_e is high (Figure 12.9). The initial concentration in soil solution is usually much higher than 0.1 μmol cm^{-3}, a concentration that allows a maximum rate of uptake (Burns, 1980; Steingrobe and Schenk, 1991). Therefore, plants can take up NO_3^- at a maximum rate until the soil is nearly depleted, that is, depletion curves are similar to the right-hand curves of Figure 12.10a. Hence, I_{max} has the main influence on the net influx. The situation for P is more similar to the left curves of Figure 12.10a. As both D_e and the P concentration in soil solution are low, a high gradient is necessary, and concentration at the root surface is far too low to allow uptake at a maximum rate. Thus, I_{max} has a minor influence on net influx. Extension of the depletion zone is small due to low D_e, and the soil between two roots cannot be depleted.

Root Radius, r_0

Diffusion conditions around a root are cylindrical. Therefore, the soil volume that can be depleted is influenced by root radius (Equation 2) and is much greater for root hairs than for the same surface area of root cylinder. Assuming the same uptake rate for roots and root hairs, the depletion zone around root hairs is less extended due to the greater soil volume for nutrient supply (Δx is smaller). Hence, the concentration gradient, $\Delta C/\Delta x$, necessary for any rate of uptake, can be established with a lower ΔC (Figure 12.11). Therefore, the concentration at the surface remains higher for a root hair or a thin root compared to a thicker one. A higher concentration at root surface enables a greater decrease of this concentration by an increasing I_{max}, resulting in a greater gradient, a higher flux, and a higher uptake rate. Root hairs achieve a higher proportion of I_{max} and are therefore more efficient in nutrient uptake from soil at low solution concentration.

In summary, evaluation of soil and plant parameters has shown that transport of nutrients toward the root is the most important process for uptake. If transport is sufficient, plants can take up nutrients at a maximum rate, in which case, I_{max} is the main parameter for calculating net uptake (e.g., NO_3^-). If transport is insufficient, as for P and often for K, soil parameters become important in determining the uptake. The initial concentration in soil solution, C_{Li}, is the most important of these parameters. Poor conditions for diffusion, expressed by a low value of the effective diffusion coefficient, D_e, can be compensated for by increasing the con-

FIGURE 12.11. Potassium Concentration Around Root and Root Hairs (Depletion Curves) Calculated with Varied I_{max} over Ten Days

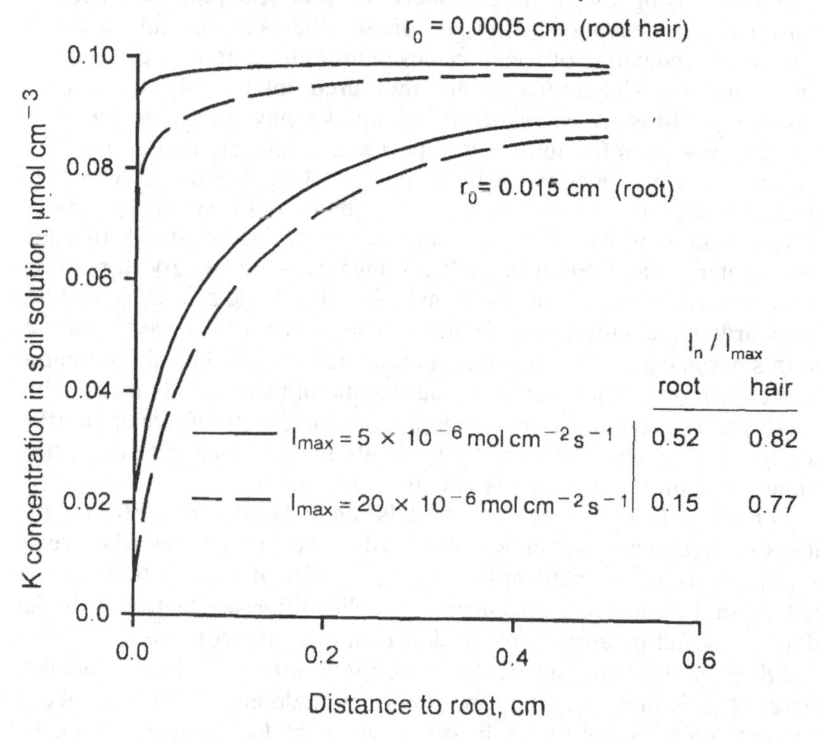

Note: The input data not specified were from Table 12.5.

centration gradient. As plants are able to lower the concentration at root surface near to zero, the highest possible concentration gradient is given by C_{Li}. Buffer power will become important only if there is competition between roots. Buffer power influences the extension of the depletion zone, and with overlapping depletion zones, the concentration between two roots decreases more quickly, thereby decreasing the potential concentration gradient toward the roots.

GENERAL CONCLUSIONS AND FINAL REMARKS

The validation of the presented models has shown a good agreement between simulated and measured uptake. In some cases, mainly at a low nutrient availability, the models underestimated uptake. This highlights the difficulty in using mechanistic models for practical purposes, such as fertilizer recommendations. However, these models are useful for developing an understanding of the processes in the rhizosphere. A close agreement between model simulation and measured uptake indicates that our ideas about diffusion, mass flow, and uptake physiology as the most important processes for nutrient transport and uptake are appropriate, and they are described adequately by Equations 3 to 7 and 11. A lack of agreement does not necessarily mean that the model is wrong. It may be due to several reasons: (1) the assumptions regarding constancy of volumetric water content, Θ, of the half distance between two roots, r_1, or of the parameters of the Michaelis-Menten kinetics, I_{max}, K_m, C_{min}, and the one regarding the uniform distribution of roots and root hairs are incorrect, and this has had a major influence on model results; (2) the values of input parameters used are not correct, because some of them are not measurable directly; and (3) the model is incomplete, as the ability of plants to influence the rhizosphere and to make nutrients more available through root exudates and microorganisms is not taken into account.

Even with a lack of agreement, models can be helpful in identifying the causes of divergence. As shown previously, a sensitivity analysis reveals the importance of a parameter for a specific nutrient and can therefore be used to find parameters responsible for the difference between model calculations and measured uptake. For example, in most cases, transport conditions in the rhizosphere have a minor effect on NO_3^- uptake, whereas I_{max} is most important. An incorrect modeling of NO_3^- uptake is therefore often caused by an incorrect value of I_{max}, which cannot be measured in the field directly and changes greatly with growing conditions. A preliminary attempt to overcome this problem was carried out by

Steingrobe and Schenk (1993, 1994, 1997) who calculated I_{max} in relation to relative growth rates.

In the case of P and often K, the initial solution concentration, C_{Li}, and the root hair growth are of importance. Normally, it is possible to measure C_{Li} in bulk soil solution, and, hence, the use of incorrect C_{Li} values can be excluded as a reason for inadequate model results. However, under conditions of nutrient shortage, plants are able to increase nutrient availability by changing the pH of the rhizosphere or by excreting organic compounds. For example, under Fe (iron) deficiency, plants exudate citrate or phytosiderophores to complex Fe and increase its availability (Römheld and Marschner, 1986). Shortage of P results in exudation of organic acids, which increase the availability of P in the rhizosphere (Gerke, 1992; Gerke, Römer, and Jungk, 1994). This increase of nutrient availability in the rhizosphere is brought about by an increase in soil solution concentration, C_{Li}. An implementation of these processes into the model will allow verification of the hypothesis that exudation of organic compounds increases nutrient availability and, hence, is responsible for the underestimation of uptake at low P conditions. Preliminary attempts to quantify exudation rate and dissolving capacity of organic acids were conducted by Gerke (1992, 1994), Beissner and Römer (1995), and Keller and Römer (1997), using the approach of Nye (1983, 1984).

The importance of C_{Li} for nutrient transport and uptake allows for more practical conclusions. Efforts to increase nutrient uptake are only as good as their effect on C_{Li}. Therefore, fertilizing a highly buffered soil has a smaller benefit on uptake than fertilizing a poorly buffered soil. Buffer power itself has no influence on nutrient transport, as the sensitivity analysis has shown, and is therefore not implemented in the steady state model (Equation 10). Buffer power affects transport only indirectly through its influence on C_{Li} (e.g., fertilization) or the extension of the depletion zone, which finally acts through C_{Li} as well.

The sensitivity analysis has shown that a decrease of the root radius r_0, often observed as a plant reaction to P shortage, is a useful mechanism because, besides the greater surface area per root volume, it increases uptake capacity of a unit root surface. A significant enhancement of uptake capacity is achieved only in the case of a large reduction in r_0 down to the radius of root hairs or hyphae. Therefore, increased proliferation of root hairs or an infection by mycorrhizal fungi are efficient strategies to address P shortage.

The assumptions regarding constancy and homogeneity of parameters cannot easily be assessed for their significance in explaining insufficient model results. Parameters used in the model (such as Θ or I_{max}) can be assessed by a sensitivity analysis, but an evaluation of the consequence of nonhomogenous root or root hair distribution on model results is not possible. Perhaps, this can be done by comparing results of this model and of the model suggested by Hoffland and colleagues (1990), which takes into account increasing root density during the simulation.

Finally, a good agreement between the model and measured uptake is comforting, as one's ideas are supported, but a lack of agreement is much more interesting and challenging because it leads to work on improving our understanding of processes underlying nutrient uptake by roots.

REFERENCES

Baldwin, J.P., P.H. Nye, and P.B. Tinker (1973). Uptake of solutes by multiple root systems from soil. III. A model for calculating the solute uptake by a randomly dispersed root system developing in a finite volume of soil. *Plant and Soil* 38: 621-635.

Barber, S.A. (1962). A diffusion and mass-flow concept of soil nutrient availability. *Soil Science* 93: 39-49.

Barber, S.A. (1995). *Soil Nutrient Bioavailability—A Mechanistic Approach*. Second Edition. New York: John Wiley.

Barber, S.A. and J.H. Cushman (1981). Nitrogen uptake model for agronomic crops. In *Modeling Waste Water Renovation-Land Treatment*, Ed. I.K. Iskandar. New York: Wiley-Interscience, pp. 382-409.

Barber, S.A. and P.G. Ozanne (1970). Autoradiographic evidence for the differential effect of four plant species altering the calcium content of the rhizosphere soil. *Soil Science Society of America Proceedings* 34: 635-637.

Barraclough, P.B. (1986). The growth and activity of winter wheat roots in the field: Nutrient inflows of high yielding crops. *Journal of Agricultural Science* 106: 53-59.

Barraclough, P.B. and P.B. Tinker (1981). The determination of ionic diffusion coefficients in field soils. I. Diffusion coefficients in sieved soils in relation to water content and bulk density. *Journal of Soil Science* 32: 225-236.

Beissner L. and W. Römer (1995). Ausscheidung von organischen Säuren durch die Zuckerrübenwurzel und deren Bedeutung für die P-Mobilisierung im Boden. In *Pflanzliche Stoffaufnahme und mikrobielle Wechselwirkungen in der Rhizosphäre. 6. Borkheider Seminar zur Ökophysiologie des Wurzelraums*, Ed. W. Merbach. Stuttgart, Germany: Teubner, pp. 137-144.

Bhadoria, P.B.S., J. Kaselowski, N. Claassen, and A. Jungk (1991). Impedance factor for chloride diffusion in soil as affected by bulk density and water content. *Zeitschrift für Pflanzenernährung und Bodenkunde* 154: 69-72.

Burns, I.G. (1980). Influence of the spatial distribution of nitrate on the uptake of N by plants: A review and a model for rooting depth. *Journal of Soil Science* 31: 155-173.

Chanter, D.O. (1981). The use and misuse of linear regression methods in crop modelling. In *Mathematics and Plant Physiology*, Eds. D.A. Rose and D.A. Charles-Edwards. London, UK: Academic Press, pp. 253-267.

Claassen, N. (1990a). *Die Aufnahme von Nährstoffen aus dem Boden durch die höhere Pflanze als Ergebnis von Verfügbarkeit und Aneignungsvermögen*. Habilitation thesis. Göttingen, Germany: Severin Verlag.

Claassen, N. (1990b). Fundamentals of soil-plant interactions as derived from nutrient diffusion in soil, uptake kinetics and morphology of roots. *14th Int. Congress of Soil Science, Kyoto, Japan*, II: 118-123.

Claassen, N. and S.A. Barber (1976). Simulation model for nutrient uptake from soil by a growing plant root system. *Agronomy Journal* 68: 961-964.

Claassen, N., K.M. Syring, and A. Jungk (1986). Verification of a mathematical model by simulating potassium uptake from soil. *Plant and Soil* 95: 209-220.

Cushman, J.H. (1979). An analytical solution to solute transport near root surfaces for low initial concentration. I. Equation development. *Soil Science Society of America Journal* 43: 1087-1090.

Cushman, J.H. (1980). Analytical study of the effect of ion depletion (replenishment) caused by microbial activity near a root. *Soil Science* 129: 69-87.

Fick, A. (1855). Über Diffusion. *Annalen der Physik und Chemie* 94: 59-86.

Föhse, D., N. Claassen, and A. Jungk (1991). Phosphorus efficiency of plants. II. Significance of root radius, root hairs and cation-anion balance for phosphorus influx in seven plant species. *Plant and Soil* 132: 261-272.

Gerke, J. (1992). Phosphate, aluminium and iron in the soil solution of three different soils in relation to varying concentrations of citric acids. *Zeitschrift für Pflanzenernährung und Bodenkunde* 155: 339-343.

Gerke, J. (1994). Kinetics of soil phosphate desorption as affected by citric acid. *Zeitschrift für Pflanzenernährung und Bodenkunde* 157: 17-22.

Gerke, J., W. Römer, and A. Jungk (1994). The excretion of citric and malic acid by proteoid roots of *Lupinus albus* L.: Effects on soil solution concentration of phosphate, iron and aluminium in the proteoid rhizosphere in samples of an oxisol and a luvisol. *Zeitschrift für Pflanzenernährung und Bodenkunde* 157: 289-294.

Gregory, P.J., D.V. Crawford, and M. McGowan (1979). Nutrient relations of winter wheat. 2. Movement of nutrients to the root and their uptake. *Journal of Agricultural Science* 93: 495-504.

Heins, B. (1989). Bedeutung von Wurzeleigenschaften für die Nutzung des Nitratangebotes durch Spinat und Kohlrabi. Doctoral thesis. University of Hannover, Hannover, Germany.

Hendriks, L., N. Claassen, and A. Jungk (1981). Phosphatverarmung des wurzelnahen Bodens und Phosphataufnahme von Mais und Raps. *Zeitschrift für Pflanzenernährung und Bodenkunde* 144: 486-499.

Hendriks, L. and A. Jungk (1981). Erfassung der Mineralstoffverteilung in Wurzelnähe durch getrennte Analyse von Rhizo- und Restboden. *Zeitschrift für Pflanzenernährung und Bodenkunde* 144: 276-282.

Hoffland, E., H.S. Bloemhof, P.A. Leffelaar, G.R. Findenegg, and J.A. Nelemans (1990). Simulation of nutrient uptake by a growing root system considering increasing root density and inter-root competition. *Plant and Soil* 124: 149-155.

Itoh, S. and S.A. Barber (1983). A numerical solution of whole plant nutrient uptake for soil-root systems with root hairs. *Plant and Soil* 70: 403-413.

Jones, D.L., P.R. Darrah, and L.V. Kochian (1996). Critical evaluation of organic acid mediated iron dissolution in the rhizosphere and its potential role in root iron uptake. *Plant and Soil* 180: 57-66.

Keller H. and W. Römer (1997). Ausscheidung organischer Säuren bei Spinat in Abhängigkeit von der P-Ernährung und deren Einfluß auf die Löslichkeit von Cu, Zn und Cd im Boden. In *Pflanzenernährung, Wurzelleistung und Exsudation. 8. Borkheider Seminar zur Ökophysiologie des Wurzelraumes.* Ed. W. Merbach. Stuttgart, Germany: Teubner, pp. 187-195.

Kuchenbuch, R. (1983). Die Bedeutung von Ionenaustauschprozessen im wurzelnahen Boden für die Pflanzenverfügbarkeit von Kalium. Doctoral thesis. University of Göttingen, Göttingen, Germany.

Kuchenbuch, R. and S.A. Barber (1988). Significance of temperature and precipitation for maize root distribution in the field. *Plant and Soil* 106: 9-14.

Kuchenbuch, R. and A. Jungk (1982). A method for determining concentration profiles at the soil-root interface by thin slicing rhizospheric soil. *Plant and Soil* 68: 391-394.

Kuhlmann, H. (1988). Ursachen und Ausmaß der N-, P-, K- und Mg-Ernährung der Pflanzen aus dem Unterboden. Habilitation thesis. University of Hannover, Hannover, Germany.

Müller, R. (1988). Bedeutung des Wurzelwachstums und der Phosphatmobilität im Boden für die Phosphaternährung von Winterweizen, Wintergerste und Zuckerrüben. Doctoral thesis. University of Göttingen, Göttingen, Germany.

Nielsen, N.E. (1972). A transport kinetic concept of ion uptake from soil by plants. II. The concept and some theoretic considerations. *Plant and Soil* 37: 561-576.

Nye, P.H. (1983). The diffusion of two interacting solutes in soil. *Journal of Soil Science* 34: 677-691.

Nye, P.H. (1984). On estimating the uptake of nutrients solubilized near roots or other surfaces. *Journal of Soil Science* 35: 439-446.

Nye, P.H. and F.C.H. Marriott (1969). A theoretical study of the distribution of substances around roots resulting from simultaneous diffusion and mass flow. *Plant and Soil* 30: 459-472.

Nye, P.H. and P.B. Tinker (1977). *Solute Movement in Soil-Root System.* Oxford, UK: Blackwell.

Rengel, Z. (1993). Mechanistic simulation models of nutrient uptake: A review. *Plant and Soil* 152: 161-173.

Retzer, F. (1995). Untersuchungen zur Stickstoffverwertung von Weizenbeständen. Doctoral thesis. Technical University of München, München, Germany.

Römheld, V. and H. Marschner (1986). Mobilization of iron in the rhizosphere of different plant species. *Advances in Plant Nutrition* 2: 155-204.

Ross, G.J.S. (1981). The use of non-linear regression methods in crop modelling. In *Mathematics and Plant Physiology*, Eds. D.A. Rose and D.A. Charles-Edwards. London, UK: Academic Press, pp. 269-282.

Seeling, B. and N. Claassen (1990). A method for determining Michaelis-Menten kinetic parameters of nutrient uptake for plants growing in soil. *Zeitschrift für Pflanzenernährung und Bodenkunde* 153: 301-303.

Seiffert, S., J. Kaselowsky, A. Jungk, and N. Claassen (1995). Observed and calculated potassium uptake by maize as affected by soil water content and bulk density. *Agronomy Journal* 87: 1070-1077.

Steingrobe, B. and M.K. Schenk (1991). Influence of nitrate concentration at the root surface on yield and nitrate uptake of kohlrabi (*Brassica oleracea gongyloides* L.) and spinach (*Spinacia oleracea* L.). *Plant and Soil* 135: 205-211.

Steingrobe, B. and M.K. Schenk (1993). Simulation of the maximum nitrate inflow (I_{max}) of lettuce (*Lactuca sativa* L.) grown under fluctuating climatic conditions in the greenhouse. *Plant and Soil* 155/156: 163-166.

Steingrobe, B. and M.K. Schenk (1994). A model relating the maximum nitrate inflow of lettuce (*Lactuca sativa* L.) to the growth of roots and shoots. *Plant and Soil* 162: 249-257.

Steingrobe, B. and M.K. Schenk (1997). Calculation of the total nitrate uptake of lettuce (*Lactuca sativa* L.) by use of a mathematical model to simulate nitrate inflow. *Zeitschrift für Pflanzenernährung und Bodenkunde* 160: 73-79.

Strebel, O. and W.H.M. Duynisveld (1989). Nitrogen supply to cereals and sugar beet by mass flow and diffusion on a silty loam soil. *Zeitschrift für Pflanzenernährung und Bodenkunde* 152: 135-141.

Syring, K. M. and N. Claassen (1995). Estimation of the influx and the radius of the depletion zone developing around a root during nutrient uptake. *Plant and Soil* 175: 115-123.

Wehrmann, J. and H.C. Scharpf (1986). The N_{min}-method—An aid to integrating various objectives of nitrogen fertilization. *Zeitschrift für Pflanzenernährung und Bodenkunde* 149: 428-440.

Wiesler, F. (1991). Sortentypische Unterschiede im Wurzelwachstum und in der Nutzung des Nitratangebotes des Bodens bei Mais. Doctoral thesis. University of Hohenheim, Stuttgart, Germany.

Index

Page numbers followed by the letter "t" indicate tables; those followed by the letter "f" indicate figures.

Printed in the United States
by Baker & Taylor Publisher Services

Printed in the United States
by Baker & Taylor Publisher Services